Inverse Problems, Geometric and Topological Aspects of Representation Theory, and Low-Dimensional Topology

CONTEMPORARY MATHEMATICS

834

Inverse Problems, Geometric and Topological Aspects of Representation Theory, and Low-Dimensional Topology

AMS Special Session
Inverse Problems
October 22–23, 2022
University of Utah, Salt Lake City, UT
AMS Special Session
Geometry and Representation Theory of Quantum Algebras and Related Topics
October 22–23, 2022
University of Utah, Salt Lake City, UT

Mee Seong Im
Hanna E. Makaruk
Bach Nguyen
Robert Owczarek
Editors

AMERICAN MATHEMATICAL SOCIETY

2020 *Mathematics Subject Classification.* Primary 57K10, 57K16, 57K45, 57K99, 57K18, 41A50, 14L24, 81P40, 81P42.

Library of Congress Cataloging-in-Publication Data

Cataloging-in-Publication Data has been applied for by the AMS.
See http://www.loc.gov/publish/cip/.

∞ The paper used in this book is acid-free and falls within the guidelines
established to ensure permanence and durability.
Visit the AMS home page at `https://www.ams.org/`

10 9 8 7 6 5 4 3 2 1 31 30 29 28 27 26

Contents

Preface

Hanna E. Makaruk and Robert M. Owczarek have been organizing special sessions at many AMS sectional meetings for over 20 years, beginning in 1999. By design, the sessions have been interdisciplinary, and their goal has been on one hand to present high quality research in the different disciplines and on the other hand encourage interactions between pure mathematicians, mathematical physicists, applied mathematicians, computer scientists, and occasionally representatives of other disciplines.

Parallel to the highly visible mathematicians and researchers in the special session Inverse Problems, noncommutative algebras and noncommutative algebraic geometry have been an active field of research for the past decades, with many important applications in mathematical physics, representation theory, number theory, combinatorics, and geometry. Mee Seong Im, Bach Nguyen, and Arik Wilbert's purpose for their special session is to bring together junior and experienced researchers on noncommutative algebras, especially quantum algebras, noncommutative algebraic geometry, low-dimensional topology, and related topics to celebrate and share their contributions in the area. This event helped to initiate collaborations among the junior and senior mathematicians and create connections to further advance their careers.

Both special sessions had wide ranges of participants giving talks, from undergraduate and graduate students to senior mathematicians and researchers. Many of our speakers reacted enthusiastically to the possibility of publishing their contributions in a volume of proceedings in American Mathematical Society's Contemporary Mathematics series. While all papers in the proceedings contain original results and research, a few are additionally expository, discussing new concepts with detailed examples and current trends with novel and important results, all of which are important to students, junior and senior researchers.

All manuscripts in this Contemporary Mathematics volume contain original research, written by the speakers, participants, and their collaborators in the two special sessions. The papers in this volume also discuss novel ideas with completely worked out examples not found in the mathematics literature, all of which yield as an invaluable contribution to the mathematics community.

Mee Seong Im
Hanna E. Makaruk
Bach Nguyen
Robert M. Owczarek

List of participants

Special Session: Geometry and Representation Theory of Quantum Algebras and Related Topics

Organizers:
Mee Seong Im
Bach Nguyen
Arik Wilbert

Prerna Agarwal
Louisiana State University
Baton Rouge, LA

Eric Bucher
Xavier University
Cincinnati, OH

Tamanna Chatterjee
University of Georgia
Athens, GA

Anne Dranowski
University of Southern California
Los Angeles, CA

Nicolle González
University of Berkeley
Berkeley, CA
University of California
Los Angeles, CA

Tekin Karadağ
College of Charleston
Charleston, SC
University of Georgia
Athens, GA

Scott Larson
University of Georgia
Athens, GA

John Machacek
University of Oregon
Eugene, OR

Daniel K. Nakano
University of Georgia
Athens, GA

Victor Ostrik
University of Oregon
Eugene, OR

Joseph Pappe
University of California
Davis, CA

Manuel Reyes
University of California
Irvine, CA

Adam Robertson
Utah State University
Logan, UT

Eric Samperton
Purdue University
West Lafayette, IN

Yaolong Shen
University of Virginia
Charlottesville, VA

Yun Shi
Brandeis University
Waltham, MA

Jan-Luca Spellmann
Utah State University
Logan, UT

Daping Weng
University of California
Davis, CA

Milen Yakimov
Northeastern University
Boston, MA

Matthew Young
Utah State University
Logan, UT

Weinan Zhang
University of Virginia
Charlottesville, VA

Special Session: Inverse Problems

Organizers:
Hanna E. Makaruk
Robert M. Owczarek

Rhea Palak Bakshi
ETH Zurich
Zürich, Switzerland

Mieczyslaw K. Dabkowski
University of Texas at Dallas
Richardson, TX

Valentina Harizanov
George Washington University
Washington, DC

Louis H. Kauffman
University of Illinois Chicago
Chicago, IL

Eric A. Machorro
Pacific Northwest National Laboratory
Richland, WA

Hanna E. Makaruk
Los Alamos National Laboratory
Los Alamos, NM

Gabriel Montoya-Vega
NSF-Ascend CUNY Graduate Center
New York, NY

Sujoy Mukherjee
University of Denver
Denver, CO

Robert M. Owczarek
University of New Mexico
Albuquerque, NM

Józef Henryk Przytycki
George Washington University
Washington, DC

Sushmita Sinha Roy
University of Texas at Dallas
Richardson, TX

Daniel Scofield
Francis Marion University
Florence, SC

Keshav Srinivasan
George Washington University
Washington, DC

Dario Verta
George Washington University
Washington, DC

Philip White
George Washington University
Washington, DC

Contemporary Mathematics
Volume **834**, 2026
https://doi.org/10.1090/conm/834/16681

Renormalized Reshetikhin–Turaev invariants for the unrolled quantum group of $\mathfrak{sl}_2(\mathbb{C})$

Nathan Geer, Adam Robertson, Jan-Luca Spellmann, and Matthew B. Young

ABSTRACT. This paper is a self-contained introduction to the theory of renormalized Reshetikhin–Turaev invariants of links defined by Geer, Patureau-Mirand and Turaev. Renormalized Reshetikhin–Turaev theory is a cornerstone of non-semisimple quantum topology and is the focal point of many on-going interactions between low dimensional topology, representation theory and high energy physics. Whereas the standard Reshetikhin–Turaev theory of a $\mathbb{C}$-linear ribbon category assigns the trivial invariant to any link with a component colored by a simple object of vanishing quantum dimension, the renormalized theory does not. We give a streamlined development of the renormalized theory for the category of weight modules over the restricted unrolled quantum group of $\mathfrak{sl}_2(\mathbb{C})$ at an even root of unity. In this way, we illustrate many features of the general renormalized theory in a concrete representation theoretic setting.

Introduction

Background and motivation. Let $L \subset S^3$ be a framed link and M_L the closed orientable 3-manifold obtained from S^3 by surgery along L. By a theorem of Lickorish and Wallace, any closed connected orientable 3-manifold arises in this way [**Wal60, Lic62**]. Moreover, the 3-manifolds M_L and $M_{L'}$ are homeomorphic if and only if the framed links L and L' are related by a finite sequence of Kirby moves [**Kir78**]. These results are the starting point for a knot theoretic approach to problems and constructions in 3-manifold topology. For example, it follows from the previous two results that an isotopy invariant of framed links which is also invariant under Kirby moves defines an invariant of 3-manifolds.

Reshetikhin and Turaev constructed a large class of link invariants using the theory of ribbon categories [**RT90**]. Associated to each ribbon category $\mathcal{D}$ is a ribbon functor $\mathcal{F}_{\mathcal{D}} : \mathbf{Rib}_{\mathcal{D}} \to \mathcal{D}$ with domain the category of $\mathcal{D}$-colored ribbon graphs. Interpreting (an isotopy class of) a $\mathcal{D}$-colored framed link L as a $(0,0)$-tangle, and so an endomorphism of the unit object $\mathbb{I} \in \mathbf{Rib}_{\mathcal{D}}$, produces an invariant

2020 *Mathematics Subject Classification.* Primary 81R50; Secondary 57K16.
Key words and phrases. Reshetikhin–Turaev theory, Representation theory of quantum groups.
The first author is partially supported by NSF grants DMS-1664387 and DMS-2104497.
The fourth author is partially supported by NSF grant DMS-2302363 and a Simons Foundation Collaboration Grant for Mathematicians (Award ID 853541).

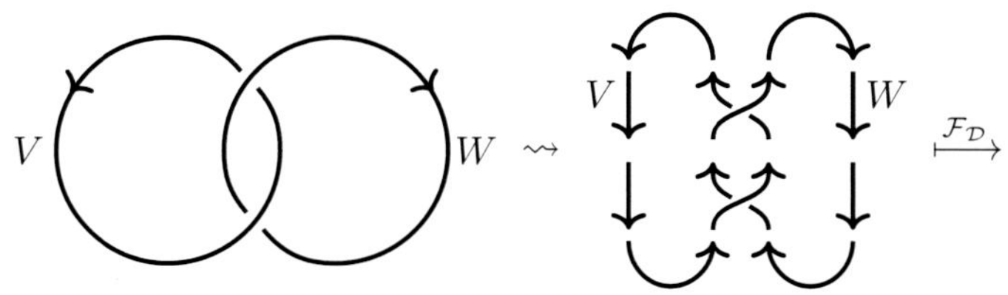

$$\begin{array}{c} \mathrm{ev}_V \otimes \widehat{\mathrm{ev}}_W \\ \circ \\ \mathrm{id}_{V^\vee} \otimes c_{W,V} \otimes \mathrm{id}_{W^\vee} \\ \circ \\ \mathrm{id}_{V^\vee} \otimes c_{V,W} \otimes \mathrm{id}_{W^\vee} \\ \circ \\ \widehat{\mathrm{coev}}_V \otimes \mathrm{coev}_W \end{array}$$

FIGURE 1. The Reshetikhin–Turaev invariant of a Hopf link colored by objects V and W of $\mathcal{D}$. Composition in $\mathcal{D}$ is read from bottom to top.

$\mathcal{F}_{\mathcal{D}}(L) \in \mathrm{End}_{\mathcal{D}}(\mathbb{I})$ of L, the *Reshetikhin–Turaev invariant*. The invariant $\mathcal{F}_{\mathcal{D}}(L)$ is computed as follows. Choose a regular diagram D for L. Decompose D into elementary pieces consisting of cups, caps, simple crossings and twists and assign to these pieces the corresponding coevaluations, evaluations, braidings and twists, respectively, of $\mathcal{D}$. The composition of these morphisms in $\mathcal{D}$ is $\mathcal{F}_{\mathcal{D}}(L)$. Figure 1 illustrates this procedure for the Hopf link.

The Reshetikhin–Turaev construction highlights the topological significance of ribbon categories. Classical representation theory produces many examples of symmetric monoidal categories: representations of groups and Lie algebras and, more generally, cocommutative Hopf algebras. Unfortunately, Reshetikhin–Turaev invariants associated to a symmetric monoidal category are uninteresting since they retain information only about the number of components of a link. On the other hand, many categories arising in quantum representation theory, such as representations of quantum groups and, more generally, quasi-triangular Hopf algebras, have naturally (non-symmetric) ribbon structures [**Jim85, Dri86, Dri90, CP94**]. The resulting quantum invariants of links, which include the Jones and HOMFLYPT polynomials, are at the foundation of quantum topology [**Jon87, FYH$^+$85, PT88, RT90, Tur16**].

Many ribbon categories arising in representation theory have the following properties:

(P1) The category has simple objects with vanishing quantum dimension.
(P2) The category is non-semisimple, that is, not every short exact sequence splits.
(P3) The category has infinitely many isomorphism classes of simple objects.

For example, the category $U_q(\mathfrak{g})$-mod of finite dimensional representations of the quantum group $U_q(\mathfrak{g})$ associated to a complex simple Lie (super)algebra $\mathfrak{g}$ at a root of unity has Properties (P1)–(P3). It is well-known that the Reshetikhin–Turaev invariant of a link colored by a simple object of vanishing quantum dimension is zero. For this reason, the Reshetikhin–Turaev construction is not well-suited to extracting the full topological content of categories having Property (P1). Properties (P2) and (P3) do not cause problems for Reshetikhin–Turaev invariants of links but are serious obstructions to extending these invariants to 3-manifolds. For example, these properties obstruct the definition of the Kirby color, a weighted sum of isomorphism classes of simple objects which is crucial to the construction of 3-manifold invariants from links [**RT91**].

A standard approach to simultaneously eliminating Properties (P1)–(P3) for the category $U_q(\mathfrak{g})$-mod, with $\mathfrak{g}$ a simple Lie algebra, is semisimplification [**And92**], whereby simple objects of vanishing quantum dimension are formally set to zero. The semisimplified categories are, for particular roots of unity, modular tensor categories. The resulting 3-manifold invariants comprise the top level of a three dimensional topological quantum field theory which is a mathematical model for Chern–Simons theory with gauge group the simply connected compact Lie group associated to $\mathfrak{g}$ [**Wit89, RT91**]. On the other hand, for the category $U_q(\mathfrak{g})$-mod, with $\mathfrak{g}$ a type I Lie superalgebra, typical representations have vanishing quantum dimension and semisimplification eliminates most interesting content of the category.

Early examples of link invariants constructed from ribbon categories with Properties (P1)–(P3) include the work of Akutsu, Deguchi and Ohtsuki [**ADO92**] and Murakami and Murakami [**MM01**], who defined (framed) link invariants from typical representations of the unrolled quantum group $\overline{U}_q^H(\mathfrak{sl}_2(\mathbb{C}))$ at an even root of unity. A systematic program to define and study quantum invariants from ribbon categories with Properties (P1)–(P3), called *renormalized Reshetikhin–Turaev theory*, was developed by Blanchet, Costantino, Geer, Patureau-Mirand and Turaev [**BCGPM16, GPMT09, CGPM14**]. In particular, the renormalized invariant of a link colored by objects with vanishing quantum dimensions need not vanish.

Ribbon categories with Properties (P1)–(P3) also arise naturally in quantum field theory. For example, such categories arise as line operators in Chern–Simons theories with non-compact gauge groups [**Wit91, BNW91, RS94, Mik15**] and topological twists of supersymmetric quantum field theories [**KS09, CDGG21**]. Another (related) source of examples are categories of modules over vertex operator algebras in non-rational (or logarithmic) conformal field theories [**RS93, CR13a, CR13b**]. The physical construction of renormalized Reshetikhin–Turaev invariants in the above settings is an active area of research [**CGP21, CDGG21, GY22**].

The goal of this paper is to give a self-contained introduction to the theory of renormalized Reshetikhin–Turaev invariants of links in the simplest case of the category of modules over $\overline{U}_q^H(\mathfrak{sl}_2(\mathbb{C}))$, following [**GPMT09, CGPM15b**]. While this paper contains no new results, we do offer a number of new proofs of known results and give complete details where they are often not available in the literature. Some familiarity with the representation theory of $U_q(\mathfrak{sl}_2(\mathbb{C}))$, at the level of [**Jan96**], and its associated Reshetikhin–Turaev invariants would be beneficial, but is not strictly necessary. We assume basic knowledge of Hopf algebras and monoidal categories.

Contents of this paper. Fix an integer $r \geq 2$ and set $q = e^{\frac{\pi\sqrt{-1}}{r}}$. The De Concini–Kac quantum group $U_q(\mathfrak{sl}_2(\mathbb{C}))$ has generators $K^{\pm 1}$, E and F with relations $KK^{-1} = 1 = K^{-1}K$ and

$$KE = q^2 EK, \qquad KF = q^{-2} FK, \qquad EF - FE = \frac{K - K^{-1}}{q - q^{-1}}. \tag{0.1}$$

The *unrolled quantum group* $U_q^H(\mathfrak{sl}_2(\mathbb{C}))$, introduced in [**GPMT09**], is defined similarly to the De Concini–Kac quantum group but with an additional generator H, thought of as a logarithm of K, which commutes with K and satisfies the classical limit of the first two relations (0.1):

$$[H, E] = 2E, \qquad [H, F] = -2F.$$

The *restricted unrolled quantum group* $\overline{U}_q^H(\mathfrak{sl}_2(\mathbb{C}))$ is the quotient of $U_q^H(\mathfrak{sl}_2(\mathbb{C}))$ by the relations $E^r = F^r = 0$. A $\overline{U}_q^H(\mathfrak{sl}_2(\mathbb{C}))$-module V is called a *weight module* if it is a direct sum of H-eigenspaces and $K = q^H$ as operators on V. The category $\mathcal{C}$ of finite dimensional weight modules over $\overline{U}_q^H(\mathfrak{sl}_2(\mathbb{C}))$ is the central algebraic object of this paper.

Section 1 is devoted to a detailed study of the category $\mathcal{C}$. A natural Hopf algebra structure on $\overline{U}_q^H(\mathfrak{sl}_2(\mathbb{C}))$ gives $\mathcal{C}$ the structure of a rigid monoidal abelian category. We use Verma modules, which are finite dimensional due to the relations $E^r = F^r = 0$, to classify simple objects of $\mathcal{C}$ in Proposition 1.7. The result is that there is a discrete family of simple modules S_n^{lr} of highest (H-)weight $n + lr$ and dimension $n+1$, $l \in \mathbb{Z}$, $0 \leq n \leq r-2$, and a continuous family of simple Verma modules V_α of highest weight $\alpha + r - 1$, $\alpha \in \mathbb{C} \setminus \mathbb{Z} \cup r\mathbb{Z}$, and dimension r.

Tracking H-weights modulo $2\mathbb{Z}$ defines a $\mathbb{C}/2\mathbb{Z}$-grading $\mathcal{C} = \bigoplus_{\overline{\alpha} \in \mathbb{C}/2\mathbb{Z}} \mathcal{C}_{\overline{\alpha}}$ which is compatible with the rigid monoidal structure. While the category $\mathcal{C}$ is not semisimple, it is *generically semisimple* in the sense that most homogeneous subcategories $\mathcal{C}_{\overline{\alpha}} \subset \mathcal{C}$ are semisimple. More precisely, we prove in Theorem 1.13 that $\mathcal{C}_{\overline{\alpha}}$ is semisimple unless $\overline{\alpha} \in \mathbb{Z}/2\mathbb{Z} \subset \mathbb{C}/2\mathbb{Z}$. Proposition 1.17 asserts that $\mathcal{C}$ is braided, a complete proof of which does not seem to be in the literature. The proof we present is elementary and self-contained. The form of the braiding is motivated by the well-known universal R-matrix of the $\hbar$-adic quantum group of $\mathfrak{sl}_2(\mathbb{C})$ [**Dri86, Oht02**]. In Theorem 1.19 we prove that $\mathcal{C}$ is ribbon. The candidate ribbon structure is based on the twist associated to the rigid monoidal structure, namely the right partial trace of the braiding. We use the generic semisimplicity of $\mathcal{C}$ to prove that this twist is compatible with duality by checking that this is so generically and concluding, via a general result of [**GPM18**], that this compatibility holds for the entirety of $\mathcal{C}$. The results of Section 1 can be summarized as follows.

THEOREM A. *The category $\mathcal{C}$ is a $\mathbb{C}/2\mathbb{Z}$-graded generically semisimple ribbon category.*

In Section 2 we recall standard material related to the Reshetikhin–Turaev functor $\mathcal{F}_{\mathcal{D}} : \mathbf{Rib}_{\mathcal{D}} \to \mathcal{D}$ associated to a ribbon category $\mathcal{D}$. Central to the renormalized theory is the well-known statement, proved in this paper as Lemma 2.4, that if $V \in \mathcal{D}$ is a simple object of a $\mathbb{C}$-linear ribbon category, L is a $\mathcal{D}$-colored link and T is a $(1,1)$-tangle whose closure is L and whose open strand colored by V, then

$$\mathcal{F}_{\mathcal{D}}(L) = \operatorname{qdim}_{\mathcal{D}}(V)\mathcal{F}_{\mathcal{D}}(T).$$

Here both sides of the equation are identified with the scalar by which they act. In particular, if $\operatorname{qdim}_{\mathcal{D}}(V) = 0$, then $\mathcal{F}_{\mathcal{D}}(L)$ vanishes, while $\mathcal{F}_{\mathcal{D}}(T)$ need not. We prove in Proposition 2.6 that it, if L is a knot, then $\mathcal{F}_{\mathcal{D}}(T)$ is an isotopy invariant of L.

In Section 3 we extend the invariant $L \mapsto \mathcal{F}_{\mathcal{D}}(T)$ from framed knots to framed links. To clarify the exposition, we restrict attention to $\mathcal{D} = \mathcal{C}$, the category of weight $\overline{U}_q^H(\mathfrak{sl}_2(\mathbb{C}))$-modules. The main obstacle in this extension is that cutting a link L with multiple components produces a $(1,1)$-tangle whose isotopy type depends on the component cut. Ambidextrous modules are the key to overcoming this obstacle. A simple $\overline{U}_q^H(\mathfrak{sl}_2(\mathbb{C}))$-module V is called *ambidextrous* if the equality

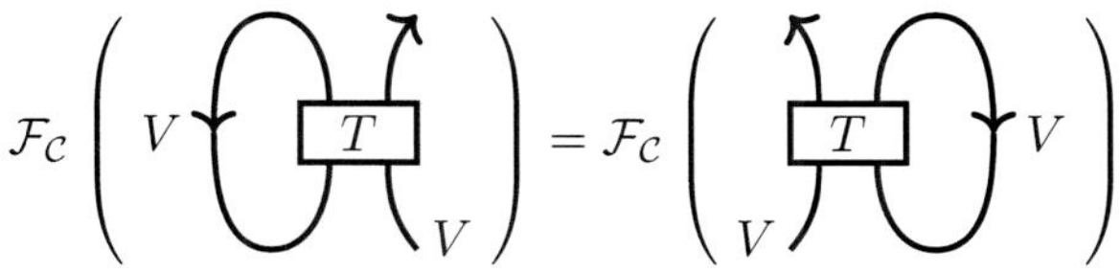

of endomorphisms of V holds for all (2,2)-tangles T whose open strands are colored by V. We prove in Corollary 3.12 that all simple $\overline{U}_q^H(\mathfrak{sl}_2(\mathbb{C}))$-modules are ambidextrous.

Define a function $S' : \mathbb{C} \times \mathbb{C} \to \mathbb{C}$ by

$$S'(\beta,\alpha) = \mathcal{F}_{\mathcal{C}}$$

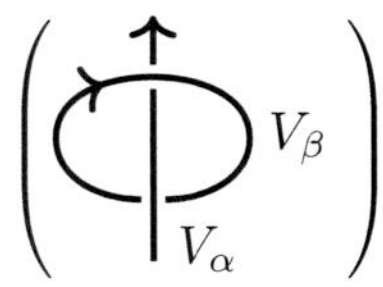

$$\in \mathrm{End}_{\mathcal{C}}(V_\alpha) \simeq \mathbb{C}.$$

For a fixed ambidextrous Verma module V_η, the *modified quantum dimension* of a simple Verma module V_α is defined to be $\mathbf{d}_\eta(\alpha) = \frac{S'(\alpha,\eta)}{S'(\eta,\alpha)}$. The main result of this paper can be stated as follows.

THEOREM B ((Theorem 3.13)). *Let $V_\eta \in \mathcal{C}$ be an ambidextrous Verma module and L a framed link with at least one component colored by V_α for some $\alpha \in \mathbb{C}\backslash\mathbb{Z}\cup r\mathbb{Z}$. Then the assignment*

$$L \mapsto \mathcal{F}'_\eta(L) := \mathbf{d}_\eta(\alpha)\mathcal{F}_{\mathcal{C}}(T),$$

where T is a $(1,1)$-tangle whose closure is L and whose open strand is colored by V_α, is a well-defined isotopy invariant of colored framed links.

In Section 3.4 we discuss some basic properties of the renormalized invariant $\mathcal{F}'_\eta$, such as its behavior under connect sum and its associated skein relations, and compute some renormalized invariants for knots with few crossings. We also show that renormalizations with respect to different ambidextrous Verma modules lead to invariants which differ by a global scalar.

Finally, in Section 4 we present a brief guide to further mathematical and physical applications of renormalized Reshetikhin–Turaev theory.

Conventions. The ground field is $\mathbb{C}$. Write $\otimes$ for $\otimes_{\mathbb{C}}$. All modules are left modules and finite dimensional over $\mathbb{C}$. Any categorical notion regarding monoidal categories is in congruence with [**EGNO15**]. Given a scalar endomorphism e of a vector space V, define $\langle e\rangle \in \mathbb{C}$ by $e = \langle e\rangle \cdot \mathrm{id}_V$.

1. The unrolled quantum group $\overline{U}_q^H(\mathfrak{sl}_2(\mathbb{C}))$ and its weight modules

Fix an integer $r \geq 2$. Set $q = e^{\frac{\pi\sqrt{-1}}{r}}$. For $z \in \mathbb{C}$, define

$$q^z = e^{\frac{\pi\sqrt{-1}z}{r}}, \qquad \{z\} = q^z - q^{-z}, \qquad [z] = \frac{\{z\}}{\{1\}}.$$

Set $\{0\}! = 1$ and $\{n\}! = \prod_{i=1}^n \{i\}$ for $n \in \mathbb{Z}_{>0}$, and similarly for $[n]!$. For $0 \leq k \leq l$, set $\begin{bmatrix} l \\ k \end{bmatrix} = \frac{[l]!}{[k]![l-k]!}$.

1.1. The unrolled quantum group of $\mathfrak{sl}_2(\mathbb{C})$. We recall the definition of the unrolled quantum group of $\mathfrak{sl}_2(\mathbb{C})$, as introduced in [**GPMT09, CGPM15b**]. Pre-cursors of the unrolled quantum group appear in work of Ohtuski [**Oht02**].

DEFINITION 1.1. The *unrolled quantum group of* $\mathfrak{sl}_2(\mathbb{C})$ is the unital associative algebra $U_q^H(\mathfrak{sl}_2(\mathbb{C}))$ generated by K, K^{-1}, H, E and F with relations

$$KK^{-1} = K^{-1}K = 1, \qquad HK = KH,$$

$$HE - EH = 2E, \qquad HF - FH = -2F,$$

$$KE = q^2EK, \qquad KF = q^{-2}FK,$$

$$EF - FE = \frac{K - K^{-1}}{q - q^{-1}}.$$

The *restricted unrolled quantum group* $\overline{U}_q^H(\mathfrak{sl}_2(\mathbb{C}))$ is the quotient of $U_q^H(\mathfrak{sl}_2(\mathbb{C}))$ by the relations $E^r = F^r = 0$.

Informally, the generator H should be viewed as a logarithm of K. While this constraint is not imposed at the level of algebras, it is imposed on the modules of interest in this paper. See Section 1.2.

Both $U_q^H(\mathfrak{sl}_2(\mathbb{C}))$ and $\overline{U}_q^H(\mathfrak{sl}_2(\mathbb{C}))$ are Hopf algebras with coproduct, counit and antipode defined by

$$\begin{aligned}
\Delta(E) &= 1 \otimes E + E \otimes K, & \varepsilon(E) &= 0, & S(E) &= -EK^{-1},\\
\Delta(F) &= F \otimes 1 + K^{-1} \otimes F, & \varepsilon(F) &= 0, & S(F) &= -KF,\\
\Delta(K) &= K \otimes K, & \varepsilon(K) &= 1, & S(K) &= K^{-1},\\
\Delta(H) &= H \otimes 1 + 1 \otimes H, & \varepsilon(H) &= 0, & S(H) &= -H.
\end{aligned}$$

The De Concini–Kac quantum group $U_q(\mathfrak{sl}_2(\mathbb{C}))$ is isomorphic to the Hopf subalgebra of $U_q^H(\mathfrak{sl}_2(\mathbb{C}))$ generated by $K^{\pm 1}$, E and F. Similarly, the restricted quantum group $\overline{U}_q(\mathfrak{sl}_2(\mathbb{C}))$ is isomorphic to the Hopf subalgebra of $\overline{U}_q^H(\mathfrak{sl}_2(\mathbb{C}))$ generated by $K^{\pm 1}$, E and F. The algebra $\overline{U}_q^H(\mathfrak{sl}_2(\mathbb{C}))$ shares many properties with $\overline{U}_q(\mathfrak{sl}_2(\mathbb{C}))$. For example, $\overline{U}_q^H(\mathfrak{sl}_2(\mathbb{C}))$ has a Poincaré–Birkhoff–Witt basis

$$\{F^aH^bK^cE^d \mid 0 \leq a, d \leq r-1,\ b \in \mathbb{Z}_{\geq 0},\ c \in \mathbb{Z}\}$$

and admits a triangular decomposition

$$\overline{U}_q^{H,-}(\mathfrak{sl}_2(\mathbb{C})) \otimes \overline{U}_q^{H,0}(\mathfrak{sl}_2(\mathbb{C})) \otimes \overline{U}_q^{H,+}(\mathfrak{sl}_2(\mathbb{C})) \xrightarrow{\sim} \overline{U}_q^H(\mathfrak{sl}_2(\mathbb{C}))$$

where

$$\overline{U}_q^{H,-}(\mathfrak{sl}_2(\mathbb{C})), \overline{U}_q^{H,0}(\mathfrak{sl}_2(\mathbb{C})) \text{and} \overline{U}_q^{H,+}(\mathfrak{sl}_2(\mathbb{C}))$$

are the subalgebras of $\overline{U}_q^H(\mathfrak{sl}_2(\mathbb{C}))$ generated by F, H and $K^{\pm 1}$ and E, respectively. For later use, let $\overline{U}_q^H(\mathfrak{b})$ be the Hopf subalgebra of $\overline{U}_q^H(\mathfrak{sl}_2(\mathbb{C}))$ generated by H, $K^{\pm 1}$ and E.

1.2. Weight modules. Recall that all modules are assumed to be finite dimensional.

DEFINITION 1.2. Let V be a $\overline{U}_q^H(\mathfrak{sl}_2(\mathbb{C}))$-module.

(1) A *weight vector of weight* $\lambda \in \mathbb{C}$ is a non-zero vector $v \in V$ which satisfies $Hv = \lambda v$. If, moreover, $Ev = 0$, then v is called a *highest weight vector*. The subspace $V[\lambda] = \{v \in V \mid Hv = \lambda v\}$ is called the *weight space of weight* λ.
(2) The module V is called a *weight module* if it is the direct sum of its weight spaces, $V = \bigoplus_{\lambda \in \mathbb{C}} V[\lambda]$, and $Kv = q^{\lambda} v$ for all $v \in V[\lambda]$.
(3) The module V is called a *highest weight module* if it is generated by a highest weight vector.

All $\overline{U}_q^H(\mathfrak{sl}_2(\mathbb{C}))$-modules considered in this paper are assumed to be weight modules. The second condition in Definition 1.2(2) can be written as the equality as operators $K = q^H$ on V. In view of this, when speaking of weight modules we often give the action of H and omit that of K. Finally, note that a highest weight module is necessarily a weight module.

Let $\mathcal{C}$ be the category of weight $\overline{U}_q^H(\mathfrak{sl}_2(\mathbb{C}))$-modules and their $\overline{U}_q^H(\mathfrak{sl}_2(\mathbb{C}))$-linear maps. The category $\mathcal{C}$ is $\mathbb{C}$-linear, locally finite and abelian. The bialgebra structure of $\overline{U}_q^H(\mathfrak{sl}_2(\mathbb{C}))$ makes $\mathcal{C}$ into a monoidal category with unit object the one dimensional module $\mathbb{C}$ on which H, E and F act by zero. The associators and unitors of the monoidal structure are as for the category of complex vector spaces and are henceforth suppressed from the notation.

Let $V \in \mathcal{C}$. Denote by $V^{\vee} \in \mathcal{C}$ the dual vector space $\mathrm{Hom}_{\mathbb{C}}(V, \mathbb{C})$ with $\overline{U}_q^H(\mathfrak{sl}_2(\mathbb{C}))$-module structure given by

$$(x \cdot f)(v) = f(S(x)v), \qquad x \in \overline{U}_q^H(\mathfrak{sl}_2(\mathbb{C})), \qquad f \in V^{\vee}, \qquad v \in V.$$

Given a basis $\{v_i\}_{i=1}^n$ of V with dual basis $\{v_i^{\vee}\}_{i=1}^n$ of $V^{\vee}$, define

$$\widehat{\mathrm{ev}}_V : V \otimes V^{\vee} \to \mathbb{C}, \qquad v \otimes f \mapsto f(K^{1-r} v)$$

and

$$\widehat{\mathrm{coev}}_V : \mathbb{C} \to V^{\vee} \otimes V, \qquad 1 \mapsto \sum_{i=1}^{n} K^{r-1} v_i^{\vee} \otimes v_i.$$

Despite its definition, $\widehat{\mathrm{coev}}_V$ is independent of the choice of basis. A direct check shows that $\widehat{\mathrm{ev}}_V$ and $\widehat{\mathrm{coev}}_V$ are $\overline{U}_q^H(\mathfrak{sl}_2(\mathbb{C}))$-linear and satisfy the snake relations: the compositions

$$V \xrightarrow{\mathrm{id}_V \otimes \widehat{\mathrm{coev}}_V} V \otimes V^{\vee} \otimes V \xrightarrow{\widehat{\mathrm{ev}}_V \otimes \mathrm{id}_V} V$$

and

$$V^{\vee} \xrightarrow{\widehat{\mathrm{coev}}_V \otimes \mathrm{id}_{V^{\vee}}} V^{\vee} \otimes V \otimes V^{\vee} \xrightarrow{\mathrm{id}_{V^{\vee}} \otimes \widehat{\mathrm{ev}}_V} V^{\vee} \tag{1.1}$$

are the respective identities. It follows that $\widehat{\mathrm{ev}}_V$ and $\widehat{\mathrm{coev}}_V$ are right duality morphisms. Define also

$$\mathrm{ev}_V : V^{\vee} \otimes V \to \mathbb{C}, \qquad f \otimes v \mapsto f(v)$$

and

$$\mathrm{coev}_V : \mathbb{C} \to V \otimes V^{\vee}, \qquad 1 \mapsto \sum_{i=1}^{n} v_i \otimes v_i^{\vee}.$$

These are the usual left duality morphisms in the category of finite dimensional vector spaces and are easily verified to be $\overline{U}_q^H(\mathfrak{sl}_2(\mathbb{C}))$-linear. It follows that the category $\mathcal{C}$ is rigid. Hence, $\mathcal{C}$ is tensor in the sense of [**EGNO15**, Definition 4.1.1].

Given a finite dimensional vector space V, write $V \to V^{\vee\vee}$, $v \mapsto \langle -, v\rangle$, for the canonical evaluation isomorphism. Explicitly, $\langle f, v\rangle = f(v)$ for $f \in V^{\vee}$ and $v \in V$.

LEMMA 1.3. *The maps* $\{p_V : V \to V^{\vee\vee}\}_{V\in\mathcal{C}}$ *given by* $p_V(v) = K^{1-r}\langle -, v\rangle$ *define a pivotal structure on* $\mathcal{C}$.

PROOF. We need to verify that $\{p_V\}_{V\in\mathcal{C}}$ are the components of a monoidal natural isomorphism $p : \mathrm{id}_{\mathcal{C}} \Rightarrow (-)^{\vee} \circ (-)^{\vee}$. Naturality is immediate and a direct check shows that p_V is $\overline{U}_q^H(\mathfrak{sl}_2(\mathbb{C}))$-linear. The relation $\Delta(K) = K \otimes K$ implies by induction the relation $\Delta(K^{1-r}) = K^{1-r} \otimes K^{1-r}$. This in turn implies the equality $p_{V\otimes W} = p_V \otimes p_W$, $V, W \in \mathcal{C}$, which is the required monoidality. □

One can readily see that the left and right duality structures are compatible with the pivotal structure, in the sense that the equalities $\mathrm{id}_{V^\vee} \otimes (p_V \circ \widehat{\mathrm{coev}}_V) = \mathrm{coev}_{V^\vee}$ and $\widehat{\mathrm{ev}}_V = (\mathrm{ev}_{V^\vee} \circ p_V) \otimes \mathrm{id}_{V^\vee}$ hold for all $V \in \mathcal{C}$.

1.3. Simple modules. Recall that a non-zero module $V \in \mathcal{C}$ is called *simple* (or *irreducible*) if it has no non-zero proper submodules. In this section, we classify simple objects of $\mathcal{C}$. The results of this section are contained in [**CGPM15b**, §5], although we give different proofs.

LEMMA 1.4. *Every simple object of* $\mathcal{C}$ *is a highest weight module.*

PROOF. Let $V \in \mathcal{C}$ be simple and $v \in V$ a weight vector. Since $E^r = 0$, there exists a minimal integer $l > 0$ such that $E^l v = 0$. Then $E^{l-1}v$ is a highest weight vector and $\overline{U}_q^H(\mathfrak{sl}_2(\mathbb{C})) \cdot E^{l-1}v \subset V$ is a non-zero submodule which, by simplicity, is equal to V. □

For each $\alpha \in \mathbb{C}$, let $\mathbb{C}_{\alpha+r-1}$ be the one dimensional weight $\overline{U}_q^H(\mathfrak{b})$-module of H-weight $\alpha + r - 1$ on which E acts by zero.

DEFINITION 1.5. The *Verma module of highest weight* $\alpha+r-1$ is the $\overline{U}_q^H(\mathfrak{sl}_2(\mathbb{C}))$-module $V_\alpha = \overline{U}_q^H(\mathfrak{sl}_2(\mathbb{C})) \otimes_{\overline{U}_q^H(\mathfrak{b})} \mathbb{C}_{\alpha+r-1}$.

Write v_i for the vector $F^i \otimes 1 \in V_\alpha$. The Poincaré–Birkhoff–Witt basis of $\overline{U}_q^H(\mathfrak{sl}_2(\mathbb{C}))$ shows that $\{v_0, \dots, v_{r-1}\}$ is a basis of V_α. Direct calculations show that the $\overline{U}_q^H(\mathfrak{sl}_2(\mathbb{C}))$-action on V_α is given by

$$Hv_i = (\alpha + r - 1 - 2i)v_i, \quad Ev_i = \frac{\{i\}\{i-\alpha\}}{\{1\}^2} v_{i-1}, \quad Fv_i = v_{i+1},$$

where by convention $v_{-1} = v_r = 0$. In particular, $V_\alpha \in \mathcal{C}$ is a highest weight module generated by v_0. The structure of V_α is summarized by the diagram

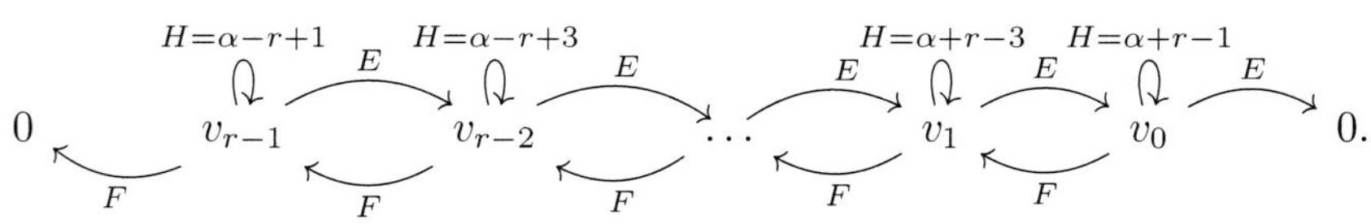

LEMMA 1.6. *If* V *is a highest weight module of highest weight* $\alpha + r - 1$, *then there exists a surjection* $V_\alpha \twoheadrightarrow V$.

PROOF. By Hom-tensor adjunction, there is an isomorphism

$$\mathrm{Hom}_{\mathcal{C}}(V_\alpha, V) \simeq \mathrm{Hom}_{\overline{U}_q^H(\mathfrak{b})}(\mathbb{C}_{\alpha+r-1}, V_{|\overline{U}_q^H(\mathfrak{b})}).$$

It follows that $\mathrm{Hom}_{\mathcal{C}}(V_\alpha, V)$ is isomorphic to the subspace of highest weight vectors of weight $\alpha+r-1$ in V. In particular, if $v \in V$ is a generating highest weight vector of weight $\alpha+r-1$, then the assignment $v_0 \mapsto v$ extends to a surjective morphism $V_\alpha \to V$ in $\mathcal{C}$. □

Using Lemma 1.6, it is straightforward to verify that the map $v^\vee_{\alpha,r-1} \mapsto v_{-\alpha,0}$ extends to a $\overline{U}_q^H(\mathfrak{sl}_2(\mathbb{C}))$-module isomorphism

$$V_\alpha^\vee \xrightarrow{\sim} V_{-\alpha}. \tag{1.2}$$

PROPOSITION 1.7. *Let $\alpha \in \mathbb{C}$.*

(1) *If $\alpha \notin \mathbb{Z} \setminus r\mathbb{Z}$, then V_α is simple.*
(2) *If $\alpha \in \mathbb{Z} \setminus r\mathbb{Z}$ is written in its unique form as $\alpha = (l-1)r+n+1$ with $0 \leq n \leq r-2$ and $l \in \mathbb{Z}$, then there exists a non-split short exact sequence*

$$0 \to S^{(l-1)r}_{r-n-2} \to V_\alpha \to S^{lr}_n \to 0 \tag{1.3}$$

which is a Jordan–Hölder filtration of V_α.
(3) *Any simple object of $\mathcal{C}$ is isomorphic to a unique module of the form V_α, $\alpha \in \mathbb{C} \setminus \mathbb{Z} \cup r\mathbb{Z}$, or S^{lr}_n, $l \in \mathbb{Z}$, $0 \leq n \leq r-2$.*

PROOF. If $\alpha \notin \mathbb{Z} \setminus r\mathbb{Z}$, then $\frac{\{i\}\{i-\alpha\}}{\{1\}^2} \neq 0$ for $i = 1, \ldots, r-1$, as follows from the assumption that q is a primitive $2r^{\text{th}}$ root of unity. It follows from the explicit form of the action of E on V_α that $Ev_i \neq 0$ for $i = 1, \ldots, r-1$, whence V_α is simple.

If instead $\alpha \in \mathbb{Z} \setminus r\mathbb{Z}$, then V_α has exactly one proper submodule. Indeed, write $\alpha = (l-1)r+n+1$ as in the statement of the proposition, so that V_α is of highest weight $lr+n$. Examining the action of E on V_α shows that $Ev_{n+1} = 0$ and $Ev_i \neq 0$ if $i \neq 0, n+1$. Hence, $S := \mathrm{span}\{v_{n+1}, \cdots, v_{r-1}\}$ is the unique proper submodule of V_α. The module S has dimension $r-n-1$ and its quotient $S^{lr}_n := V_\alpha/S$ is a simple highest weight module of highest weight $lr+n$ and dimension $n+1$. By Lemma 1.6, there exists a surjection $V_{(l-1)r-n-1} \to S$ which, by the argument of this paragraph, descends to an isomorphism $S^{(l-1)r}_{r-n-2} \xrightarrow{\sim} S$. Finally, the uniqueness of S implies that the sequence (1.3) is non-split.

By Lemmas 1.4 and 1.6, any simple object of $\mathcal{C}$ is a quotient of a Verma module. Thus, the third statement of the proposition follows from the first two. □

REMARK 1.8. Since $\overline{U}_q(\mathfrak{sl}_2(\mathbb{C}))$ is a Hopf subalgebra of $\overline{U}_q^H(\mathfrak{sl}_2(\mathbb{C}))$, there is a monoidal forgetful functor $\mathcal{C} \to \overline{U}_q(\mathfrak{sl}_2(\mathbb{C}))$-mod. In the notation of [**Jan96**, §2.11], this functor sends the simple objects S^{lr}_n and V_α of $\mathcal{C}$ to $L(n, (-1)^l)$ and $Z_0(q^{\alpha+r-1})$, respectively.

Proposition 1.7 implies that the category $\mathcal{C}$ has Properties (P2) and (P3) from the introduction. Moreover, a simple object is determined up to isomorphism by its highest weight and the simple objects S^{lr}_n are neither injective nor projective.

PROPOSITION 1.9. *If $\alpha \in \mathbb{C} \setminus \mathbb{Z} \cup r\mathbb{Z}$, then the Verma module $V_\alpha \in \mathcal{C}$ is projective and injective.*

PROOF. Let $f : V \twoheadrightarrow W$ be a surjection in $\mathcal{C}$ and $\phi : V_\alpha \to W$ a non-zero morphism. By the proof of Lemma 1.6, the map ϕ is determined by a highest weight vector $\phi(v_0) = w \in W$ of weight $\alpha + r - 1$. Surjectivity of f implies that w has a preimage under f, say v, which is of weight $\alpha + r - 1$ and satisfies $Ev \in \ker f$.

Let $\xi = Ev$, which is of weight $\alpha + r + 1$ and satisfies $E^{r-1}\xi = 0$. For any $a_0, \dots, a_{r-2} \in \mathbb{C}$, the vector

$$v' = v + \sum_{i=0}^{r-2} a_i F^{i+1} E^i \xi$$

is of weight $\alpha + r - 1$ and satisfies $f(v') = w$. Using [**Jan96**, §1.3], we compute

$$Ev' = \xi + \sum_{i=0}^{r-2} a_i (F^{i+1}E^{i+1}\xi + [i+1][\alpha + r + 1 + i] F^i E^i \xi).$$

Then $Ev' = 0$ if and only if the recursive system of equations

$$a_i = -[i+2][\alpha + r + 2 + i] a_{i+1} \qquad i = -1, \dots, r-3,$$

holds, with $a_{-1} = 1$. This recursive system determines $\{a_i\}_i$ if and only if $\alpha \in \mathbb{C} \setminus \mathbb{Z} \cup r\mathbb{Z}$, as otherwise the coefficient of some a_{i+1} vanishes. Arguing as in the beginning of the proof, the assignment $v_0 \mapsto v'$ determines a $\overline{U}_q^H(\mathfrak{sl}_2(\mathbb{C}))$-linear map $\tilde{\phi} : V_\alpha \to V$ which satisfies $f \circ \tilde{\phi} = \phi$. This establishes the projectivity of V_α.

In view of the isomorphism (1.2) and the previous paragraph, the module $V_\alpha^\vee$ is projective. Standard adjunction isomorphisms (see [**EGNO15**, Proposition 2.10.8]) give a natural isomorphism of contravariant functors

$$\mathrm{Hom}_{\mathcal{C}}(-, V_\alpha) \simeq \mathrm{Hom}_{\mathcal{C}}(V_\alpha^\vee, -) \circ (-)^\vee.$$

Because $V_\alpha^\vee$ is projective, $\mathrm{Hom}_{\mathcal{C}}(V_\alpha^\vee, -)$ is an exact functor. Because $(-)^\vee$ is an exact functor at the level of complex vector spaces, it is also exact on $\mathcal{C}$. Hence, the functor $\mathrm{Hom}_{\mathcal{C}}(-, V_\alpha)$ is exact and V_α is injective. □

1.4. Generic semisimplicity. Recall that an abelian category is called *semisimple* if every object is a direct sum of simple objects. In view of Proposition 1.7(2), the category $\mathcal{C}$ is not semisimple. However, $\mathcal{C}$ fails to be semisimple in a controlled manner. The goal of this section is to make the latter statement precise. To do so, we begin with some general definitions from [**GPM18**].

Let G be an additive abelian group.

DEFINITION 1.10. A *G-grading* on a rigid monoidal category $\mathcal{D}$ is the data of non-empty full subcategories $\mathcal{D}_g \subset \mathcal{D}$, $g \in G$, such that $\mathcal{D} = \bigoplus_{g \in G} \mathcal{D}_g$ and $V^\vee \in \mathcal{D}_{-g}$ and $V \otimes V' \in \mathcal{D}_{g+g'}$ whenever $V \in \mathcal{D}_g$ and $V' \in \mathcal{D}_{g'}$.

DEFINITION 1.11. A subset $X \subset G$ is called *symmetric* if $-X = X$ and *small* if $G \neq \bigcup_{i=1}^n (g_i + X)$ for all $g_1, \dots, g_n \in G$.

DEFINITION 1.12. A G-graded category $\mathcal{D}$ is called *generically semisimple with small symmetric subset* $X \subset G$ if $\mathcal{D}_g$ is semisimple whenever $g \in G \setminus X$. In this case, a simple object $V \in \mathcal{D}_g$ in degree $g \in G \setminus X$ is called *generic simple*.

Consider again $\mathcal{C}$, the category of weight modules over $\overline{U}_q^H(\mathfrak{sl}_2(\mathbb{C}))$. Let G be the additive group $\mathbb{C}/2\mathbb{Z}$. For each $\overline{\alpha} \in \mathbb{C}/2\mathbb{Z}$, let $\mathcal{C}_{\overline{\alpha}}$ be the full subcategory of $\mathcal{C}$ consisting of modules whose weights are in the class $\overline{\alpha}$. The explicit form of the Hopf algebra structure of $\overline{U}_q^H(\mathfrak{sl}_2(\mathbb{C}))$ shows that $\mathcal{C} = \bigoplus_{\overline{\alpha} \in \mathbb{C}/2\mathbb{Z}} \mathcal{C}_{\overline{\alpha}}$ is a $\mathbb{C}/2\mathbb{Z}$-grading.

THEOREM 1.13. *The $\mathbb{C}/2\mathbb{Z}$-graded category $\mathcal{C}$ is generically semisimple with small symmetric subset $\mathbb{Z}/2\mathbb{Z} \subset \mathbb{C}/2\mathbb{Z}$.*

PROOF. Let $\overline{\alpha} \in (\mathbb{C}/2\mathbb{Z}) \setminus (\mathbb{Z}/2\mathbb{Z})$ and $V \in \mathcal{C}_{\overline{\alpha}}$ a non-zero object. Then V contains a highest weight vector v of weight $\alpha \in \mathbb{C}$ for some α in the class $\overline{\alpha}$. The assumption on $\overline{\alpha}$ implies that the submodule generated by v is isomorphic to $V_{\alpha-r+1}$; see the proof of Proposition 1.7. The module $V_{\alpha-r+1}$ is injective by Proposition 1.9, whence there is a splitting $V \simeq V_{\alpha-r+1} \oplus V'$ for some $V' \in \mathcal{C}_{\overline{\alpha}}$ of dimension strictly less than that of V. An induction argument on the dimension of V then completes the proof. □

In view of Proposition 1.7, the generic simple objects of $\mathcal{C}$ are the Verma modules V_α with $\alpha \in \mathbb{C} \setminus \mathbb{Z}$.

1.5. Braiding. In this section, we construct a braiding on $\mathcal{C}$. The form of the braiding is motivated by the universal R-matrix for the $\hbar$-adic quantum group $U_\hbar(\mathfrak{sl}_2(\mathbb{C}))$, as described in [**Dri86**, §10], [**Oht02**, §§4.5 and A.2].

DEFINITION 1.14. The *r-truncated q-exponential map* is

$$\exp_q^{<r}(x) = \sum_{l=0}^{r-1} \frac{q^{l(l-1)/2}}{[l]!} x^l.$$

Let $V, W \in \mathcal{C}$ with weight bases $\{v_i\}_i$ and $\{w_j\}_j$ of weights $\{\lambda_i^v\}_i$ and $\{\lambda_j^w\}_j$, respectively. Define $q^{H\otimes H/2} \in \operatorname{End}_{\mathbb{C}}(V \otimes W)$ by

$$q^{H\otimes H/2}(v_i \otimes w_j) = q^{\lambda_i^v \lambda_j^w/2} v_i \otimes w_j$$

and $R \in \operatorname{End}_{\mathbb{C}}(V \otimes W)$ by

$$R = q^{H\otimes H/2} \circ \exp_q^{<r}(\{1\}E \otimes F) = q^{H\otimes H/2} \circ \sum_{l=0}^{r-1} \frac{\{1\}^{2l}}{\{l\}!} q^{l(l-1)/2} E^l \otimes F^l,$$

where $\exp_q^{<r}(\{1\}E \otimes F)$ is viewed as a $\mathbb{C}$-linear map via left multiplication. Finally, define $c_{V,W} \in \operatorname{Hom}_{\mathbb{C}}(V \otimes W, W \otimes V)$ by

$$c_{V,W}(v \otimes w) = \tau R(v \otimes w),$$

where τ is the swap map $V \otimes W \to W \otimes V$, $v \otimes w \mapsto w \otimes v$.

LEMMA 1.15. *The map $c_{V,W}$ is $\overline{U}_q^H(\mathfrak{sl}_2(\mathbb{C}))$-linear.*

PROOF. It suffices to check linearity of $c_{V,W}$ on the generators H, E and F. Let $v \in V$ and $w \in W$ of weight λ^v and λ^w, respectively. We have

$$\begin{aligned} H \cdot c_{V,W}(v \otimes w) &= \tau q^{H\otimes H/2} \sum_{l=0}^{r-1} \frac{q^{l(l-1)/2}}{[l]!} E^l \otimes F^l((H+2l)v \otimes w + v \otimes (H-2l)w) \\ &= (\lambda^v + \lambda^w) c_{V,W}(v \otimes w) \\ &= c_{V,W}(H \cdot v \otimes w). \end{aligned}$$

We prove E-linearity. We have an equality

$$K \otimes E \circ q^{H\otimes H/2} = q^{H\otimes H/2} \circ 1 \otimes E$$

in $\operatorname{End}_{\mathbb{C}}(V \otimes W)$. Indeed, we compute

$$K \otimes E \circ q^{H\otimes H/2}(v_i \otimes w_j) = q^{\lambda_i^v} q^{\lambda_i^v \lambda_j^w/2} v_i \otimes E w_j = q^{\lambda_i^v(\lambda_j^w+2)/2} v_i \otimes E w_j$$

and

$$q^{H\otimes H/2}\circ 1\otimes E(v_i\otimes w_j)=q^{H\otimes H/2}v_i\otimes Ew_j=q^{\lambda_i^v(\lambda_j^w+2)/2}v_i\otimes Ew_j.$$

A similar calculation shows that $E\otimes 1\circ q^{H\otimes H/2}=q^{H\otimes H/2}\circ E\otimes K^{-1}$. Using these two equalities, E-linearity of $c_{V,W}$ reduces to the equality

$$(E\otimes K^{-1}+1\otimes E)\exp_q^{<r}(\{1\}E\otimes F)=\exp_q^{<r}(\{1\}E\otimes F)(E\otimes K+1\otimes E)$$

in $\overline{U}_q^H(\mathfrak{sl}_2(\mathbb{C}))\otimes\overline{U}_q^H(\mathfrak{sl}_2(\mathbb{C}))$, which is proved in [**Oht02**, Equation A.10]. Linearity of F is proved similarly. □

LEMMA 1.16. *The map $c_{V,W}$ is invertible.*

PROOF. Clearly τ is invertible. We claim that the inverse of R is

$$R^{-1}=\exp_{q^{-1}}^{<r}(-\{1\}E\otimes F)\circ q^{-H\otimes H/2}.$$

Compare with [**Oht02**, §A.2]. We have $q^{H\otimes H/2}\circ q^{-H\otimes H/2}=1$. By definition,

$$\exp_q^{<r}(\{1\}E\otimes F)\exp_{q^{-1}}^{<r}(-\{1\}E\otimes F)$$
$$=\sum_{l=0}^{r-1}\sum_{k=0}^{r-1}\frac{q^{l(l-1)/2}q^{-k(k-1)/2}}{[l]![k]!}(-1)^k(\{1\}E\otimes F)^{l+k}.$$

Since $(E\otimes F)^r=0$, the double sum is

$$\sum_{i=0}^{r-1}\frac{q^{-i(i-1)/2}}{[i]!}(-\{1\}E\otimes F)^i\sum_{l=0}^{i}(-1)^l\begin{bmatrix}i\\l\end{bmatrix}q^{l(i-1)}.$$

The sum $\sum_{l=0}^{i}(-1)^l\left[\begin{smallmatrix}i\\l\end{smallmatrix}\right]q^{l(i-1)}$ is 0 for $i>0$ and 1 if $i=0$; see [**Jan96**, §0.2]. The inverse of R is thus as stated. □

PROPOSITION 1.17. *The maps $\{c_{V,W}:V\otimes W\to W\otimes V\}_{V,W}$ define a braiding on $\mathcal{C}$.*

PROOF. Lemmas 1.15 and 1.16 yield that the maps $c_{V,W}$ give a family of isomorphisms in $\mathcal{C}$. Naturality of $c_{V,W}$ follows from the fact that the endomorphism $q^{H\otimes H/2}$ commutes with $\overline{U}_q^H(\mathfrak{sl}_2(\mathbb{C}))$-linear maps. It remains to verify the hexagon axioms [**EGNO15**, Definition 8.1.1]. Let $V,W,U\in\mathcal{C}$. We prove that

$$c_{V,W\otimes U}=(\mathrm{id}_W\otimes c_{V,U})\circ(c_{V,W}\otimes\mathrm{id}_U)$$

and leave the verification of the equality $c_{V\otimes W,U}=(c_{V,U}\otimes\mathrm{id}_W)\circ(\mathrm{id}_V\otimes c_{W,U})$ to the reader. Let $v\in V,w\in W,u\in U$ be weight vectors of weights λ^v,λ^w, λ^u, respectively. We compute

$$c_{V,W\otimes U}(v\otimes w\otimes u)=q^{H\otimes H/2}\sum_{l=0}^{r-1}\frac{\{1\}^{2l}}{\{l\}!}q^{l(l-1)/2}\Delta(F^l)(w\otimes u)\otimes E^lv.$$

A straightforward induction argument shows that

$$\Delta(F^l)=\sum_{i=0}^{l}q^{i(l-i)}\begin{bmatrix}l\\i\end{bmatrix}F^iK^{-(l-i)}\otimes F^{l-i}.$$

Compare with [**Jan96**, §3.1], where slightly different conventions are used. Using this, we find

$$\begin{aligned} & c_{V,W\otimes U}(v\otimes w\otimes u) \\ &= q^{H\otimes H/2}\sum_{l=0}^{r-1}\frac{\{1\}^{2l}}{\{l\}!}q^{l(l-1)/2}\sum_{i=0}^{l}q^{(i-\lambda^w)(l-i)}\begin{bmatrix} l \\ i \end{bmatrix} F^i\otimes F^{l-i}\otimes E^l(w\otimes u\otimes v) \end{aligned}$$

which evaluates to

$$\sum_{l=0}^{r-1}\frac{\{1\}^{2l}}{\{l\}!}q^{l(l-1)/2}\sum_{i=0}^{l}q^{(\lambda^w+\lambda^u-2l)(\lambda^v+2l)/2}q^{(i-\lambda^w)(l-i)}\begin{bmatrix} l \\ i \end{bmatrix} F^i\otimes F^{l-i}\otimes E^l(w\otimes u\otimes v).$$

On the other hand, we compute

$$\begin{aligned} c_{V,W}\otimes \mathrm{id}_U(v\otimes w\otimes u) &= q^{H\otimes H/2}\sum_{l=0}^{r-1}\frac{\{1\}^{2l}}{\{l\}!}q^{l(l-1)/2}F^n w\otimes E^n v\otimes u \\ &= \sum_{l=0}^{r-1}\frac{\{1\}^{2l}}{\{l\}!}q^{l(l-1)/2}q^{(\lambda^w-2l)(\lambda^v+2l)/2}F^l w\otimes E^l v\otimes u. \end{aligned}$$

Applying $\mathrm{id}_W\otimes c_{V,U}$ then gives

$$\begin{aligned} \mathrm{id}_W\otimes c_{V,U}&(\sum_{l=0}^{r-1}\frac{\{1\}^{2l}}{\{l\}!}q^{l(l-1)/2}q^{(\lambda^w-2l)(\lambda^v+2l)/2}F^l w\otimes E^l v\otimes u) \\ &= \sum_{l=0}^{r-1}\sum_{k=0}^{r-1}\frac{\{1\}^{2l+2k}}{\{l\}!\{k\}!}q^{l(l-1)/2}q^{k(k-1)/2}q^{(\lambda^w-2l)(\lambda^v+2l)/2} \\ &\qquad\cdot q^{(\lambda^u-2k)(\lambda^v+2l+2k)/2}(F^l w\otimes F^k u\otimes E^k E^l v). \end{aligned}$$

To check the equality $c_{V,W\otimes U}(v\otimes w\otimes u) = c_{V,W}\otimes \mathrm{id}_U(v\otimes w\otimes u)$, we compare the coefficients of $F^a w\otimes F^b u\otimes E^{a+b}v$. The coefficients on the left- and right-hand sides of the desired equality are

$$\frac{\{1\}^{2(a+b)}}{\{a+b\}!}q^{(a+b)((a+b)-1)/2}q^{(\lambda^w+\lambda^u-2(a+b))(\lambda^v+2(a+b))/2}q^{(a-\lambda^w)((a+b)-a)}\begin{bmatrix} a+b \\ a \end{bmatrix}$$

and

$$\frac{\{1\}^{2a+2b}}{\{a\}!\{b\}!}q^{a(a-1)/2}q^{b(b-1)/2}q^{(\lambda^w-2a)(\lambda^v+2a)/2}q^{(\lambda^u-2b)(\lambda^v+2a+2b)/2},$$

respectively, which are equal by direct verification. □

1.6. Ribbon structure. In this section, we construct a ribbon structure on $\mathcal{C}$. Having already established that $\mathcal{C}$ is braided (Proposition 1.17), a ribbon structure is the additional data of a *twist*, that is, a natural automorphism of the identity functor $\theta : \mathrm{id}_{\mathcal{C}} \Rightarrow \mathrm{id}_{\mathcal{C}}$ which satisfies the *balancing condition*

$$\theta_{V\otimes W} = (\theta_V\otimes\theta_W)\circ c_{W,V}\circ c_{V,W}$$

and the *ribbon condition*

$$(\theta_V)^\vee = \theta_{V^\vee}$$

for all $V, W\in\mathcal{C}$.

Recall that the right partial trace of $f \in \mathrm{End}_{\mathcal{C}}(V \otimes W)$ is the endomorphism $\mathrm{ptr}_R(f) \in \mathrm{End}_{\mathcal{C}}(V)$ defined by

$$V \xrightarrow{\mathrm{id}_V \otimes \mathrm{coev}_W} V \otimes W \otimes W^\vee \xrightarrow{f \otimes \mathrm{id}_{W^\vee}} V \otimes W \otimes W^\vee \xrightarrow{\mathrm{id}_V \otimes \widetilde{\mathrm{ev}}_W} V.$$

Define a natural automorphism $\theta : \mathrm{id}_{\mathcal{C}} \Rightarrow \mathrm{id}_{\mathcal{C}}$ by

$$\theta_V := \mathrm{ptr}_R(c_{V,V}), \qquad V \in \mathcal{C} \tag{1.4}$$

where c is the braiding of $\mathcal{C}$. The hexagon axioms of the braiding ensure that θ satisfies the balancing condition. To verify that θ satisfies the ribbon condition, we use the following generic extension result.

THEOREM 1.18 ([**GPM18**, Theorem 9]). *Let $\mathcal{D}$ be a generically semisimple pivotal braided category. Define a natural automorphism $\theta : \mathrm{id}_{\mathcal{D}} \Rightarrow \mathrm{id}_{\mathcal{D}}$ so that its components are given by Equation (1.4). If $\theta_V^\vee = \theta_{V^\vee}$ for any generic simple object $V \in \mathcal{D}$, then θ is a twist on $\mathcal{D}$.*

We can now prove the main result of this section.

THEOREM 1.19. *The natural transformations c and θ equip $\mathcal{C}$ with the structure of a ribbon category.*

PROOF. Lemma 1.3, Theorem 1.13, and Proposition 1.17 imply that $\mathcal{C}$ satisfies the categorical assumptions of Theorem 1.18. Recall that the generic simple objects of $\mathcal{C}$ are the Verma modules V_α with $\alpha \in \mathbb{C} \setminus \mathbb{Z}$. For any $\alpha \in \mathbb{C}$ the right partial trace of c_{V_α, V_α} is

$$V_\alpha \xrightarrow{\mathrm{id}_{V_\alpha} \otimes \mathrm{coev}_{V_\alpha}} V_\alpha \otimes V_\alpha \otimes V_\alpha^\vee \xrightarrow{c_{V_\alpha, V_\alpha} \otimes \mathrm{id}_{V_\alpha^\vee}} V_\alpha \otimes V_\alpha \otimes V_\alpha^\vee \xrightarrow{\mathrm{id}_{V_\alpha} \otimes \widetilde{\mathrm{ev}}_{V_\alpha}} V_\alpha.$$

Since $\mathrm{End}_{\mathcal{C}}(V_\alpha) \simeq \mathbb{C}$ (see the proof of Lemma 1.6), in order describe θ_{V_α} it suffices to compute the image of the highest weight vector $v_0 \in V_\alpha$ under this composition. We have

$$\begin{aligned} v_0 \mapsto \sum_{i=0}^{r-1} v_0 \otimes v_i \otimes v_i^\vee \mapsto \sum_{i=0}^{r-1} q^{(\alpha+r-1)(\alpha+r-1-2i)/2} v_i \otimes v_0 \otimes v_i^\vee \\ \mapsto q^{(\alpha+r-1)(\alpha+r-1)/2} q^{(\alpha+r-1)(1-r)} v_0 = q^{(\alpha+r-1)(\alpha-r+1)/2} v_0. \end{aligned}$$

Since $V_\alpha^\vee \simeq V_{-\alpha}$ and the scalar $q^{(\alpha+r-1)(\alpha-r+1)/2}$ is unchanged under the substitution $\alpha \mapsto -\alpha$, it follows that $\theta_{V_\alpha^\vee} = \theta_{V_\alpha}^\vee$ for all $\alpha \in \mathbb{C}$. Theorem 1.18 therefore applies in the present setting, allowing the conclusion that the maps $\{\theta_V\}_{V \in \mathcal{C}}$ define a twist on $\mathcal{C}$. □

2. Reshetikhin–Turaev invariants

We recall background material on Reshetikhin–Turaev invariants of links [**RT90**]. For a detailed introduction to the theory, the reader is referred to [**Oht02, Tur16**]. We end this section by modifying the Reshetikhin–Turaev construction to produce a non-zero invariant for knots colored by simple objects of vanishing quantum dimension. Readers who are well-versed in Reshetikhin–Turaev theory could remind themselves of Lemma 2.4 and proceed directly to Section 3.

2.1. Reshetikhin–Turaev invariants of links. Let $\mathcal{D}$ be a ribbon category. Associated to $\mathcal{D}$ is the ribbon category $\mathbf{Rib}_{\mathcal{D}}$ of $\mathcal{D}$-colored ribbon graphs [**Tur16**, §I.I.2]. Objects of $\mathbf{Rib}_{\mathcal{D}}$ are finite sequences of pairs (V, ϵ), where $V \in \mathcal{D}$ and $\epsilon \in \{\pm\}$. Morphisms in $\mathbf{Rib}_{\mathcal{D}}$ are isotopy classes of $\mathcal{D}$-colored ribbon graphs bordering two such sequences of objects. The colorings of the ribbon graphs are required to be compatible with the domain and codomain objects in the obvious sense. Composition of morphisms is defined by concatenation of ribbon graphs. The monoidal structure of $\mathbf{Rib}_{\mathcal{D}}$ is defined on objects by concatenation of sequences and on morphisms by disjoint union.

THEOREM 2.1 ([**Tur16**, Theorem 2.5]). *There exists a unique ribbon functor* $\mathcal{F}_{\mathcal{D}} : \mathbf{Rib}_{\mathcal{D}} \to \mathcal{D}$ *such that* $\mathcal{F}_{\mathcal{D}}(V,+) = V$ *and* $\mathcal{F}_{\mathcal{D}}(V,-) = V^{\vee}$ *for all* $V \in \mathcal{D}$.

The functor $\mathcal{F}_{\mathcal{D}}$ is called the *Reshetikhin–Turaev functor*. The definition of the ribbon structure of $\mathbf{Rib}_{\mathcal{D}}$ and the fact that $\mathcal{F}_{\mathcal{D}}$ is ribbon implies that $\mathcal{F}_{\mathcal{D}}$ takes the following values on morphisms in $\mathbf{Rib}_{\mathcal{D}}$:

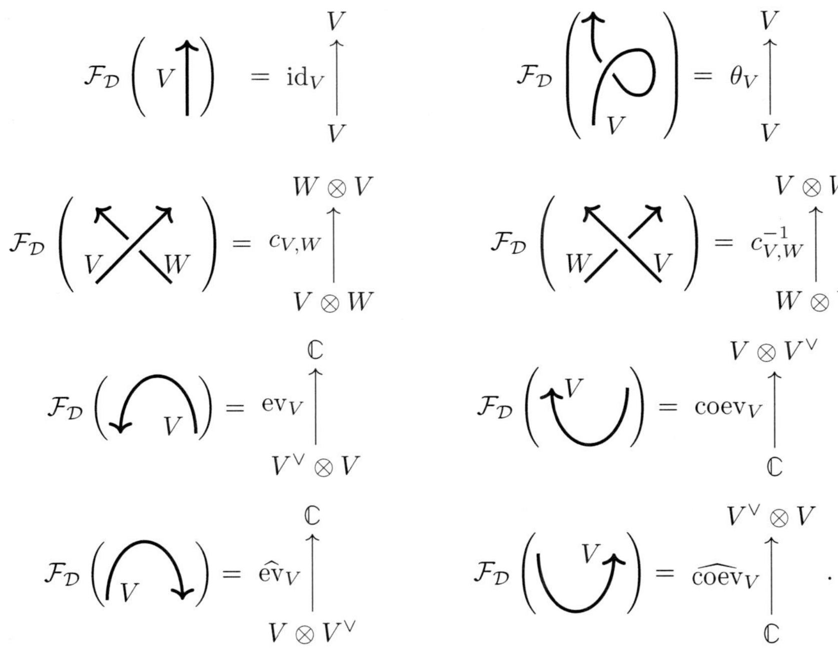

The above eight morphisms in $\mathbf{Rib}_{\mathcal{D}}$ generate all morphisms [**Tur16**, §I.3-4]. In particular, the value $\mathcal{F}_{\mathcal{D}}$ on any morphism of $\mathbf{Rib}_{\mathcal{D}}$ can be computed as an iterated composition of (co)evaluations, (inverse) braidings and (inverse) twists in $\mathcal{D}$. Isotopy classes of colored framed links are particular examples of morphisms in $\mathbf{Rib}_{\mathcal{D}}$- they are endomorphisms of the empty sequence. Thus, the assignment $L \mapsto \langle \mathcal{F}_{\mathcal{D}}(L) \rangle$ is an isotopy invariant of colored framed links.

We record the following result for later use.

LEMMA 2.2. *For any* $V, W \in \mathcal{D}$, *the following equality of morphisms in* $\mathcal{D}$ *holds:*

(2.1)

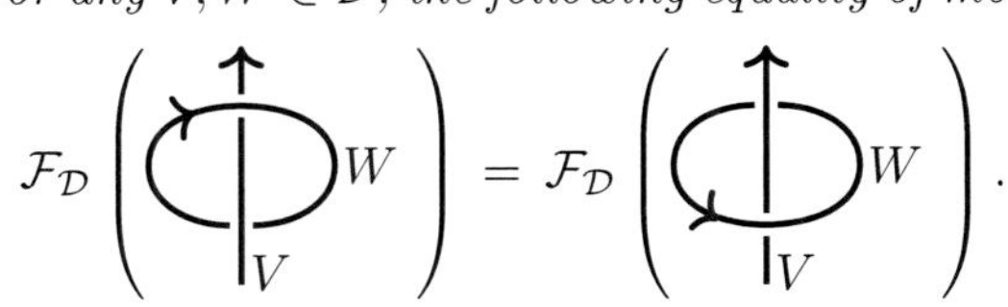

Moreover, if V is simple, then the following equality of scalars holds:

(2.2) 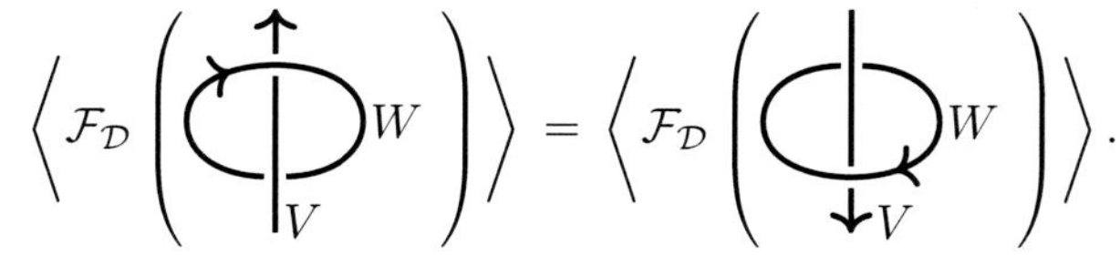

PROOF. Equation (2.1) holds by the following indicated combination of framed Reidemeister moves:

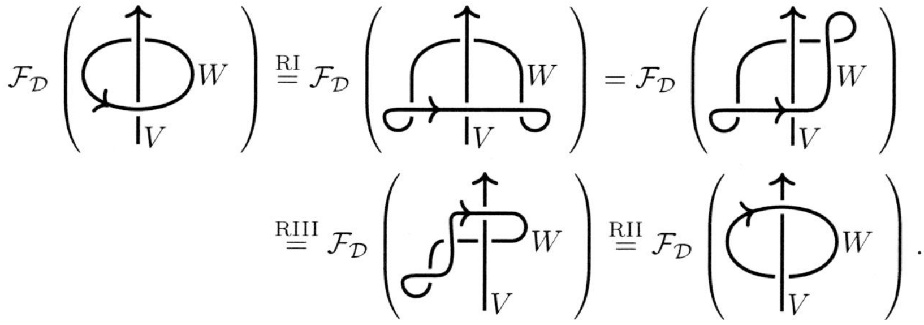

If V is simple then Equation (2.2) holds by planar isotopy:

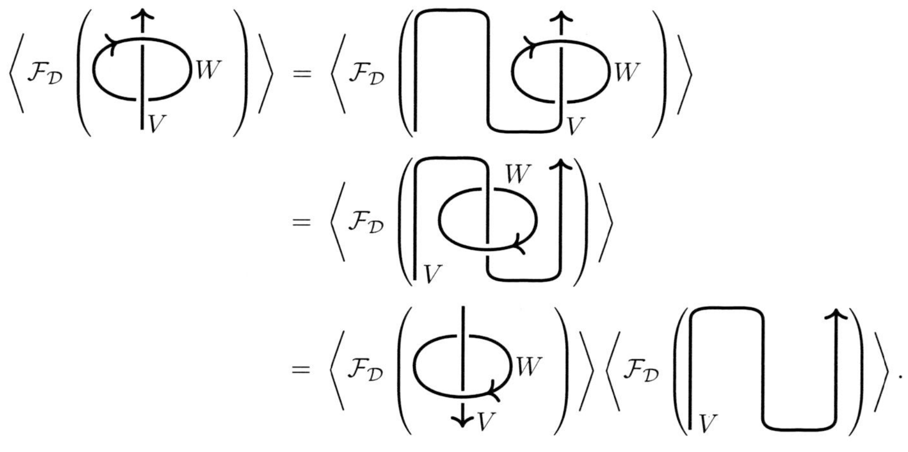

The snake relation (1.1) implies that the second scalar in the final line is equal to 1. □

2.2. Reshetikhin–Turaev invariants and quantum dimension. Let K be the unknot. Color K by an object V of a $\mathbb{C}$-linear ribbon category $\mathcal{D}$. The scalar $\langle \mathcal{F}_{\mathcal{D}}(K) \rangle$ associated to the linear map $\mathcal{F}_{\mathcal{D}}(K) : \mathbb{C} \to \mathbb{C}$ is called the *quantum dimension of* V and is denoted by $\mathrm{qdim}_{\mathcal{D}}(V)$. Explicitly, we have $\mathrm{qdim}_{\mathcal{D}}(V) = \langle \widehat{\mathrm{ev}}_V \circ \mathrm{coev}_V \rangle$.

EXAMPLE 2.3. Consider again the category $\mathcal{C}$ of weight $\overline{U}_q^H(\mathfrak{sl}_2(\mathbb{C}))$-modules. Let K be the unknot colored by V_α, $\alpha \in \mathbb{C}$. Let $\{v_i \mid 0 \leq i \leq r-1\}$ be the weight basis of V_α described in Section 1.3 with $\{v_i^\vee \mid 0 \leq i \leq r-1\}$ its dual basis. Then $\mathcal{F}_{\mathcal{C}}(K)$ is the composition

$$1 \xmapsto{\mathrm{coev}_{V_\alpha}} \sum_{i=0}^{r-1} v_i \otimes v_i^\vee \xmapsto{\widehat{\mathrm{ev}}_{V_\alpha}} \sum_{i=0}^{r-1} q^{(\alpha+r-1-2i)(1-r)} = q^{(\alpha+r-1)(1-r)} \sum_{i=0}^{r-1} q^{-2i+2ir}.$$

$T =$ 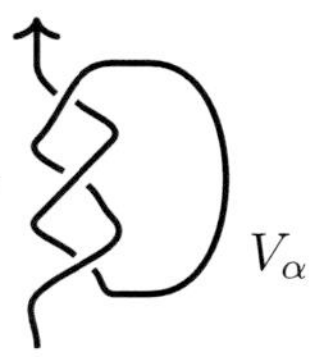

FIGURE 2. A $(1,1)$-tangle T whose closure is the right-handed trefoil.

As q is a primitive $2r^{\text{th}}$ root of unity, we have $\sum_{i=0}^{r-1} q^{-2i+2ir} = \sum_{i=0}^{r-1} q^{-2i} = 0$. Hence, $\mathcal{F}_{\mathcal{C}}(K) = 0$ and $\operatorname{qdim}_{\mathcal{C}}(V_\alpha) = 0$. The category $\mathcal{C}$ therefore has Property (P1) from the introduction. If instead K is colored by the simple module S_n^{lr}, $0 \leq n \leq r-2$ and $l \in \mathbb{Z}$, then

$$\langle \mathcal{F}_{\mathcal{C}}(K) \rangle = \sum_{j=0}^{n} v_j^{\vee}(K^{1-r} v_j) = q^{(1-r)(lr+n)} \sum_{j=0}^{n} q^{-2j} = (-1)^{n+l+lr}[n+1],$$

whence $\operatorname{qdim}_{\mathcal{C}}(S_n^{lr}) \neq 0$.

LEMMA 2.4. *Let $\mathcal{D}$ be a $\mathbb{C}$-linear ribbon category, $V \in \mathcal{D}$ a simple object, L a $\mathcal{D}$-colored link and T a $(1,1)$-tangle whose closure is L and whose open strand is colored by V. Then*

$$\langle \mathcal{F}_{\mathcal{D}}(L) \rangle = \operatorname{qdim}_{\mathcal{D}}(V) \langle \mathcal{F}_{\mathcal{D}}(T) \rangle. \tag{2.3}$$

PROOF. Using isotopy invariance we can draw a diagram of L of the form

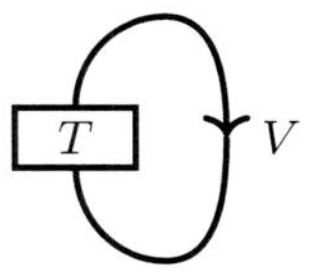

Since V is simple, the endomorphism $\mathcal{F}_{\mathcal{D}}(T)$ is a scalar and Equation (2.3) follows. □

Thus, whenever a knot is colored by a simple object of vanishing quantum dimension, the Reshetikhin–Turaev invariant is trivial. In particular, in view of Example 2.3, the Reshetikhin–Turaev invariants of $\mathcal{C}$-colored links with at least one component colored by a simple Verma module are zero.

2.3. Knot invariants via cutting. Equation (2.3) is the starting point of the theory of renormalized quantum invariants of [**GPMT09**]. The main idea is that even though $\operatorname{qdim}_{\mathcal{D}}(V)$, and hence $\mathcal{F}_{\mathcal{D}}(K)$, vanish, $\mathcal{F}_{\mathcal{D}}(T)$ need not and may provide an interesting invariant of K. In graphical language, to get a non-trivial invariant of a knot K we cut it to obtain a $(1,1)$-tangle T and apply the standard Reshetikhin–Turaev functor to T.

EXAMPLE 2.5. Let K be the right-handed trefoil knot colored by a Verma module $V_\alpha \in \mathcal{C}$. Example 2.3 shows that $\operatorname{qdim}_{\mathcal{C}}(V_\alpha) = 0$. It follows from Lemma 2.4 that $\mathcal{F}_{\mathcal{C}}(K) = 0$.

Let T be the $(1,1)$-tangle pictured in Figure 2. The closure of T is K. The endomorphism $\mathcal{F}_{\mathcal{C}}(T) \in \operatorname{End}_{\mathcal{C}}(V_\alpha)$ is the composition

$$V_\alpha \xrightarrow{\operatorname{id}_{V_\alpha} \otimes \operatorname{coev}_{V_\alpha}} V_\alpha \otimes V_\alpha \otimes V_\alpha^{\vee} \xrightarrow{(c_{V_\alpha, V_\alpha} \otimes \operatorname{id}_{V_\alpha^{\vee}})^{\circ 3}} V_\alpha \otimes V_\alpha \otimes V_\alpha^{\vee} \xrightarrow{\operatorname{id}_{V_\alpha} \otimes \widetilde{\operatorname{ev}}_{V_\alpha}} V_\alpha.$$

Since V_α is highest weight, $\mathcal{F}_{\mathcal{C}}(T)$ is determined by its value on a highest weight vector $v_0 \in V_\alpha$. Using the explicit form of the braiding, we compute

$$\langle \mathcal{F}_{\mathcal{C}}(T) \rangle = q^{\frac{3}{2}(\alpha+r-1)^2+(\alpha+r-1)(1-r)} \sum_{i=0}^{r-1} q^{i(-3\alpha-r+1)} \prod_{j=0}^{i-1} \{i-j-\alpha\}.$$

For example, when $r=2$ and $\alpha=2$, this specializes to $\langle \mathcal{F}_{\mathcal{C}}(T) \rangle = -3e^{\frac{\pi\sqrt{-1}}{4}} \neq 0$.

PROPOSITION 2.6. *The assignment $K \mapsto \langle \mathcal{F}_{\mathcal{D}}(T) \rangle$, where K is a colored framed knot and T is a $(1,1)$-tangle whose closure is K, is a well-defined invariant of colored framed knots.*

PROOF. This follows from Theorem 2.1 and the standard fact that two connected $(1,1)$-tangles T and T' are isotopic if and only if their closures are framed isotopic knots. $\square$

3. Renormalized Reshetikhin–Turaev invariants

For the remainder of the paper we restrict attention to the ribbon category $\mathcal{C}$ of weight $\overline{U}_q^H(\mathfrak{sl}_2(\mathbb{C}))$-modules and write $\mathcal{F}$ for the Reshetikhin–Turaev functor $\mathcal{F}_{\mathcal{C}}$.

3.1. Ambidextrous modules. The construction of a non-zero invariant from a knot by cutting to obtain a $(1,1)$-tangle does not immediately extend to links, as the following example shows.

EXAMPLE 3.1. Let $\alpha, \beta \in \mathbb{C}$ and L the Hopf link with components colored by V_α and V_β. Up to isotopy, there are two choices of how to cut L. Cutting the component colored by V_α gives

$\mathcal{F}$ 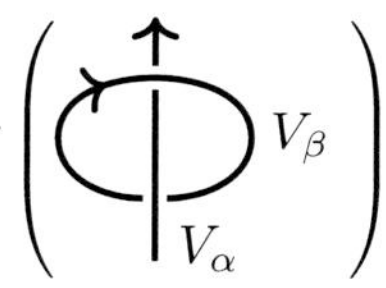

$$= \begin{cases} q^{\beta\alpha} \frac{\{\alpha r\}}{\{\alpha\}} \cdot \mathrm{id}_{V_\alpha} & \text{if } \alpha \in \mathbb{C} \setminus r\mathbb{Z}, \\ q^{\beta r z} \cdot (-1)^{(r+1)z} r \cdot \mathrm{id}_{V_\alpha} & \text{if } \alpha = rz \in r\mathbb{Z}. \end{cases}$$

Indeed, the map defined by the above tangle is the composition

$$V_\alpha \xrightarrow{\mathrm{coev}_{V_\beta}} V_\alpha \otimes V_\beta \otimes V_\beta^\vee \xrightarrow{c_{V_\alpha,V_\beta}} V_\beta \otimes V_\alpha \otimes V_\beta^\vee \xrightarrow{c_{V_\beta,V_\alpha}} V_\alpha \otimes V_\beta \otimes V_\beta^\vee \xrightarrow{\widetilde{\mathrm{ev}}_{V_\beta}} V_\alpha.$$

As in the proof of Theorem 1.19, it suffices to compute the image under this map of a highest weight vector $v_0 \in V_\alpha$. Let $\{w_i \mid 0 \le i \le r-1\}$ be a weight basis of V_β with dual basis $\{w_i^\vee \mid 0 \le i \le r-1\}$. Then we have under the above composition

$$v_0 \xmapsto{\mathrm{id}\otimes \mathrm{coev}_{V_\beta}} \sum_{i=0}^{r-1} v_0 \otimes w_i \otimes w_i^\vee \xmapsto{c_{V_\alpha,V_\beta}\otimes \mathrm{id}} \sum_{i=0}^{r-1} q^{(\alpha+r-1)(\beta+r-1-2i)/2} w_i \otimes v_0 \otimes w_i^\vee$$

$$\xmapsto{c_{V_\beta,V_\alpha}\otimes \mathrm{id}} \sum_{i=0}^{r-1} q^{(\alpha+r-1)(\beta+r-1-2i)} v_0 \otimes w_i \otimes w_i^\vee + \cdots$$

$$\xmapsto{\mathrm{id}\otimes \widetilde{\mathrm{ev}}_{V_\beta}} \sum_{i=0}^{r-1} q^{(\alpha+r-1)(\beta+r-1-2i)} q^{(\beta+r-1-2i)(1-r)} v_0,$$

where the omitted quantity $\cdots$ is a linear combination of terms of the form $E^j v_0 \otimes w_{i+j} \otimes w_i^\vee$, $j > 0$, and so is in the kernel of $\mathrm{id}_{V_\alpha} \otimes \widehat{\mathrm{ev}}_{V_\beta}$. We have

$$\sum_{i=0}^{r-1} q^{(\alpha+r-1)(\beta+r-1-2i)} q^{(\beta+r-1-2i)(1-r)} = q^{\alpha(\beta+r-1)} \sum_{i=0}^{r-1} q^{-2\alpha i}.$$

If $\alpha \notin r\mathbb{Z}$, then $q^{-2\alpha} \neq 1$ and the previous line evaluates to

$$q^{\alpha(\beta+r-1)} \frac{1-q^{-2\alpha r}}{1-q^{-2\alpha}} = q^{\alpha\beta} \frac{q^{\alpha r} - q^{-\alpha r}}{q^{\alpha} - q^{-\alpha}} = q^{\alpha\beta} \frac{\{\alpha r\}}{\{\alpha\}}.$$

If instead $\alpha = rz \in r\mathbb{Z}$, then

$$q^{rz(\beta+r-1)} \sum_{i=0}^{r-1} q^{-2rzi} = q^{rz(\beta+r-1)} r = (-1)^{rz+z} q^{\beta rz} r.$$

In particular, taking $r = 2$ with $\alpha = 0$ and $\beta = 2$, we obtain

$\mathcal{F}$ (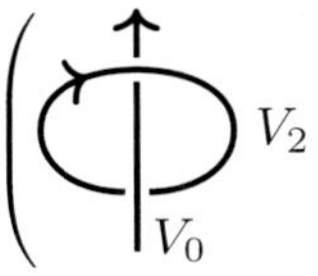) $= 2\mathrm{id}_{V_0}$, $\mathcal{F}$ (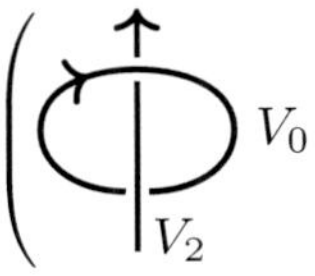) $= -2\mathrm{id}_{V_2}$.

In view of Equation (2.3), we want to attach to the Hopf link colored by V_0 and V_2 a scalar given by cutting the Hopf link open to a $(1,1)$-tangle. However, we see that the scalar depends non-trivially on which component is cut.

The following notion is the key to resolving the cutting ambiguity illustrated by the previous example.

DEFINITION 3.2 ([**GPMT09**, §3]). A module $V \in \mathcal{C}$ is called *ambidextrous* if the equality

(3.1) $\mathcal{F}$ (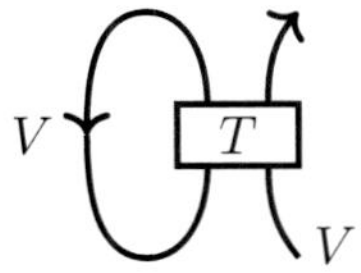) $= \mathcal{F}$ (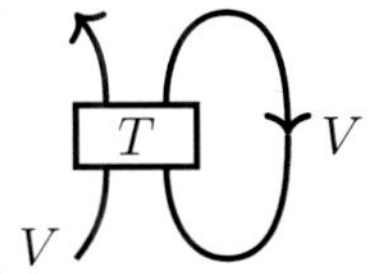)

holds for all (2,2)-tangles T whose open strands are colored by V.

LEMMA 3.3. *If $V \in \mathcal{C}$ is simple with non-vanishing quantum dimension, then V is ambidextrous.*

PROOF. Let T be a $(2,2)$-tangle whose open strands are colored by V and let L be the diagram

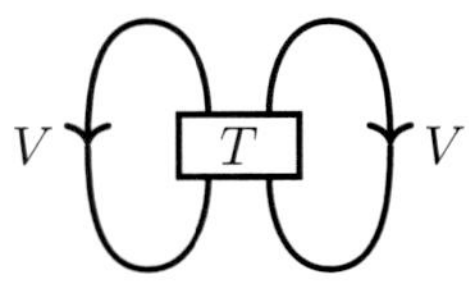

Since L is obtained by taking the right and left partial traces of the $(1,1)$-tangles appearing in Equation (3.1), we find that both sides of the desired equality are equal to $\frac{\langle \mathcal{F}(L) \rangle}{\mathrm{qdim}_{\mathcal{C}}(V)} \mathrm{id}_V$. □

When V has vanishing quantum dimension, we need to investigate further.

LEMMA 3.4. *Let $V, W \in \mathcal{C}$ be simple objects such that $V \otimes W$ is semisimple and multiplicity free. Then the algebra $\mathrm{End}_{\mathcal{C}}(V \otimes W)$ is isomorphic to a direct sum of copies of $\mathbb{C}$.*

PROOF. Let $U_1, \ldots, U_n$ be pairwise non-isomorphic simples such that $V \otimes W \simeq U_1 \oplus \cdots \oplus U_n$. Schur's Lemma implies algebra isomorphisms $\mathrm{End}_{\mathcal{C}}(V \otimes W) \simeq \bigoplus_{i=1}^n \mathrm{End}_{\mathcal{C}}(U_i) \simeq \mathbb{C}^n$. □

LEMMA 3.5. *Let $\eta \in \mathbb{C} \setminus \frac{1}{2}\mathbb{Z}$. Then $V_\eta \otimes V_\eta \in \mathcal{C}$ is semisimple and multiplicity free.*

PROOF. Since $2\eta \notin \mathbb{Z}$, the object $V_\eta \otimes V_\eta \in \mathcal{C}_{\overline{2\eta}}$ is semisimple by Theorem 1.13. Hence, there exist unique integers $m_{2\eta_r - r+2i+1} \in \mathbb{Z}_{\geq 0}$ such that

$$V_\eta \otimes V_\eta \simeq \bigoplus_{i=0}^{r-1} V_{2\eta-r+2i+1}^{\oplus m_{2\eta_r - r+2i+1}}.$$

Consider $\mathbb{Q}[\mathbb{C}]$, the group algebra of $\mathbb{C}$, with basis $\{x^\lambda\}_{\lambda \in \mathbb{C}}$. The *character* of $V \in \mathcal{C}$ is $\mathrm{ch}(V) = \sum_{\lambda \in \mathbb{C}} \dim_{\mathbb{C}}(V[\lambda]) x^\lambda$. The explicit description of V_α gives $\mathrm{ch}(V_\alpha) = x^\alpha [r]_x$, where $[r]_x = \sum_{i=0}^{r-1} x^{r-1-2i}$. We claim that

$$\mathcal{S} = \{\mathrm{ch}(V_{2\eta-r+2i+1}) \mid 0 \leq i \leq r-1\}$$

is a linearly independent subset of $\mathbb{Q}[\mathbb{C}]$. Suppose that $\sum_{i=0}^{r-1} a_i \mathrm{ch}(V_{2\eta-r+2i+1}) = 0$ for some $a_i \in \mathbb{Q}$. Since all powers of x which appear in this equation lie on the same affine real line in $\mathbb{C}$, they are naturally ordered. The largest such power is $x^{2\eta+2(r-1)}$ with coefficient a_{r-1}, resulting from $\mathrm{ch}(V_{2\eta-r+2(r-1)+1})$. Hence, $a_{r-1} = 0$. Continuing in this way shows that $a_{r-2} = \cdots = a_0 = 0$ and $\mathcal{S}$ is linearly independent.

The character of $V_\eta \otimes V_\eta$ is $\mathrm{ch}(V_\eta)^2 = x^{2\eta}[r]_x^2$. On the other hand,

$$\mathrm{ch}(V_\eta)^2 = \Big(\sum_{i=0}^{r-1} m_{2\eta-r+2i+1} x^{2\eta-r+2i}\Big)[r]_x.$$

Setting $m_{2\eta-r+1+2i} = 1$ for each i, the right-hand side of the previous equation becomes

$$\left(\sum_{i=0}^{r-1} x^{2\eta-r+2i}\right)[r]_x = x^{2\eta}[r]_x^2 = \mathrm{ch}(V_\eta)^2.$$

In view of the linear independence of $\mathcal{S}$, this completes the proof. □

REMARK 3.6. A direct modification of the proof of Lemma 3.5 shows that if $\alpha, \beta, \alpha+\beta \in \mathbb{C} \setminus \mathbb{Z}$, then

$$V_\alpha \otimes V_\beta \simeq \bigoplus_{i=0}^{r-1} V_{\alpha+\beta-r+2i+1}.$$

THEOREM 3.7. *Let $\eta \in \mathbb{C} \setminus \frac{1}{2}\mathbb{Z}$. Then V_η is ambidextrous.*

PROOF. By Lemmas 3.4 and 3.5, the algebra $\mathrm{End}_{\mathcal{C}}(V_\eta \otimes V_\eta)$ is commutative. Let T be a $(2,2)$-tangle whose open strands are colored by V_η. Then we have the following sequence of equalities, where we implicitly apply $\mathcal{F}$ to each tangle and the coupons are colored by T:

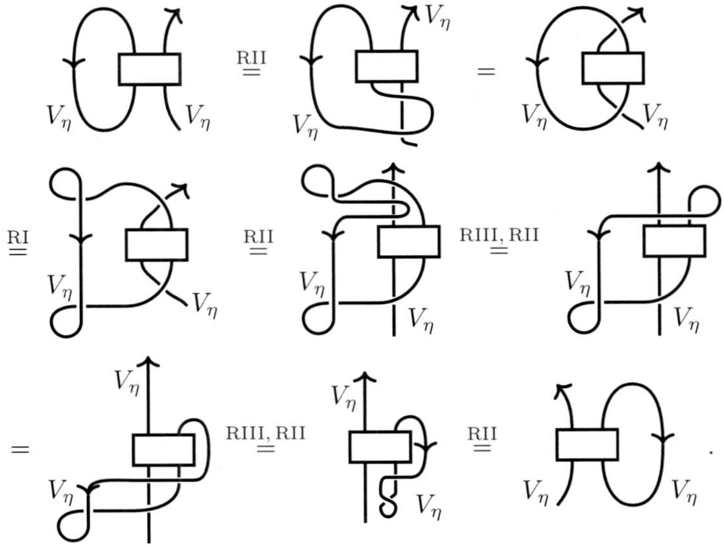

The second equality is implied by the commutativity of $\mathrm{End}_{\mathcal{C}}(V_\eta \otimes V_\eta)$. The sixth equality holds by a combination of framed Reidemeister moves that depends on T. The other equalities hold by the indicated framed Reidemeister moves. □

3.2. Modified quantum dimensions. Define a function $S' : \mathbb{C} \times \mathbb{C} \to \mathbb{C}$ by

$$S'(\beta,\alpha) = \left\langle \mathcal{F}\left(\right) \right\rangle.$$

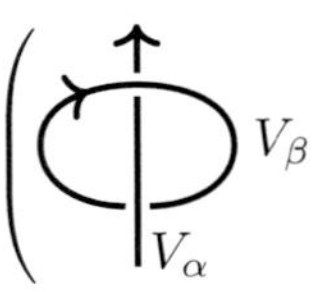

PROPOSITION 3.8. *The equality*

$$S'(\beta,\alpha) = \begin{cases} q^{\beta\alpha}\frac{\{\alpha r\}}{\{\alpha\}} & \text{if } \alpha \in \mathbb{C}\setminus r\mathbb{Z}, \\ q^{\beta r z}\cdot(-1)^{(r+1)z} r & \text{if } \alpha = rz \in r\mathbb{Z} \end{cases}$$

holds. In particular, $S'(\beta,\alpha)$ is nonzero for all $\alpha \in \mathbb{C}\setminus \mathbb{Z}\cup r\mathbb{Z}$.

PROOF. This was computed in Example 3.1. □

DEFINITION 3.9. Let $\eta \in \mathbb{C}$. The *modified quantum dimension with respect to* η is the function $\mathbf{d}_\eta : \mathbb{C}\setminus\mathbb{Z}\cup r\mathbb{Z} \to \mathbb{C}$ given by $\mathbf{d}_\eta(\alpha) = \frac{S'(\alpha,\eta)}{S'(\eta,\alpha)}$.

By Proposition 3.8, the modified quantum dimension $\mathbf{d}_\eta$ is nowhere zero. Modified quantum dimensions associated to different parameters η are related as follows.

PROPOSITION 3.10. *For $\eta, \eta' \in \mathbb{C}\setminus\mathbb{Z}$ and $\alpha \in \mathbb{C}\setminus\mathbb{Z}\cup r\mathbb{Z}$, the following equality holds:*

$$\mathbf{d}_\eta(\alpha) = \frac{\sin(\pi\frac{\eta}{r})\sin(\eta'\pi)}{\sin(\eta\pi)\sin(\pi\frac{\eta'}{r})}\mathbf{d}_{\eta'}(\alpha).$$

PROOF. This follows immediately from Proposition 3.8 and the definition of $\mathbf{d}_{(-)}$. □

THEOREM 3.11 ([**GPMT09**, Lemma 2]). *Let $\eta \in \mathbb{C}$ be such that V_η is ambidextrous and $\alpha, \beta \in \mathbb{C} \setminus \mathbb{Z} \cup r\mathbb{Z}$. Then for all $(2,2)$-tangles T, the following equality holds:*

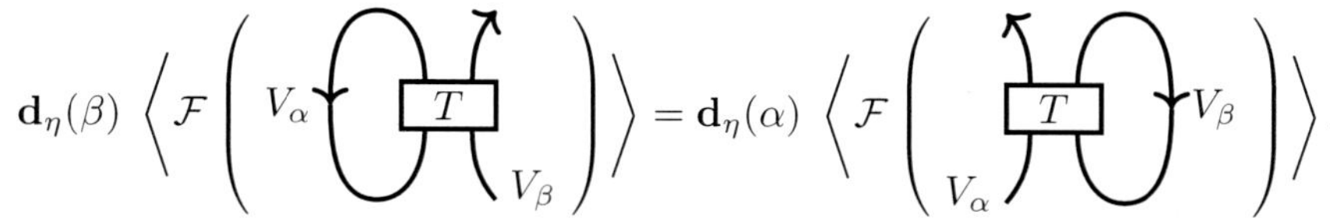

PROOF. Because V_η is ambidextrous, there is an equality

(3.2) 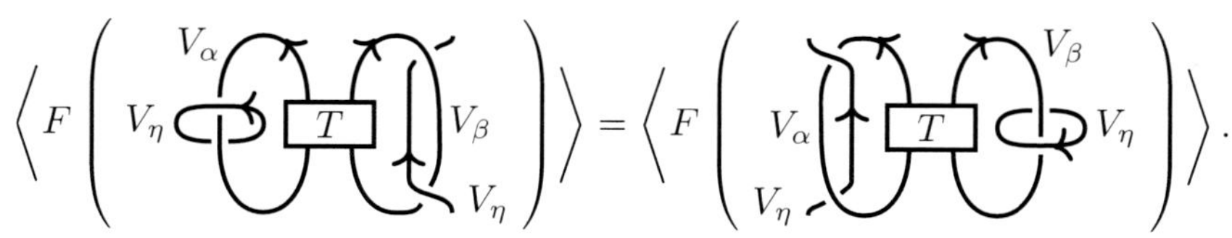

We expand both sides of this equality. The left-hand side becomes

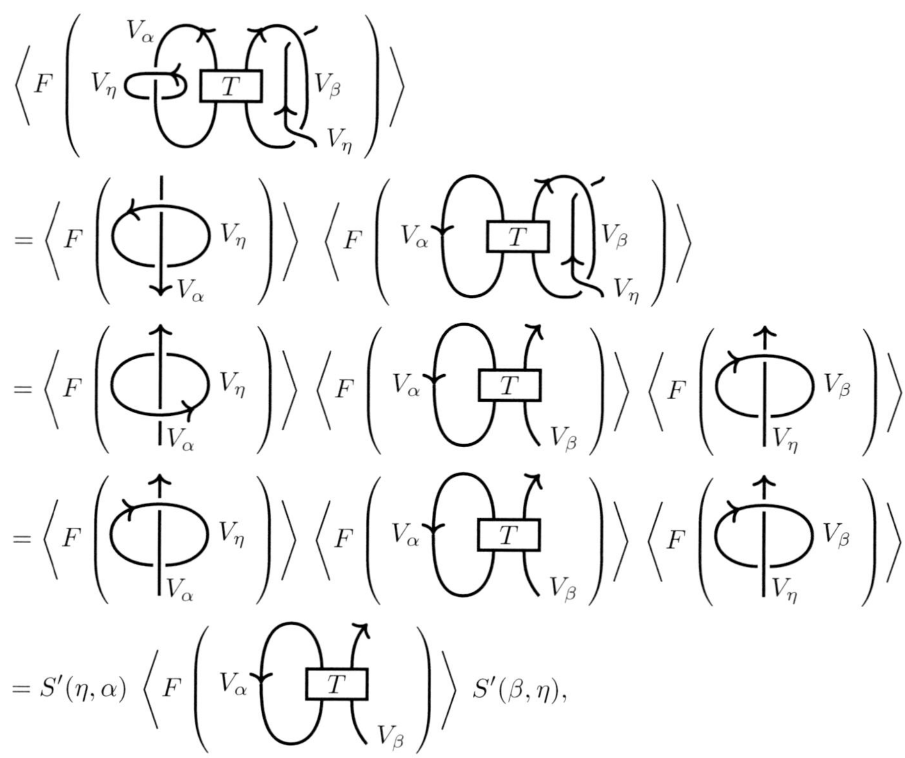

where we have repeatedly applied Lemma 2.2. The right-hand side becomes

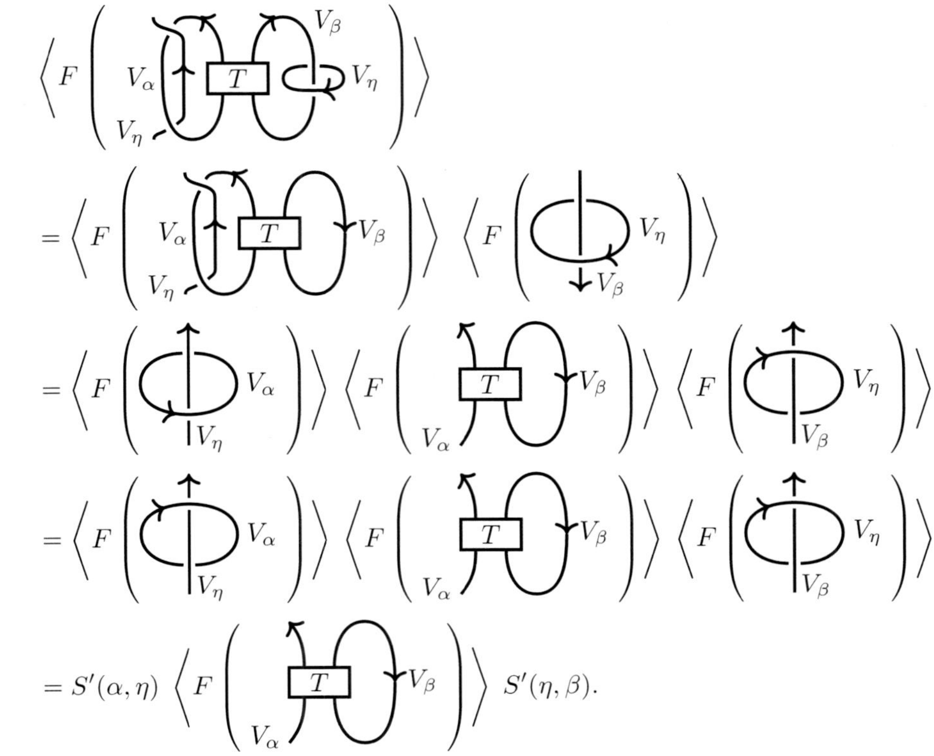

By Proposition 3.8, $S'(\eta,\alpha)$ and $S'(\eta,\beta)$ are non-zero. We can therefore divide both sides of Equation (3.2) by $S'(\eta,\alpha)S'(\eta,\beta)$ to complete the proof. □

COROLLARY 3.12. *For each $\alpha \in \mathbb{C}\setminus \mathbb{Z}\cup r\mathbb{Z}$, the module V_α is ambidextrous. In particular, any simple module in $\mathcal{C}$ is ambidextrous.*

PROOF. Let $\eta \in \mathbb{C}\setminus \frac{1}{2}\mathbb{Z}$. By Theorem 3.7, the module V_η is ambidextrous. By Proposition 3.8, the scalar $\mathbf{d}_\eta(\alpha)$ is non-zero. The first statement now follows from taking $\alpha=\beta$ in Theorem 3.11. Using the classification of simple objects of $\mathcal{C}$ given in Proposition 1.7, the second claim follows from the above and Lemma 3.3. □

3.3. Renormalized Reshetikhin–Turaev invariants of links. Denote by $\mathfrak{L}$ the set of all colored framed links for which at least one of its colors is of the form V_α for some $\alpha \in \mathbb{C}\setminus \mathbb{Z}\cup r\mathbb{Z}$. We view $\mathfrak{L}$ as a subset of morphisms of $\mathbf{Rib}_{\mathcal{C}}$.

THEOREM 3.13 ([**GPMT09**, Theorem 3]). *Let $\eta \in \mathbb{C}\setminus \mathbb{Z}\cup r\mathbb{Z}$. Then the map $\mathcal{F}'_\eta : \mathfrak{L} \to \mathbb{C}$ given by*

$$\mathcal{F}'_\eta(L) = \mathbf{d}_\eta(\alpha)\langle \mathcal{F}(T)\rangle,$$

where T is a $(1,1)$-tangle whose closure is L and whose open strand is colored by V_α for some $\alpha \in \mathbb{C}\setminus \mathbb{Z}\cup r\mathbb{Z}$, is a well-defined isotopy invariant of links in $\mathfrak{L}$.

PROOF. Well-definedness of $\mathcal{F}'_\eta$ is the statement that $\mathcal{F}'_\eta(L)$ is independent of the choice of $(1,1)$-tangle T used in its definition. If T and T' are $(1,1)$-tangles constructed from L by cutting along components colored by V_α and V_β, respectively, then, in view of Corollary 3.12, Theorem 3.11 shows that $\mathbf{d}_\eta(\alpha)\langle\mathcal{F}(T)\rangle = \mathbf{d}_\eta(\beta)\langle\mathcal{F}(T')\rangle$. Isotopy invariance of $\mathcal{F}'_\eta$ follows from Theorem 2.1. □

Comparing the definition of $\mathcal{F}'_\eta(L)$ with Equation (2.3) shows that $\mathbf{d}_\eta(-)$ plays the role of $\mathrm{qdim}_\mathcal{C}(-)$ in the standard theory. This justifies the term *modified quantum dimension* for the function $\mathbf{d}_\eta$.

3.4. Basic properties and examples. We begin by discussing the dependence of the renormalized invariants of Theorem 3.13 on the parameter η. By Proposition 3.10, the modified quantum dimension functions associated to two such parameters differ by an explicit global scalar. In fact, more is true.

LEMMA 3.14. *Let $D : \mathbb{C} \setminus \mathbb{Z} \cup r\mathbb{Z} \to \mathbb{C}$ be a function such that the assignment $L \mapsto D(\alpha)\langle \mathcal{F}(T)\rangle$, where T is any $(1,1)$-tangle whose closure is isotopic to L and whose open strand is colored by V_α for some $\alpha \in \mathbb{C} \setminus \mathbb{Z} \cup r\mathbb{Z}$, is a well-defined invariant of colored framed links in $\mathfrak{L}$. Then, for any $\eta \in \mathbb{C}\setminus\mathbb{Z}$, there exists a scalar $d \in \mathbb{C}$ such that $D = d \cdot \mathbf{d}_\eta$.*

PROOF. Let L be the Hopf link with components colored by V_α and V_η for some $\alpha \in \mathbb{C} \setminus \mathbb{Z} \cup r\mathbb{Z}$ and $\eta \in \mathbb{C} \setminus \mathbb{Z}$. The assumption of D implies that $D(\alpha)S'(\eta,\alpha) = D(\eta)S'(\alpha,\eta)$, whence $D(\alpha) = D(\eta)\mathbf{d}_\eta(\alpha)$. Taking $d = D(\eta)$ proves the lemma. □

It follows from Lemma 3.14 that renormalized invariants arising from a modified quantum dimension $\mathbf{d}_\eta$ are effectively the only invariants that incorporate the cutting process described above.

REMARK 3.15. When $r = 2$, Lemma 3.14 can be strengthened. By Proposition 3.10, any function D that gives an invariant incorporating the cutting procedure is equal to $\pm\mathbf{d}_\eta$ for some $\eta \in \mathbb{C}$, that is, the scalar d in Lemma 3.14 can be taken to be ± 1.

REMARK 3.16. By Example 2.3, the simple modules S_n^{lr} have non-zero quantum dimension and so, by Lemma 3.3, are ambidextrous. Hence, there is a modification of Theorem 3.13 in which the standard Reshetikhin–Turaev invariant is renormalized with respect to S_n^{lr} instead of V_η. The resulting invariant $\mathcal{F}'_{n,lr}$ is defined on $\mathfrak{L}$ as well as those knots $\tilde{\mathfrak{L}}$ colored by at least one simple module of the form S_m^{kr}. A calculation as in Example 3.1 shows that the S_n^{lr}-renormalized quantum dimension $\mathbf{d}_n^{lr}$ vanishes on V_α, $\alpha \in \mathbb{C} \setminus \mathbb{Z} \cup r\mathbb{Z}$. In particular, $\mathcal{F}'_{n,lr}$ is zero on $\mathfrak{L}$. On the other hand, a direct calculation gives

$$\mathrm{qdim}_\mathcal{C}(S_m^{kr}) = \mathrm{qdim}_\mathcal{C}(S_n^{lr}) \cdot \mathbf{d}_n^{lr}(S_m^{kr}).$$

It follows that the restriction of $\mathcal{F}'_{n,lr}$ to $\tilde{\mathfrak{L}}$ is equal to $\mathrm{qdim}_\mathcal{C}(S_n^{lr})^{-1}$ times the standard Reshetikhin–Turaev invariant $\mathcal{F}$. The renormalized theory therefore recovers the standard theory.

Next, we describe the behavior of renormalized invariants under connect sum.

PROPOSITION 3.17. *Let $L, L' \in \mathfrak{L}$, each of which have at least one component colored by V_α for some $\alpha \in \mathbb{C}\setminus\mathbb{Z}\cup r\mathbb{Z}$. Then the η-renormalized Reshetikhin–Turaev invariant of the connect sum $L\#L'$ along components colored by V_α satisfies*

$$\mathbf{d}_\eta(\alpha)\mathcal{F}'_\eta(L\#L') = \mathcal{F}'_\eta(L) \cdot \mathcal{F}'_\eta(L').$$

PROOF. Consider a knot diagram for $L\#L'$ of the form

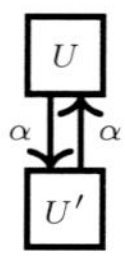

where the obvious closures of the $(2,0)$-tangle U and $(0,2)$-tangle U' are knot diagrams for L and L', respectively, and we have written α for the color V_α. Cutting this diagram along one of the connecting strands gives

$$\mathcal{F}'_\eta(L\#L') = \mathbf{d}_\eta(\alpha)\left\langle \mathcal{F}\right.$$
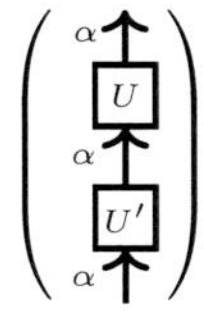

$$\left.\right\rangle$$

where, by a slight abuse of notation, we now denote by U the $(1,1)$-tangle obtained from U by pulling one open strand to the top, and similarly for U'. Since V_α is simple, the right hand side of this equation is equal to

$$\mathbf{d}_\eta(\alpha)\left\langle \mathcal{F}\right.$$
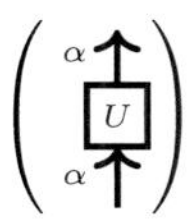

$$\left.\right\rangle\left\langle \mathcal{F}\right.$$
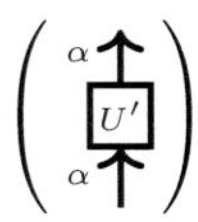

$$\left.\right\rangle = \mathbf{d}_\eta(\alpha)^{-1}\mathcal{F}'_\eta(L)\mathcal{F}'_\eta(L').$$

This gives the desired expression for $\mathcal{F}'_\eta(L\#L')$. $\square$

EXAMPLE 3.18. Let $r=2$. Set $V=V_\alpha$ for some $\alpha\in\mathbb{C}$ with standard weight basis $\{v_0,v_1\}$. Consider $V\otimes V$ with basis $\{v_0\otimes v_0, v_0\otimes v_1, v_1\otimes v_0, v_1\otimes v_1\}$. A direct computation gives

$$c_{V,V} = \begin{pmatrix} q^{\frac{(\alpha+1)^2}{2}} & 0 & 0 & 0 \\ 0 & 0 & q^{\frac{(\alpha+1)(\alpha-1)}{2}} & 0 \\ 0 & q^{\frac{(\alpha+1)(\alpha-1)}{2}} & q^{\frac{(\alpha+1)(\alpha-1)}{2}}\{1-\alpha\} & 0 \\ 0 & 0 & 0 & q^{\frac{(\alpha-1)^2}{2}} \end{pmatrix}.$$

For example, the $(3,2)$ and $(3,3)$ entries of $c_{V,V}$ are the coefficients of $c_{V,V}(v_1\otimes v_0)$:

$$\begin{aligned} c_{V,V}(v_1\otimes v_0) &= \tau(q^{H\otimes H/2}(1+\frac{\{1\}^2}{\{1\}}E\otimes F)(v_1\otimes v_0) \\ &= \tau(q^{H\otimes H/2}(v_1\otimes v_0+\frac{\{1\}^2}{\{1\}}Ev_1\otimes Fv_0) \\ &= q^{\frac{(\alpha-1)(\alpha+1)}{2}}v_0\otimes v_1 + q^{\frac{(\alpha-1)(\alpha+1)}{2}}\{1-\alpha\}v_1\otimes v_0. \end{aligned}$$

Using the explicit formula for $c_{V,V}$, we verify the equality

$$q^{-\frac{(\alpha+1)(\alpha-1)}{2}}c_{V,V} - q^{\frac{(\alpha+1)(\alpha-1)}{2}}c_{V,V}^{-1} = \{\alpha+1\}\mathrm{id}_{V\otimes V}. \tag{3.3}$$

Recall from the proof of Theorem 1.19 that

$$\theta_V = q^{-\frac{(\alpha+1)(\alpha-1)}{2}}\mathrm{id}_V.$$

Define

$$\mathcal{F}'(L) = q^{\frac{(\alpha+1)(\alpha-1)}{2}\mathrm{wr}(L)}\mathcal{F}'(L),$$

where $\mathrm{wr}(L)$ is the writhe of L. Then $\mathcal{F}'$ is an invariant of oriented links colored by V_α. This is an instance of the deframing procedure, explained, for example, in [**JM19**, §3.3]. Relation (3.3) implies that $\mathcal{F}'$ satisfies the Alexander skein relation with $t=q^{\alpha+1}$. For further connections between the renormalized Reshetikhin–Turaev theory of $\mathcal{C}$ at $r=2$, the Alexander polynomial and Reidemeister torsion, see [**BCGPM16**, §6.7].

We end this section with some calculations of renormalized invariants for knots with few crossings.

EXAMPLE 3.19. Using the result of Example 2.5, the η-renormalized invariant of the right-handed trefoil $K = 3_1$ colored by a simple module V_α is

$$\mathcal{F}'_\eta(K) = \frac{\{\eta r\}\{\alpha\}}{\{\eta\}\{\alpha r\}} q^{\frac{3}{2}(\alpha+r-1)^2+(\alpha+r-1)(1-r)} \sum_{i=0}^{r-1} q^{i(-3\alpha-r+i)} \prod_{j=0}^{i-1}\{i-j-\alpha\}.$$

EXAMPLE 3.20. Proceeding analogously to Example 3.19, the η-renormalized invariant of the figure eight knot $K = 4_1$ colored by a simple module V_α is

$$\mathcal{F}'_\eta(K) = \frac{\{\eta r\}\{\alpha\}}{\{\eta\}\{\alpha r\}} \sum_{i=0}^{r-1}\sum_{j=0}^{i}\sum_{k=0}^{r-i} q^{(-\alpha+r-1-2i)(\alpha-r+1)/2} q^{-(-\alpha+r-1-2(i-j))(i+r-1-j)}$$
$$\cdot q^{-(-\alpha+r-1-2(i-j+k))(\alpha+r-1-2(i+k))/2} q^{j(j-1)/2} q^{k(k-1)/2}$$
$$\cdot q^{(i+k-r+1)(i+k-r)/2}$$
$$\cdot \prod_{x=r-1}^{r-1-j} \frac{\{x\}\{x-\alpha\}}{\{j\}!} \prod_{y=i-j}^{i-j+k} \frac{\{y\}\{y+\alpha\}}{\{k\}!}) \prod_{z=i+k}^{r-1} \frac{\{z\}\{z-\alpha\}}{\{l\}!}.$$

4. Further reading

In this final section, we give a brief sample of recent developments and on-going research directions in renormalized Reshetikhin–Turaev theory.

The construction of renormalized Reshetikhin–Turaev invariants of links, given in Theorem 3.11 for the category of weight modules over $\overline{U}_q^H(\mathfrak{sl}_2(\mathbb{C}))$, applies more generally to certain non-semisimple ribbon categories [**GPMT09**]. The key new notions in this more general context are *modified traces* (and so modified quantum dimensions) and ambidextrous objects [**GKPM11, GKPM13, GKPM22, GPM18, FOG18, GR20, BBG21**]. Renormalized Reshetikhin–Turaev invariants of links have been studied for categories of weight modules over unrolled quantum groups of complex simple Lie algebras [**GPM13**] and Lie superalgebras [**GPM07, GPM10**]. The renormalized invariants are motivated by and recover previous *ad hoc* renormalized invariants [**KS91, ADO92, Vir06, GPM10**].

The extension of the renormalized theory from links to 3-manifolds was achieved in [**CGPM14**] for the category of weight modules over $\overline{U}_q^H(\mathfrak{sl}_2(\mathbb{C}))$, where properties (P2) and (P3) are serious obstructions. In general, the additional data and constraints on the input ribbon category required to obtain a 3-manifold invariant are termed *non-degenerate relative pre-modularity*. The resulting 3-manifold invariants have novel properties, including the ability to distinguish homotopy classes of lens spaces and connections with the Volume Conjecture, and are related to earlier non-semisimple 3-manifold invariants, including those of Hennings [**Hen96**] and Kerler–Lyubashenko [**KL01**]. See [**DRGPM18, DRGG$^+$23**]. Further examples of 3-manifold invariants associated to non-degenerate relative pre-modular categories are studied in [**AGPM21, BH21, Ha18, BI22**]. For connections between standard and renormalized Reshetikhin–Turaev invariants of links and 3-manifolds, see [**CGPM15a, CGP21, DRGPM20, Mor22**].

The further extension of renormalized 3-manifold invariants to three dimensional topological quantum field theories (TQFTs) was first accomplished in the case of weight modules over $\overline{U}_q^H(\mathfrak{sl}_2(\mathbb{C}))$ [**BCGPM16**]. The TQFTs have interesting features, including the extension of Reidemeister torsion to a TQFT and the possibility of producing representations of mapping class groups that are faithful modulo their centers. A general framework for the construction of (extended) TQFTs from *relative modular categories* was given by De Renzi [**DR22**], generalizing the so-called universal construction of semisimple TQFTs [**BHMV95**]. See [**BDR21**] for an overview of this circle of ideas. Further examples of TQFTs from renormalized invariants are constructed and studied in [**DRGG$^+$22, GY22**].

At present, the main source of relative modular categories is the representation theory of unrolled quantum groups, thereby making this class of quantum groups central to non-semisimple topology. Motivated by the success of rational conformal field theoretic techniques in semisimple topology, a number of authors have pursued a conjectural logarithmic variant of the Kazhdan–Lusztig correspondence, which asserts an equivalence between categories of weight modules over unrolled quantum groups and modules over non-rational, or logarithmic, vertex operator algebras. Much progress has been made in the case of $\overline{U}_q^H(\mathfrak{sl}_2(\mathbb{C}))$, where connections between weight modules and the singlet, triplet and Feigin–Tipunin algebras have been found [**CMR18, CR22**].

Finally, there has been exciting progress in connecting non-semisimple mathematical TQFTs to physical quantum field theories. This can be seen as a non-semisimple generalization of the celebrated connection between compact Chern–Simons theory and Reshetikhin–Turaev TQFTs [**Wit89, RT91**]. The case of TQFTs arising from $\overline{U}_q^H(\mathfrak{sl}_2(\mathbb{C}))$ is studied in [**CDGG21**], where it is connected to a topological twist of 3d $\mathcal{N} = 4$ Chern–Simons matter theory with gauge group $SU(2)$. Similarly, TQFTs arising from an unrolled quantization of the Lie superalgebra $\mathfrak{gl}(1|1)$ were shown in [**GY22**] to be related to supergroup Chern–Simons theories with gauge group $\mathfrak{psl}(1|1)$ and $U(1|1)$. A key feature in both physical realizations is the presence of global symmetry groups, allowing the quantum field theories to be coupled to background flat connections. Further physical studies of quantum field theories coupled to background flat connections can be found in [**GHN$^+$21, Jag22**]. In condensed matter physics, Levin and Wen used unitary spherical fusion categories to give a mathematical foundation of topological order and string-net condensation [**LW05**]. Recently, this construction was extended to the setting of the non-semisimple category of weight modules over $\overline{U}_q^H(\mathfrak{sl}_2(\mathbb{C}))$ [**GLPMS22a, GLPMS22b**].

Acknowledgments

The authors thank an anonymous referee for helpful comments and suggestions. The authors also thank M. S. Im, B. Nguyen and A. Wilbert for organizing the Special Session on Geometry and Representation Theory of Quantum Algebras and Related Topics, where Robertson and Spellmann spoke about the contents of this paper.

References

[ADO92] Y. Akutsu, T. Deguchi, and T. Ohtsuki, *Invariants of colored links*, J. Knot Theory Ramifications **1** (1992), no. 2, 161–184, DOI 10.1142/S0218216592000094. MR1164114

[AGPM21] C. A.-M. Anghel, N. Geer, and B. Patureau-Mirand, *Relative (pre)-modular categories from special linear Lie superalgebras*, J. Algebra **586** (2021), 479–525, DOI 10.1016/j.jalgebra.2021.06.035. MR4291491

[And92] H. H. Andersen, *Tensor products of quantized tilting modules*, Comm. Math. Phys. **149** (1992), no. 1, 149–159. MR1182414

[BBG21] A. Beliakova, C. Blanchet, and A. M. Gainutdinov, *Modified trace is a symmetrised integral*, Selecta Math. (N.S.) **27** (2021), no. 3, Paper No. 31, 51, DOI 10.1007/s00029-021-00626-5. MR4257848

[BCGPM16] C. Blanchet, F. Costantino, N. Geer, and B. Patureau-Mirand, *Non-semi-simple TQFTs, Reidemeister torsion and Kashaev's invariants*, Adv. Math. **301** (2016), 1–78, DOI 10.1016/j.aim.2016.06.003. MR3539369

[BDR21] C. Blanchet and M. De Renzi, *Modular categories and TQFTs beyond semisimplicity*, Topology and geometry—a collection of essays dedicated to Vladimir G. Turaev, IRMA Lect. Math. Theor. Phys., vol. 33, Eur. Math. Soc., Zürich, [2021] ©2021, pp. 175–208, DOI 10.4171/IRMA/33-1/11. MR4394509

[BH21] A. Beliakova and K. Hikami, *Non-semisimple invariants and Habiro's series*, Topology and geometry—a collection of essays dedicated to Vladimir G. Turaev, IRMA Lect. Math. Theor. Phys., vol. 33, Eur. Math. Soc., Zürich, [2021] ©2021, pp. 161–174, DOI 10.4171/IRMA/33-1/10. MR4394508

[BHMV95] C. Blanchet, N. Habegger, G. Masbaum, and P. Vogel, *Topological quantum field theories derived from the Kauffman bracket*, Topology **34** (1995), no. 4, 883–927, DOI 10.1016/0040-9383(94)00051-4. MR1362791

[BI22] Y. Bao and N. Ito, $\mathfrak{gl}(1|1)$*-Alexander polynomial for 3-manifolds*, Internat. J. Math. **34** (2023), no. 4, Paper No. 2350016, 23, DOI 10.1142/S0129167X23500167. MR4576428

[BNW91] D. Bar-Natan and E. Witten, *Perturbative expansion of Chern-Simons theory with noncompact gauge group*, Comm. Math. Phys. **141** (1991), no. 2, 423–440. MR1133274

[CDGG21] T. Creutzig, T. Dimofte, N. Garner, and N. Geer, *A QFT for non-semisimple TQFT*, Adv. Theor. Math. Phys. **28** (2024), no. 1, 161–405, DOI 10.4310/atmp.2024.v28.n1.a4. MR4788097

[CGP21] F. Costantino, S. Gukov, and P. Putrov, *Non-semisimple TQFT's and BPS q-series*, SIGMA Symmetry Integrability Geom. Methods Appl. **19** (2023), Paper No. 010, 71, DOI 10.3842/SIGMA.2023.010. MR4560896

[CGPM14] F. Costantino, N. Geer, and B. Patureau-Mirand, *Quantum invariants of 3-manifolds via link surgery presentations and non-semi-simple categories*, J. Topol. **7** (2014), no. 4, 1005–1053, DOI 10.1112/jtopol/jtu006. MR3286896

[CGPM15a] F. Costantino, N. Geer, and B. Patureau-Mirand, *Relations between Witten-Reshetikhin-Turaev and nonsemisimple* $\mathfrak{sl}(2)$ *3-manifold invariants*, Algebr. Geom. Topol. **15** (2015), no. 3, 1363–1386, DOI 10.2140/agt.2015.15.1363. MR3361139

[CGPM15b] F. Costantino, N. Geer, and B. Patureau-Mirand, *Some remarks on the unrolled quantum group of* $\mathfrak{sl}(2)$, J. Pure Appl. Algebra **219** (2015), no. 8, 3238–3262, DOI 10.1016/j.jpaa.2014.10.012. MR3320217

[CMR18] T. Creutzig, A. Milas, and M. Rupert, *Logarithmic link invariants of* $\overline{U}_q^H(\mathfrak{sl}_2)$ *and asymptotic dimensions of singlet vertex algebras*, J. Pure Appl. Algebra **222** (2018), no. 10, 3224–3247, DOI 10.1016/j.jpaa.2017.12.004. MR3795642

[CP94] V. Chari and A. Pressley, *A guide to quantum groups*, Cambridge University Press, Cambridge, 1994. MR1300632

[CR13a] T. Creutzig and D. Ridout, *Logarithmic conformal field theory: beyond an introduction*, J. Phys. A **46** (2013), no. 49, 494006, 72, DOI 10.1088/1751-8113/46/49/494006. MR3146012

[CR13b] T. Creutzig and D. Ridout, *Relating the archetypes of logarithmic conformal field theory*, Nuclear Phys. B **872** (2013), no. 3, 348–391, DOI 10.1016/j.nuclphysb.2013.04.007. MR3048507

[CR22] T. Creutzig and M. Rupert, *Uprolling unrolled quantum groups*, Commun. Contemp. Math. **24** (2022), no. 4, Paper No. 2150023, 27, DOI 10.1142/S0219199721500231. MR4414165

[DR22] M. De Renzi, *Non-semisimple extended topological quantum field theories*, Mem. Amer. Math. Soc. **277** (2022), no. 1364, v+161, DOI 10.1090/memo/1364. MR4403930

[DRGG+22] M. De Renzi, A. M. Gainutdinov, N. Geer, B. Patureau-Mirand, and I. Runkel, *3-Dimensional TQFTs from non-semisimple modular categories*, Selecta Math. (N.S.) **28** (2022), no. 2, Paper No. 42, 60, DOI 10.1007/s00029-021-00737-z. MR4370375

[DRGG+23] M. De Renzi, A. M. Gainutdinov, N. Geer, B. Patureau-Mirand, and I. Runkel, *Mapping class group representations from non-semisimple TQFTs*, Commun. Contemp. Math. **25** (2023), no. 1, Paper No. 2150091, 52, DOI 10.1142/S0219199721500917. MR4523145

[DRGPM18] M. De Renzi, N. Geer, and B. Patureau-Mirand, *Renormalized Hennings invariants and 2 + 1-TQFTs*, Comm. Math. Phys. **362** (2018), no. 3, 855–907, DOI 10.1007/s00220-018-3187-8. MR3845290

[DRGPM20] M. De Renzi, N. Geer, and B. Patureau-Mirand, *Nonsemisimple quantum invariants and TQFTs from small and unrolled quantum groups*, Algebr. Geom. Topol. **20** (2020), no. 7, 3377–3422, DOI 10.2140/agt.2020.20.3377. MR4194285

[Dri86] V. G. Drinfel′d, *Quantum groups* (Russian, with English summary), Zap. Nauchn. Sem. Leningrad. Otdel. Mat. Inst. Steklov. (LOMI) **155** (1986), no. Differentsial′naya Geometriya, Gruppy Li i Mekh. VIII, 18–49, 193, DOI 10.1007/BF01247086; English transl., J. Soviet Math. **41** (1988), no. 2, 898–915. MR869575

[Dri90] V. Drinfeld. Hopf algebras and the quantum Yang–Baxter equation. In *Yang-Baxter Equation in Integrable Systems*, pages 264–268. World Scientific, 1990.

[EGNO15] P. Etingof, S. Gelaki, D. Nikshych, and V. Ostrik, *Tensor categories*, Mathematical Surveys and Monographs, vol. 205, American Mathematical Society, Providence, RI, 2015, DOI 10.1090/surv/205. MR3242743

[FOG18] A. Fontalvo Orzco and A. Gainutdinov, *Module traces and Hopf group coalgebras*, `arXiv:1809.01122`, 2018.

[FYH+85] P. Freyd, D. Yetter, J. Hoste, W. B. R. Lickorish, K. Millett, and A. Ocneanu, *A new polynomial invariant of knots and links*, Bull. Amer. Math. Soc. (N.S.) **12** (1985), no. 2, 239–246, DOI 10.1090/S0273-0979-1985-15361-3. MR776477

[GHN+21] S. Gukov, P.-S. Hsin, H. Nakajima, S. Park, D. Pei, and N. Sopenko, *Rozansky-Witten geometry of Coulomb branches and logarithmic knot invariants*, J. Geom. Phys. **168** (2021), Paper No. 104311, 22, DOI 10.1016/j.geomphys.2021.104311. MR4281915

[GKPM11] N. Geer, J. Kujawa, and B. Patureau-Mirand, *Generalized trace and modified dimension functions on ribbon categories*, Selecta Math. (N.S.) **17** (2011), no. 2, 453–504, DOI 10.1007/s00029-010-0046-7. MR2803849

[GKPM13] N. Geer, J. Kujawa, and B. Patureau-Mirand, *Ambidextrous objects and trace functions for nonsemisimple categories*, Proc. Amer. Math. Soc. **141** (2013), no. 9, 2963–2978, DOI 10.1090/S0002-9939-2013-11563-7. MR3068949

[GKPM22] N. Geer, J. Kujawa, and B. Patureau-Mirand, *M-traces in (non-unimodular) pivotal categories*, Algebr. Represent. Theory **25** (2022), no. 3, 759–776, DOI 10.1007/s10468-021-10044-y. MR4421818

[GLPMS22a] N. Geer, A. D. Lauda, B. Patureau-Mirand, and J. Sussan, *A Hermitian TQFT from a non-semisimple category of quantum* $\mathfrak{sl}(2)$*-modules*, Lett. Math. Phys. **112** (2022), no. 4, Paper No. 74, 27, DOI 10.1007/s11005-022-01570-x. MR4458638

[GLPMS22b] N. Geer, A. D. Lauda, B. Patureau-Mirand, and J. Sussan, *Pseudo-Hermitian Levin-Wen models from non-semisimple TQFTs*, Ann. Physics **442** (2022), Paper No. 168937, 33, DOI 10.1016/j.aop.2022.168937. MR4438067

[GPM07] N. Geer and B. Patureau-Mirand, *Multivariable link invariants arising from* $\mathfrak{sl}(2|1)$ *and the Alexander polynomial*, J. Pure Appl. Algebra **210** (2007), no. 1, 283–298, DOI 10.1016/j.jpaa.2006.09.015. MR2311186

[GPM10] N. Geer and B. Patureau-Mirand, *Multivariable link invariants arising from Lie superalgebras of type I*, J. Knot Theory Ramifications **19** (2010), no. 1, 93–115, DOI 10.1142/S0218216510007784. MR2640994

[GPM13] N. Geer and B. Patureau-Mirand, *Topological invariants from nonrestricted quantum groups*, Algebr. Geom. Topol. **13** (2013), no. 6, 3305–3363, DOI 10.2140/agt.2013.13.3305. MR3248736

[GPM18] N. Geer and B. Patureau-Mirand, *The trace on projective representations of quantum groups*, Lett. Math. Phys. **108** (2018), no. 1, 117–140, DOI 10.1007/s11005-017-0993-4. MR3740431

[GPMT09] N. Geer, B. Patureau-Mirand, and V. Turaev, *Modified quantum dimensions and re-normalized link invariants*, Compos. Math. **145** (2009), no. 1, 196–212, DOI 10.1112/S0010437X08003795. MR2480500

[GR20] A. M. Gainutdinov and I. Runkel, *Projective objects and the modified trace in factorisable finite tensor categories*, Compos. Math. **156** (2020), no. 4, 770–821, DOI 10.1112/s0010437x20007034. MR4079628

[GY22] N. Geer and M. B. Young, *Three dimensional topological quantum field theory from $U_q(\mathfrak{gl}(1|1))$ and $U(1|1)$ Chern-Simons theory*, Adv. Math. **460** (2025), Paper No. 110044, 69, DOI 10.1016/j.aim.2024.110044. MR4831843

[Ha18] N. P. Ha, *Topological invariants from quantum group $\mathcal{U}_\xi\mathfrak{sl}(2|1)$ at roots of unity*, Abh. Math. Semin. Univ. Hambg. **88** (2018), no. 1, 163–188, DOI 10.1007/s12188-017-0181-6. MR3785790

[Hen96] M. Hennings, *Invariants of links and 3-manifolds obtained from Hopf algebras*, J. London Math. Soc. (2) **54** (1996), no. 3, 594–624, DOI 10.1112/jlms/54.3.594. MR1413901

[Jag22] M. Jagadale, *Decorated TQFTs and their Hilbert spaces*, J. High Energy Phys. **8** (2023), Paper No. 110, 20, DOI 10.1007/jhep08(2023)110. MR4633815

[Jan96] J. C. Jantzen, *Lectures on quantum groups*, Graduate Studies in Mathematics, vol. 6, American Mathematical Society, Providence, RI, 1996, DOI 10.1090/gsm/006. MR1359532

[Jim85] M. Jimbo, *A q-difference analogue of $U(\mathfrak{g})$ and the Yang-Baxter equation*, Lett. Math. Phys. **10** (1985), no. 1, 63–69, DOI 10.1007/BF00704588. MR797001

[JM19] D. M. Jackson and I. Moffatt, *An introduction to quantum and Vassiliev knot invariants*, CMS Books in Mathematics/Ouvrages de Mathématiques de la SMC, Springer, Cham, 2019, DOI 10.1007/978-3-030-05213-3. MR3931694

[Jon87] V. F. R. Jones, *Hecke algebra representations of braid groups and link polynomials*, Ann. of Math. (2) **126** (1987), no. 2, 335–388, DOI 10.2307/1971403. MR908150

[Kir78] R. Kirby, *A calculus for framed links in S^3*, Invent. Math. **45** (1978), no. 1, 35–56, DOI 10.1007/BF01406222. MR467753

[KL01] T. Kerler and V. V. Lyubashenko, *Non-semisimple topological quantum field theories for 3-manifolds with corners*, Lecture Notes in Mathematics, vol. 1765, Springer-Verlag, Berlin, 2001. MR1862634

[KS91] L. H. Kauffman and H. Saleur, *Free fermions and the Alexander-Conway polynomial*, Comm. Math. Phys. **141** (1991), no. 2, 293–327. MR1133269

[KS09] A. Kapustin and N. Saulina, *Chern-Simons-Rozansky-Witten topological field theory*, Nuclear Phys. B **823** (2009), no. 3, 403–427, DOI 10.1016/j.nuclphysb.2009.07.006. MR2571746

[Lic62] W. B. R. Lickorish, *A representation of orientable combinatorial 3-manifolds*, Ann. of Math. (2) **76** (1962), 531–540, DOI 10.2307/1970373. MR151948

[LW05] X.-G. Wen, *An introduction to quantum order, string-net condensation, and emergence of light and fermions*, Ann. Physics **316** (2005), no. 1, 1–29, DOI 10.1016/j.aop.2004.07.001. MR2125773

[Mik15] V. Mikhaylov, *Analytic torsion 3d mirror symmetry and supergroup Chern–Simons theories*, `arXiv:1505.03130`, 2015.

[MM01] H. Murakami and J. Murakami, *The colored Jones polynomials and the simplicial volume of a knot*, Acta Math. **186** (2001), no. 1, 85–104, DOI 10.1007/BF02392716. MR1828373

[Mor22] A. Mori, *Relations between Reshetikhin–Turaev and re-normalized link invariants*, `arXiv:2209.03608`, 2022.

[Oht02] T. Ohtsuki, *Quantum invariants*, Series on Knots and Everything, vol. 29, World Scientific Publishing Co., Inc., River Edge, NJ, 2002. A study of knots, 3-manifolds, and their sets. MR1881401

[PT88] J. H. Przytycki and P. Traczyk, *Invariants of links of Conway type*, Kobe J. Math. **4** (1988), no. 2, 115–139. MR945888

[RS93] L. Rozansky and H. Saleur, *S- and T-matrices for the super* U(1, 1) WZW *model. Application to surgery and* 3*-manifolds invariants based on the Alexander-Conway polynomial*, Nuclear Phys. B **389** (1993), no. 2, 365–423, DOI 10.1016/0550-3213(93)90326-K. MR1201534

[RS94] L. Rozansky and H. Saleur, *Reidemeister torsion, the Alexander polynomial and* U(1, 1) *Chern-Simons theory*, J. Geom. Phys. **13** (1994), no. 2, 105–123, DOI 10.1016/0393-0440(94)90022-1. MR1260593

[RT90] N. Yu. Reshetikhin and V. G. Turaev, *Ribbon graphs and their invariants derived from quantum groups*, Comm. Math. Phys. **127** (1990), no. 1, 1–26. MR1036112

[RT91] N. Reshetikhin and V. G. Turaev, *Invariants of* 3*-manifolds via link polynomials and quantum groups*, Invent. Math. **103** (1991), no. 3, 547–597, DOI 10.1007/BF01239527. MR1091619

[Tur16] V. G. Turaev, *Quantum invariants of knots and 3-manifolds*, De Gruyter Studies in Mathematics, vol. 18, De Gruyter, Berlin, 2016. Third edition [of MR1292673], DOI 10.1515/9783110435221. MR3617439

[Vir06] O. Ya. Viro, *Quantum relatives of the Alexander polynomial* (Russian, with Russian summary), Algebra i Analiz **18** (2006), no. 3, 63–157, DOI 10.1090/S1061-0022-07-00956-9; English transl., St. Petersburg Math. J. **18** (2007), no. 3, 391–457. MR2255851

[Wal60] A. H. Wallace, *Modifications and cobounding manifolds*, Canadian J. Math. **12** (1960), 503–528, DOI 10.4153/CJM-1960-045-7. MR125588

[Wit89] E. Witten, *Quantum field theory and the Jones polynomial*, Comm. Math. Phys. **121** (1989), no. 3, 351–399. MR990772

[Wit91] E. Witten, *Quantization of Chern-Simons gauge theory with complex gauge group*, Comm. Math. Phys. **137** (1991), no. 1, 29–66. MR1099255

Department of Mathematics & Statistics, Utah State University, Logan, Utah 84322
Email address: nathan.geer@gmail.com

Department of Mathematics & Statistics, Utah State University, Logan, Utah 84322
Email address: adam.robertson@usu.edu

Department of Mathematics & Statistics, Utah State University, Logan, Utah 84322
Email address: jspellmann@outlook.de

Department of Mathematics & Statistics, Utah State University, Logan, Utah 84322
Email address: matthew.young@usu.edu

Contemporary Mathematics
Volume **834**, 2026
https://doi.org/10.1090/conm/834/16682

Two-dimensional topological quantum field theories of rank two over Dedekind domains

Fabian Espinoza, Mee Seong Im, and Mikhail Khovanov

ABSTRACT. We give examples of Frobenius algebras of rank two over ground Dedekind rings which are projective but not free and discuss possible applications of these algebras to link homology.

1. Introduction

An n-dimensional topological quantum field theory (TQFT) is a monoidal functor $\mathcal{F} : \mathsf{Cob}_n \longrightarrow R\text{–}\mathsf{mod}$, where Cob_n is the category of n-dimensional cobordisms between $(n-1)$-dimensional manifolds, that are either oriented or unoriented, and the target category $R\text{–}\mathsf{mod}$ is the category of R-modules, where R is a commutative ring. We write $(n-1)$-manifolds to mean $(n-1)$-dimensional manifolds.

It is easy to see that, for any $(n-1)$-manifold N, the module $\mathcal{F}(N)$ is a finitely-generated projective R-module; see [**GIK⁺23**, Proposition 2.1] for $n = 1$, and the same argument works for any n.

A one-dimensional oriented TQFT $\mathcal{F}$ is given by such a finitely generated projective R-module P. Namely, if $P = \mathcal{F}(+)$, then its dual $P^* \cong \mathcal{F}(-)$, where $+$ and $-$ are points viewed as 0-manifolds with positive and negative orientation, respectively. To *cup* and *cap* cobordisms [**GIK⁺23**], this TQFT assigns co-evaluation and evaluation maps, respectively. Consequently, the isomorphism classes of oriented one-dimensional TQFTs are in a bijection with finitely generated projective R-modules.

Topological quantum field theories are most commonly studied over a field K, with $\mathcal{F}(N)$ a finite-dimensional K-vector space for any $(n-1)$-manifold N. At the same time, integral structures are often essential in a more refined analysis of TQFTs. For example, the Witten–Reshetikhin–Turaev (WRT) 3D TQFT [**Wit89, RT91**] is defined over $\mathbb{C}$, with a root of unity parameter $q \in \mathbb{C}$. Integral structures in these TQFTs, see [**CL05**] and references therein, relate to Lusztig–Kashiwara canonical bases in irreducible representations of quantum groups [**Kas90, Kas91, Lus90, Lus10**]. The integral structure also leads to the Habiro rings [**Hab02, Hab04, GSWZ24**] and to the conjectural expansions at $q = 0$ of the analytic

2020 *Mathematics Subject Classification.* Primary 11R04, 57K16, 18M05, 16L60.

Key words and phrases. Two-dimensional TQFT, commutative Frobenius algebra, Dedekind domain, ideal class group, number field.

The third author would like to acknowledge partial support from NSF grant DMS-2204033 and Simons Collaboration Award 994328.

extension of the WRT invariants into the unit disk. These expansions are expected to have integer coefficients which lift to the Euler characteristics of an unknown homology theory for 3-manifolds [**GPV17**]. In these developments, however, one does not study modifications of the WRT TQFT with state spaces of 2-manifolds being non-free modules over the ground ring.

Another notable example of integrality is the theory of invertible TQFTs. In an invertible n-dimensional TQFT, the object $\mathcal{F}(N)$ associated to an $(n-1)$-manifold N is an invertible object of the target monoidal category, such as the category of line bundles over an algebraic variety, and the map that $\mathcal{F}$ associates to each n-cobordism is invertible [**SP24, Fre14**].

In this paper, we start to investigate two-dimensional TQFTs over a commutative ring R such that the state space A of a circle is not a free R-module. A 2D TQFT is determined by the commutative Frobenius algebra structure on the R-algebra $A = \mathcal{F}(\mathbb{S}^1)$ which is the state space of a circle $\mathbb{S}^1$. If R has no zero divisors (i.e. R is a domain), consider the field of fractions $K = K(R)$. Tensoring with K over R converts commutative Frobenius pair (A, R) to commutative Frobenius pair (A_K, K) over the field K, where $A_K := A \otimes_R K$.

We are looking for examples of commutative Frobenius R-algebras A which are projective but not free modules over R. For simplicity and convenience, assume that R is a Dedekind ring. A Dedekind ring $\mathcal{O}$ often admit projective non-free modules, and the supply of the latter is measured by the ideal class group $\mathsf{Cl}(\mathcal{O})$. Examples of Dedekind rings include the ring of integers $\mathcal{O}_K$ in a number field K and the ring of functions on a smooth irreducible algebraic curve.

Any projective $\mathcal{O}$-module M of rank n has the form $M \cong \mathcal{O}^{n-1} \oplus P$, where P is a rank one projective $\mathcal{O}$-module. The isomorphism classes of the latter are parametrized by the elements of the ideal class group $\mathsf{Cl}(\mathcal{O})$, which is an abelian group. If $\mathcal{O} = \mathcal{O}_K$ is the ring of integers in a number field K, the ideal class group is finite. For ideals $\mathfrak{a}$ and $\mathfrak{a}'$, viewed as projective $\mathcal{O}$-modules, we have $\mathfrak{a} \oplus \mathfrak{a}' \cong \mathcal{O} \oplus \mathfrak{a}\mathfrak{a}'$.

Let A be a commutative Frobenius algebra over a Dedekind ring $\mathcal{O}$. The algebra A is equipped with the unit element, given by a morphism $\iota : \mathcal{O} \longrightarrow A$, with commutative, associative and unital multiplication $m : A^{\otimes 2} \longrightarrow A$ and with a nondegenerate trace map $\varepsilon : A \longrightarrow \mathcal{O}$.

Define the dual $\mathcal{O}$-module $A^* := \mathsf{Hom}_{\mathcal{O}}(A, \mathcal{O})$. The map ε as above induces a map of $\mathcal{O}$-modules

$$(1.1) \qquad \widetilde{\varepsilon} : A \longrightarrow A^*, \quad A \ni a \mapsto \varepsilon_a \in A^*, \quad \varepsilon_a(b) := \varepsilon(ab), \quad b \in A.$$

The map ε is called *nondegenerate* if $\widetilde{\varepsilon}$ is an isomorphism of $\mathcal{O}$-modules. The trace ε is nondegenerate precisely when A is Frobenius over $\mathcal{O}$.

Dualize the multiplication to a map

$$A^* \xrightarrow{m^*} (A \otimes_{\mathcal{O}} A)^* \cong A^* \otimes_{\mathcal{O}} A^*.$$

Since A is Frobenius, ε is nondegenerate, and composing with $\widetilde{\varepsilon}$ and $\widetilde{\varepsilon}^{-1} \otimes \widetilde{\varepsilon}^{-1}$ results in the map $\Delta : A \longrightarrow A^{\otimes 2}$ dual to the multiplication m in an appropriate sense:

(1.2)
$$\begin{array}{ccc} A^* & \xrightarrow{m^*} & (A^{\otimes 2})^* \simeq A^* \otimes A^* \\ \Big\uparrow \tilde{\varepsilon} & & \Big\downarrow \tilde{\varepsilon}^{-1} \otimes \Big\downarrow \tilde{\varepsilon}^{-1}. \\ A & \xrightarrow{\Delta} & A \otimes A \end{array}$$

There is an $\mathcal{O}$-module decomposition $A \cong \mathcal{O}^{n-1} \oplus P$ for some projective rank one $\mathcal{O}$-module $P \in \mathsf{Cl}(\mathcal{O})$. Then $A^* \cong \mathcal{O}^{n-1} \oplus P^*$. An isomorphism of $\mathcal{O}$-modules $A \cong A^*$ implies that $P \cong P^*$, so that $P \otimes P \cong \mathcal{O}$.

COROLLARY 1.1. *For a commutative Frobenius $\mathcal{O}$-algebra A as above, rank one projective $\mathcal{O}$-module P has order at most two in the ideal class group $\mathsf{Cl}(\mathcal{O})$.*

In our search for projective non-free commutative Frobenius algebras A, we are thus limited to the case when $A \cong \mathcal{O}^{n-1} \oplus P$ with $P \otimes P \cong \mathcal{O}$ and $P \ncong \mathcal{O}$. Thus, P is a rank one projective of order two in the ideal class group.

2. Rank two TQFTs

2.1. Integrality conditions and equations. We now look for rank two commutative Frobenius algebras A over a Dedekind ring $\mathcal{O}$ with the field of fractions $K = K(\mathcal{O})$ such that A is not a free $\mathcal{O}$-module.

Since $\mathsf{rk}_{\mathcal{O}}(A) = 2$, we can decompose $A \cong \mathcal{O} \oplus \mu$, where projective module (or ideal) μ has order two in $\mathsf{Cl}(K)$, $\mu \ncong \mathcal{O}$, $\mu^2 \cong \mathcal{O}$. We can assume that $\mu \subset \mathcal{O}$, write $\mu\mu = (z)$ for some $z \in \mathcal{O}$, and interchangeably treat μ as a projective $\mathcal{O}$-module and as an ideal of $\mathcal{O}$. Assume that

$$A \cong \mathcal{O}1 \oplus \mu X, \tag{2.1}$$

where 1 and X are generators, and the multiplication is given by

$$X^2 = aX + b, \quad a, b \in K. \tag{2.2}$$

Computing the multiplication on μ-multiples of X,

$$(u_1X)(u_2X) = u_1u_2aX + u_1u_2b, \quad u_1, u_2 \in \mu,$$

tells us that $u_1u_2a \in \mu$ and $u_1u_2b \in \mathcal{O}$ for all $u_1, u_2 \in \mu$. Consequently,

$$a \in \mu^{-1} = z^{-1}\mu, \quad b \in z^{-1}\mathcal{O},$$

and any such pair (a, b) turns A into a commutative unital $\mathcal{O}$-algebra with the above multiplication.

We next ask for a nondegenerate trace map $\varepsilon : A \longrightarrow \mathcal{O}$. The maps ε that are $\mathcal{O}$-linear are described by the following data:

$$\varepsilon(1) \in \mathcal{O}, \quad \varepsilon|_{\mu X} : \mu \longrightarrow \mathcal{O}.$$

The map $\varepsilon : \mu X \longrightarrow \mathcal{O}$ is given by the element $\varepsilon(X) \in \mu^{-1} = z^{-1}\mu$. Thus, without the nondegeneracy condition, a trace map is given by a pair $(\varepsilon(1), \varepsilon(X))$ with $\varepsilon(1) \in \mathcal{O}$ and $\varepsilon(X) \in z^{-1}\mu$.

Together, the data of an algebra A with a trace ε is given by the quadruple

$$(a, b, \varepsilon(1), \varepsilon(X)), \quad a \in z^{-1}\mu, \quad b \in z^{-1}\mathcal{O}, \quad \varepsilon(1) \in \mathcal{O}, \quad \varepsilon(X) \in z^{-1}\mu. \tag{2.3}$$

Next, consider the nondegeneracy condition on the trace. First, write the map $\widetilde{\varepsilon}$ in (1.1) explicitly over K, using the dual basis $\{1^*, X^*\}$ for $A^*_K := A^* \otimes_{\mathcal{O}} K$. Here

$$1^*(1) = 1, \quad 1^*(X) = 0, \quad X^*(1) = 0, \quad X^*(X) = 1.$$

Integrally,

$$A \cong \mathcal{O}1 \oplus \mu X \quad \text{and} \quad A^* = \mathcal{O}1^* \oplus z^{-1}\mu X^*. \tag{2.4}$$

The map

$$\widetilde{\varepsilon}: A \longrightarrow A^* \tag{2.5}$$

in (1.1) should be an isomorphism. Let us compute this map, first over the field of fractions K:

(2.6)

$$\widetilde{\varepsilon}(1) = \varepsilon(1)1^* + \varepsilon(X)X^*, \quad \widetilde{\varepsilon}(X) = \varepsilon(X)1^* + \varepsilon(X^2)X^* = \varepsilon(X)1^* + (a\varepsilon(X) + b\varepsilon(1))X^*,$$

or, in matrix notation,

$$\begin{pmatrix} \widetilde{\varepsilon}(1) \\ \widetilde{\varepsilon}(X) \end{pmatrix} = \begin{pmatrix} \varepsilon(1) & \varepsilon(X) \\ \varepsilon(X) & \varepsilon(X^2) \end{pmatrix} \begin{pmatrix} 1^* \\ X^* \end{pmatrix}. \tag{2.7}$$

Let

$$M := \begin{pmatrix} \varepsilon(1) & \varepsilon(X) \\ \varepsilon(X) & \varepsilon(X^2) \end{pmatrix}, \quad \widetilde{\Delta} := \mathsf{det}(M) = \varepsilon(1)\varepsilon(X^2) - \varepsilon(X)^2. \tag{2.8}$$

Necessarily, $\widetilde{\Delta} \neq 0$. The inverse map $\widetilde{\varepsilon}^{-1} : A^*_K \longrightarrow A_K$ in these bases is given by the inverse matrix M^{-1}:

$$\begin{pmatrix} \widetilde{\varepsilon}^{-1}(1^*) \\ \widetilde{\varepsilon}^{-1}(X^*) \end{pmatrix} = \widetilde{\Delta}^{-1} \begin{pmatrix} \varepsilon(X^2) & -\varepsilon(X) \\ -\varepsilon(X) & \varepsilon(1) \end{pmatrix} \begin{pmatrix} 1 \\ X \end{pmatrix}. \tag{2.9}$$

Let us write down multiplication in A_K:

$$m: \ 1 \otimes 1 \mapsto 1, \quad 1 \otimes X \mapsto X, \quad X \otimes 1 \mapsto X, \quad X \otimes X \mapsto aX + b1, \tag{2.10}$$

and transpose it to comultiplication in A^*_K:

$$m^* : \ 1^* \longrightarrow 1^* \otimes 1^* + bX^* \otimes X^*, \quad X^* \longrightarrow 1^* \otimes X^* + X^* \otimes 1^* + aX^* \otimes X^*. \tag{2.11}$$

We have

$$\Delta = (\widetilde{\varepsilon}^{-1} \otimes \widetilde{\varepsilon}^{-1}) \circ m^* \circ \widetilde{\varepsilon}$$

and

$$\begin{aligned}
\Delta(1) &= (\widetilde{\varepsilon}^{-1} \otimes \widetilde{\varepsilon}^{-1}) \circ m^* \circ \widetilde{\varepsilon}(1) \\
&= (\widetilde{\varepsilon}^{-1} \otimes \widetilde{\varepsilon}^{-1}) \circ m^*(\varepsilon(1)1^* + \varepsilon(X)X^*) \\
&= (\widetilde{\varepsilon}^{-1} \otimes \widetilde{\varepsilon}^{-1})(\varepsilon(1)\,(1^* \otimes 1^* + bX^* \otimes X^*) \\
&\quad + \varepsilon(X)\,(1^* \otimes X^* + X^* \otimes 1^* + aX^* \otimes X^*) \\
&= (\widetilde{\varepsilon}^{-1} \otimes \widetilde{\varepsilon}^{-1})\,(\varepsilon(1)1^* + \varepsilon(X)X^*) \otimes 1^* + \big(\varepsilon(X)1^* + \varepsilon(X^2)X^*\big) \otimes X^*.
\end{aligned}$$

We compute

$$\widetilde{\varepsilon}^{-1}(\varepsilon(1)1^* + \varepsilon(X)X^*) = \widetilde{\Delta}^{-1}(\varepsilon(1)(\varepsilon(X^2)1 - \varepsilon(X)X) + \varepsilon(X)(-\varepsilon(X)1 + \varepsilon(1)X)) = 1,$$

and

$$\begin{aligned}
&\widetilde{\varepsilon}^{-1}(\varepsilon(X)1^* + \varepsilon(X^2)X^*) \\
&\qquad = \widetilde{\Delta}^{-1}\,\big(\varepsilon(X)(\varepsilon(X^2)1 - \varepsilon(X)X) + \varepsilon(X^2)(-\varepsilon(X)1 + \varepsilon(1)X)\big) = X.
\end{aligned}$$

Thus,

$$\begin{aligned}\Delta(1) &= 1 \otimes \widetilde{\varepsilon}^{-1}(1^*) + X \otimes \widetilde{\varepsilon}^{-1}(X^*) \\ &= \widetilde{\Delta}^{-1}\left(1 \otimes (\varepsilon(X^2)1 - \varepsilon(X)X) + X \otimes (-\varepsilon(X)1 + \varepsilon(1)X)\right)\end{aligned}$$

and we arrive at the formula

$$\Delta(1) = \widetilde{\Delta}^{-1}\left(\varepsilon(X^2)1 \otimes 1 - \varepsilon(X)(1 \otimes X + X \otimes 1) + \varepsilon(1)X \otimes X\right) \tag{2.12}$$

in the Frobenius K-algebra A_K. Rewrite this formula as

$$\Delta(1) = \widetilde{\Delta}^{-1}\left(1 \otimes (\varepsilon(X^2)1 - \varepsilon(X)X) + X \otimes (-\varepsilon(X)1 + \varepsilon(1)X)\right). \tag{2.13}$$

It tells us that the dual to the basis $(1, X)$ of A_K is the basis

$$\widetilde{\Delta}^{-1}\left(\varepsilon(X^2)1 - \varepsilon(X)X, -\varepsilon(X)1 + \varepsilon(1)X\right). \tag{2.14}$$

Map $\widetilde{\varepsilon}^{-1} : A_K^* \longrightarrow A_K$ on K-vector spaces should restrict to a map $\widetilde{\varepsilon}^{-1} : A^* \longrightarrow A$. From (2.4) this means

$$\widetilde{\varepsilon}^{-1}(1^*), \widetilde{\varepsilon}^{-1}(z^{-1}uX^*) \in \mathcal{O}1 \oplus \mu X, \quad \forall u \in \mu.$$

Equation (2.9) gives

$$\widetilde{\varepsilon}^{-1}(1^*) = \widetilde{\Delta}^{-1}\left(\varepsilon(X^2)1 - \varepsilon(X)X\right), \quad \widetilde{\varepsilon}^{-1}(z^{-1}uX^*) = \widetilde{\Delta}^{-1}z^{-1}u\left(-\varepsilon(X)1 + \varepsilon(1)X\right) \tag{2.15}$$

and the integrality conditions are

$$\widetilde{\Delta}^{-1}\varepsilon(X^2) \in \mathcal{O}, \quad \widetilde{\Delta}^{-1}\varepsilon(X) \in \mu, \quad \widetilde{\Delta}^{-1}z^{-1}u\varepsilon(X) \in \mathcal{O}, \quad \widetilde{\Delta}^{-1}z^{-1}u\varepsilon(1) \in \mu, \quad \forall u \in \mu, \tag{2.16}$$

where $\widetilde{\Delta}$ is given by (2.8). These conditions can be rewriten as

$$\varepsilon(X^2) \in \widetilde{\Delta}\mathcal{O}, \quad \varepsilon(X) \in \widetilde{\Delta}\mu, \quad \varepsilon(1) \in \widetilde{\Delta}z\mathcal{O}. \tag{2.17}$$

Let us go back to formulas (2.4) and (2.6) and continue with the integrality constraints. Integrally, $\widetilde{\varepsilon}(1) \in A^*$, and for $\widetilde{\varepsilon}(uX) \in A^*$, $u \in \mu$, one additionally needs

$$\varepsilon(uX) = u\varepsilon(X)1^* + u(a\varepsilon(X) + b\varepsilon(1))X^* \in \mathcal{O}1^* \oplus z^{-1}\mu X^*, \quad u \in \mu.$$

Note that $\varepsilon(X^2) = a\varepsilon(X) + b\varepsilon(1)$. Equivalently, we need $u\varepsilon(X) \in \mathcal{O}$, which holds due to (2.3), and $u(a\varepsilon(X) + b\varepsilon(1)) \in z^{-1}\mu$ since

$$ua\varepsilon(X) \in \mu z^{-1}\mu z^{-1}\mu = z^{-1}\mu$$

and

$$ub\,\varepsilon(1) \in \mu z^{-1}\mathcal{O} = z^{-1}\mu.$$

Next, for $\widetilde{\varepsilon} : A \longrightarrow A^*$ to be an isomorphism, one needs 1^* and $z^{-1}uX^*$ for $u \in \mu$ to be in the image of $\widetilde{\varepsilon}$.

(I) For 1^* to be in the image of $\widetilde{\varepsilon}$, one needs $\widetilde{\varepsilon}(c1 + dX) = 1^*$ for some c and d. That is,

$$\varepsilon_{c1+dX}(1) = \varepsilon(c1 + dX) = c\varepsilon(1) + d\varepsilon(X) = 1$$

and

$$\begin{aligned}\varepsilon_{c1+dX}(X) &= \varepsilon((c1+dX)X)\\ &= \varepsilon(cX+dX^2)\\ &= \varepsilon(cX+d(aX+b))\\ &= c\varepsilon(X)+ad\varepsilon(X)+bd\varepsilon(1)\\ &= 0.\end{aligned}$$

Thus, for some $c \in \mathcal{O}, d \in \mu$, the two equations

$$c\varepsilon(1)+d\varepsilon(X)=1, \quad c\varepsilon(X)+d(a\varepsilon(X)+b\varepsilon(1))=0 \tag{2.18}$$

should hold, or, in matrix form, they are written as

$$\begin{pmatrix} c & d \\ bd & ad+c \end{pmatrix}\begin{pmatrix}\varepsilon(1)\\ \varepsilon(X)\end{pmatrix}=\begin{pmatrix}1\\0\end{pmatrix}. \tag{2.19}$$

A useful parameter is $t = a\varepsilon(X)+b\varepsilon(1)=\varepsilon(X^2)$, also see (2.24).

(II) For $z^{-1}uX^*$ to be in the image of $\widetilde{\varepsilon}$, where $u \in \mu$, one needs $c'(u)=uc'(1) \in \mathcal{O}$ and $d'(u)=ud'(1) \in \mu$, where $\widetilde{\varepsilon}(c'(u)+d'(u)X)=z^{-1}uX^*$. Here, $c' : \mu \longrightarrow \mathcal{O}$ and $d' : \mu \longrightarrow \mu$ are suitable $\mathcal{O}$-linear maps that describe elements mapped to multiples of X^* by $\widetilde{\varepsilon}$. These extend to $\mathcal{O}$-linear maps $c' : \mathcal{O} \longrightarrow z^{-1}\mu$ and $d' : \mathcal{O} \longrightarrow \mathcal{O}$.

Similar to Case I, we have

$$\varepsilon_{c'(u)+d'(u)X}(1)=\varepsilon(c'(u)+d'(u)X)=c'(u)\varepsilon(1)+d'(u)\varepsilon(X)=0,$$

and

$$\begin{aligned}\varepsilon_{c'(u)+d'(u)X}(X) &= \varepsilon(c'(u)X+d'(u)X^2)\\ &= \varepsilon(c'(u)X+d'(u)(aX+b))\\ &= c'(u)\varepsilon(X)+ad'(u)\varepsilon(X)+bd'(u)\varepsilon(1)\\ &= uz^{-1}.\end{aligned}$$

These relations give us

$$\begin{pmatrix} c'(u) & d'(u) \\ bd'(u) & ad'(u)+c'(u) \end{pmatrix}\begin{pmatrix}\varepsilon(1)\\ \varepsilon(X)\end{pmatrix}=\begin{pmatrix}0\\ uz^{-1}\end{pmatrix}, \tag{2.20}$$

Since $c'(u)$ and $d'(u)$ are linear in u, we can factor u out and use shorthand notation

$$c' := c'(1) \in z^{-1}\mu, \quad d' := d'(1) \in \mathcal{O} \tag{2.21}$$

to get

$$\begin{pmatrix} c' & d' \\ bd' & ad'+c' \end{pmatrix}\begin{pmatrix}\varepsilon(1)\\ \varepsilon(X)\end{pmatrix}=\begin{pmatrix}0\\ z^{-1}\end{pmatrix}. \tag{2.22}$$

Write this as

$$c'\varepsilon(1)+d'\varepsilon(X)=0, \quad c'\varepsilon(X)+d'(a\varepsilon(X)+b\varepsilon(1))=z^{-1}.$$

Summary I. Given $a, b, \varepsilon(1), \varepsilon(X)$ as in (2.3) and copied below:

$$(a,b,\varepsilon(1),\varepsilon(X)), \quad a \in z^{-1}\mu, \quad b \in z^{-1}\mathcal{O}, \quad \varepsilon(1) \in \mathcal{O}, \quad \varepsilon(X) \in z^{-1}\mu, \tag{2.23}$$

also define

$$t := a\varepsilon(X)+b\varepsilon(1)=\varepsilon(X^2), \tag{2.24}$$

TABLE 2.1

Integrality conditions on the parameters			
$z^{-1}\mathcal{O}$	$z^{-1}\mu$	$\mathcal{O}$	μ
b	a $\varepsilon(X)$	 $\varepsilon(1)$	
	 c'	c d'	d
t			

where the second equality is obtained by applying ε to (2.2) (in particular, $t \in z^{-1}\mathcal{O}$), and look for $c \in \mathcal{O}, d \in \mu, c' \in z^{-1}\mu, d' \in \mathcal{O}$ such that matrix relations (2.19) and (2.22) hold. These are linear relations

$$c\varepsilon(1) + d\varepsilon(X) = 1, \tag{2.25}$$

$$c\varepsilon(X) + dt = 0, \tag{2.26}$$

$$c'\varepsilon(1) + d'\varepsilon(X) = 0, \tag{2.27}$$

$$c'\varepsilon(X) + d't = z^{-1}, \tag{2.28}$$

which can also be written as matrix equations (cf. (2.8))

$$\begin{pmatrix} \varepsilon(1) & \varepsilon(X) \\ \varepsilon(X) & \varepsilon(X^2) \end{pmatrix} \begin{pmatrix} c \\ d \end{pmatrix} = \begin{pmatrix} 1 \\ 0 \end{pmatrix}, \quad \begin{pmatrix} \varepsilon(1) & \varepsilon(X) \\ \varepsilon(X) & \varepsilon(X^2) \end{pmatrix} \begin{pmatrix} c' \\ d' \end{pmatrix} = \begin{pmatrix} 0 \\ z^{-1} \end{pmatrix}. \tag{2.29}$$

In Table 2.1, the ideals decrease from left to right: $z^{-1}\mathcal{O} \supset z^{-1}\mu \supset \mathcal{O} \supset \mu$. The four columns describe where corresponding parameters should lie. We have different rows for the two different quadruples of parameters. In the relations above, c and d interact only with the first four parameters (before the change of variables using t), and likewise for c' and d'.

Multiply (2.25) by $\varepsilon(X)$ and subtract from it (2.26) multiplied by $\varepsilon(1)$ to obtain

$$d\left(\varepsilon(X)^2 - \varepsilon(1)\varepsilon(X^2)\right) = \varepsilon(X). \tag{2.30}$$

(Recall that $\varepsilon(X)^2 - \varepsilon(1)\varepsilon(X^2) = -\mathsf{det}(M) = -\widetilde{\Delta}$, see (2.8).) Next, write (2.27) as $\varepsilon(X)d' = -c'\varepsilon(1)$ and insert into (2.28) multiplied by $\varepsilon(X)$ to get

$$c'\left(\varepsilon(X)^2 - \varepsilon(1)\varepsilon(X^2)\right) = z^{-1}\varepsilon(X). \tag{2.31}$$

From (2.30) and (2.31), we obtain

$$(d - zc')\left(\varepsilon(X)^2 - \varepsilon(1)\varepsilon(X^2)\right) = 0. \tag{2.32}$$

Consequently, either (I) $d = zc'$ or (II) $\varepsilon(X)^2 = \varepsilon(1)\varepsilon(X^2)$. If (II) holds, then the 2×2 matrix in (2.29) has rank 1 over the field of fractions K, and the system of equations in (2.29) cannot both hold. Hence,

$$d \;=\; zc' \tag{2.33}$$

and

$$-\widetilde{\Delta} = \varepsilon(X)^2 - \varepsilon(1)\varepsilon(X^2) \neq 0. \tag{2.34}$$

Now write (2.27) as

$$d\varepsilon(1) + d'z\varepsilon(X) = 0 \tag{2.35}$$

TABLE 2.2

Integrality conditions on rescaled parameters		
$\mathcal{O}$	μ	$\underline{\varepsilon}_X\mathcal{O}$
$\underline{b}$	$\underline{a}$	
$\varepsilon(1)$	$\underline{\varepsilon}_X$	
c, d'	d	
$\underline{t}$		$d\varepsilon(1)$

and (2.28) as

$$d\varepsilon(X) + d'z\varepsilon(X^2) = 1. \tag{2.36}$$

Rescaling (2.36) by $\varepsilon(1)$ and (2.35) by $\varepsilon(X)$ and then subtracting one from the other gives

$$d'z\left(\varepsilon(X^2)\varepsilon(1) - \varepsilon(X)^2\right) = \varepsilon(1).$$

That d' defined by this equation is in $\mathcal{O}$, see the table above, is equivalent to the third condition in (2.17).

We can express c' and essentially reduce d' to d via the equations

$$c' = z^{-1}d, \quad d'\varepsilon(X)z = -d\varepsilon(1) \tag{2.37}$$

since $z \in K$ for the former and using (2.33) and (2.27) for the latter.

(i) If $\varepsilon(X) \neq 0$, we can express both c', d' via d, and equations (2.27) and (2.28) become redundant. Integrality condition on c' reduces to that for d, integrality condition on d' becomes

$$d\varepsilon(1)\varepsilon(X)^{-1} \in (z). \tag{2.38}$$

We can then remove c', d' from the list of our parameters, remove (2.27), (2.28) and add integrality condition (2.38).

(ii) Case $\varepsilon(X) = 0$ is more straightforward, implying $d = 0$ and $c' = 0$, via (2.30) and (2.31), respectively.

For now let us consider both cases at once and rescale the parameters to reduce them to elements of ideals $\mathcal{O}$ and μ. Rescale the parameters as follows

$$b = z^{-1}\underline{b}, \quad a = z^{-1}\underline{a}, \quad \varepsilon(X) = z^{-1}\underline{\varepsilon}_X, \quad t = z^{-1}\underline{t} \tag{2.39}$$

so that

$$\underline{b}, \underline{t} \in \mathcal{O}, \qquad \underline{a}, \underline{\varepsilon}_X \in \mu, \quad d\,\varepsilon(1) \in \underline{\varepsilon}_X\mathcal{O}.$$

The corresponding table is Table 2.1. Now

$$\underline{t} = z^{-1}\underline{a}\,\underline{\varepsilon}_X + \underline{b}\varepsilon(1), \tag{2.40}$$

and the relations (2.25)–(2.28) become

$$c\,\varepsilon(1) + z^{-1}d\,\underline{\varepsilon}_X = 1, \tag{2.41}$$

$$c\,\underline{\varepsilon}_X = -d\,\underline{t}, \tag{2.42}$$

$$d'\underline{\varepsilon}_X = -d\,\varepsilon(1), \tag{2.43}$$

$$z^{-1}\,d\underline{\varepsilon}_X + d'\,\underline{t} = 1. \tag{2.44}$$

Subtracting (2.41) from (2.44), we can replace the latter by

$$d'\,\underline{t} = c\,\varepsilon(1). \tag{2.45}$$

Equations (2.41)–(2.44) can also be written in the following matrix form

$$\begin{pmatrix} \varepsilon(1) & z^{-1}\underline{\varepsilon}_X \\ \underline{\varepsilon}_X & \underline{t} \end{pmatrix} \begin{pmatrix} c \\ d \end{pmatrix} = \begin{pmatrix} 1 \\ 0 \end{pmatrix}, \qquad \begin{pmatrix} \varepsilon(1) & \underline{\varepsilon}_X \\ z^{-1}\underline{\varepsilon}_X & \underline{t} \end{pmatrix} \begin{pmatrix} d \\ d' \end{pmatrix} = \begin{pmatrix} 0 \\ 1 \end{pmatrix}. \tag{2.46}$$

While the multiplication in A_K is given by (2.2), the multiplication in A is based on multiplying two elements in μX and given by

$$u_1 X\, u_2 X = \frac{u_1 u_2}{z}\,(\underline{a}X + \underline{b}), \qquad u_1, u_2 \in \mu. \tag{2.47}$$

This shows another use for $\underline{a}, \underline{b}$. Note that $u_1 u_2 z^{-1}, \underline{b} \in \mathcal{O}$, $\underline{a} \in \mu$, so the right hand side is in $A = \mathcal{O}1 \oplus \mu X$.

Proposition 2.1. *In any example as above,* $\varepsilon(1) \neq 0$, $\underline{t} \neq 0$, $c \neq 0$, *and* $d' \neq 0$.

Proof.

(I) Assume $\varepsilon(1) = 0$. Then $\underline{t} = z^{-1}\underline{a}\,\underline{\varepsilon}_X$ and equation (2.41) becomes $d\underline{\varepsilon}_X = z$, giving us equality of principal ideals $(d)(\underline{\varepsilon}_X) = (d\underline{\varepsilon}_X) = (z)$. There are proper inclusions of ideals $(d) \subset \mu$ and $(\underline{\varepsilon}_X) \subset \mu$. These produce a chain of inclusions of ideals

$$(z) = (d)(\underline{\varepsilon}_X) \subset (d)\mu \subset \mu\mu = (z),$$

implying that all ideals in the chain are equal. But (z) is principal and $(d)\mu$ is not principal, giving a contradiction. Hence $\varepsilon(1) \neq 0$.

(II) Assume $\underline{t} = 0$. We already know from (2.34) that $t = \varepsilon(X)^2 - \varepsilon(1)\varepsilon(X^2) \neq 0$, so that $\underline{t} \neq 0$. Here is another proof: from (2.42), $c\underline{\varepsilon}_X = 0$. If $c = 0$, from Equation (2.41), $d\underline{\varepsilon}_X = z$. We get a contradiction, as in the proof of part I, since $d, \underline{\varepsilon}_X \in \mu$. Consequently, $\underline{\varepsilon}_X = 0$. Equation (2.44) then implies $d'\underline{t} = 1$, which is a contradiction since we have assumed $\underline{t} = 0$.

(III) If $c = 0$, then (2.42) results in $d = 0$, giving a contradiction with (2.41), $0 = 1$. So $c \neq 0$.

(IV) If $d' = 0$, then (2.44) reduces to $d\underline{\varepsilon}_X = z$, leading to a contradiction as in part I. □

Remark 2.2 (Twistings). Given a 2D Frobenius algebra (A, μ, ε) as above, it admits the following modifications and parameter changes:

(1) Change of variables $X \mapsto \lambda_0 X$, given an invertible $\lambda_0 \in \mathcal{O}^\times$:

$$a \mapsto \lambda_0 a, \quad b \mapsto \lambda_0^2 b, \quad \varepsilon(X) \mapsto \lambda_0 \varepsilon(X), \quad \varepsilon(1) \mapsto \varepsilon(1). \tag{2.48}$$

(2) Change of variables $X \mapsto X + \lambda_1$, for $\lambda_1 \in z^{-1}\mu$. This changes

$$a \mapsto a + 2\lambda_1, \quad b \mapsto b + \lambda_1^2. \tag{2.49}$$

(3) For an invertible $\lambda \in \mathcal{O}^\times$ rescale

$$\varepsilon \mapsto \lambda\varepsilon, \quad \Delta \mapsto \lambda^{-1}\Delta. \tag{2.50}$$

Invertible linear changes (1) and (2) of X together generate an action of the affine group $\mathsf{Aff}(\mathcal{O}, \mu)$ of symmetries $(X \mapsto aX + b)$ on the set of Frobenius algebras as above with X fixed.

3. Some solutions

Let us now consider two cases, depending on whether or not $\varepsilon(X)$ vanishes.

3.1. Case $\varepsilon(X) = 0$.

PROPOSITION 3.1. *Assume that in an example as above, $\underline{\varepsilon}_X = 0$. Then*

(3.1)
$$c' = d = 0,\ \underline{a} \in \mathcal{O},\ \underline{b}, \varepsilon(1) \in \mathcal{O}^\times,\ \underline{t} = \underline{b}\varepsilon(1) \in \mathcal{O}^\times,\ c = \varepsilon(1)^{-1} \in \mathcal{O}^\times,\ d' = \underline{b}^{-1}\varepsilon(1)^{-1} \in \mathcal{O}^\times.$$

Each of these data provides a commutative Frobenius algebra

$$A = \mathcal{O}1 \oplus \mu X, \quad \mu^2 = (z), \quad X^2 = z^{-1}\underline{a}X + z^{-1}\underline{b}, \tag{3.2}$$

uniquely determined by the ideal μ, by z, by $\underline{a} \in \mathcal{O}$ and $\underline{b}, \varepsilon(1) \in \mathcal{O}^\times$ via the above formulas.

PROOF. Assume $\underline{\varepsilon}_X = 0$. Then $d\underline{t} = 0$ by (2.42). Since $\underline{t} \neq 0$, we get $d = 0$. Then $c\varepsilon(1) = 1$ by (2.41), so that $\varepsilon(1) \in \mathcal{O}^\times$ and $c = \varepsilon(1)^{-1}$. The last two equations (2.43) and (2.44) give $\underline{c'}\varepsilon(1) = 0$, $d'\underline{t} = 1$. Consequently, $\underline{c'} = 0$ and $\underline{t} \in \mathcal{O}^\times$, $d' = \underline{t}^{-1}$. Then (2.40) gives $\underline{t} = \underline{b}\varepsilon(1)$. This requires $\underline{b} \in \mathcal{O}^\times$. Summarizing, in this case:

$$\underline{\varepsilon}_X = c' = d = 0,\ \underline{a} \in \mathcal{O},\ \varepsilon(1), \underline{b} \in \mathcal{O}^\times,\ \underline{t} = \underline{b}\varepsilon(1) \in \mathcal{O}^\times,$$
$$c = \varepsilon(1)^{-1} \in \mathcal{O}^\times,\ d' = \underline{b}^{-1}\varepsilon(1)^{-1} \in \mathcal{O}^\times.$$

This concludes the proof. □

Thus, there is a family of solutions to the problem of constructing a Frobenius $\mathcal{O}$-algebra $A = \mathcal{O}1 \oplus \mu X$, with the square of the generator X given by (2.2), with a non-principal ideal $\mu \subset \mathcal{O}$ with $\mu^2 = (z)$ and trace ε as in (2.3). With the additional condition that $\varepsilon(X) = 0$, the solutions are classified by $\underline{a} \in \mathcal{O}$ and $\varepsilon(1), \underline{b} \in \mathcal{O}^\times$, with the parameters given by (3.1) and (2.39), see also (3.2). The trace ε is 0 on μX, and the restriction of ε to $\mathcal{O}1$ is an isomorphism $\mathcal{O} \longrightarrow \mathcal{O}$, $x \mapsto \varepsilon(1)x$. Isomorphism $\widetilde{\varepsilon}: A \longrightarrow A^*$ in (1.1) is given by

$$\widetilde{\varepsilon}(1) = \varepsilon(1)1^*, \quad \widetilde{\varepsilon}(uX) = z^{-1}\underline{b}\varepsilon(1)\, uX^*, \quad u \in \mu.$$

This construction works for any Dedekind domain $\mathcal{O}$ with a non-principal ideal μ whose square is a principal ideal.

REMARK 3.2. The comultiplication map $\Delta : A \longrightarrow A \otimes A$ is dual to the multiplication via the formula in (1.2). Since it is a map of A-bimodules, it is enough to compute $\Delta(1)$. One can first compute it in the field of fractions K to get

$$\Delta(1) = \varepsilon(1)^{-1}(1 \otimes 1 + b^{-1}X \otimes X). \tag{3.3}$$

Also note that (2.13) in case $\varepsilon(X) = 0$ gives

$$\begin{aligned}\Delta(1) &= \widetilde{\Delta}^{-1}\left(1 \otimes \varepsilon(X^2)1 + X \otimes \varepsilon(1)X\right)\\ &= \varepsilon(1)^{-1}1 \otimes 1 + \varepsilon(X^2)^{-1}X \otimes X\\ &= \varepsilon(1)^{-1}\left(1 \otimes 1 + \varepsilon(1)\varepsilon(X^2)^{-1}X \otimes X\right).\end{aligned}$$

We have $b = \varepsilon(X^2)\varepsilon(1)^{-1}$ for this case. Indeed, from (3.1) we get $\underline{t} = \underline{b}\varepsilon(1)$ so that $t = b\varepsilon(1)$ and $t = \varepsilon(X^2)$, resulting in the coefficient b^{-1} next to $X \otimes X$ in (3.3).

In the $\mathcal{O}$-algebra A term X is only available when scaled by elements of μ. Pick 2 generators s_1, s_2 of the ideal μ, so that $\mu = \mathcal{O}s_1 + \mathcal{O}s_2$. Since $\mu^2 = z$, pick elements

$s_1', s_2' \in \mu$ so that $s_1 s_1' + s_2 s_2' = z$. The element $zX \otimes X$ can be replaced by the sum of two product elements in $\mu X \otimes \mu X$ as shown below (and using that $b^{-1} = \underline{b}^{-1} z$)

$$\Delta(1) = \varepsilon(1)^{-1}\left(1 \otimes 1 + \underline{b}^{-1}(s_1 X \otimes s_1' X + s_2 X \otimes s_2' X)\right). \tag{3.4}$$

REMARK 3.3. What symmetries can we apply to this class of examples, see Remark 2.2? One can rescale X by an element of $\mathcal{O}^*$. Also one can change variables $X \mapsto X + w$, so that $uX \mapsto uX + uw$ and $uw \in \mathcal{O}$ for all $u \in \mu$. Hence $w \in z^{-1}\mu$. This modifies the trace and takes us outside the case $\varepsilon(X) = 0$ since $\varepsilon(X + w) = \varepsilon(w) = w\varepsilon(1) \neq 0$ if $w \neq 0$.

3.2. Case $\varepsilon(X)$ not equal to 0.

Note that $\varepsilon(X) = z^{-1}\underline{\varepsilon}_X$ and $\underline{\varepsilon}_X \in \mu$. Recall Proposition 2.1.

PROPOSITION 3.4. *If $\underline{\varepsilon}_X \neq 0$, then $\underline{b} \neq 0$ and $d \neq 0$.*

PROOF.

(I) If $d = 0$, then (2.42) gives $c\underline{\varepsilon}_X = 0$, so that $\underline{\varepsilon}_X = 0$ in view of Proposition 2.1, which is a contradiction.

(II) Assume $\underline{b} = 0$. Then $\underline{t} = z^{-1}\underline{a}\,\underline{\varepsilon}_X$ by (2.40). Inserting this equation into (2.42) produces $(c + dz^{-1}\underline{a})\underline{\varepsilon}_X = 0$, so that $c = -d\underline{a}z^{-1}$. Inserting the equation for c in (2.41) gives $d(\underline{\varepsilon}_X - \underline{a}\varepsilon(1)) = z$. Since both factors $d, \underline{\varepsilon}_X - \underline{a}\varepsilon(1) \in \mu$, this gives a contradiction as in Part I in the proof of Proposition 2.1, i.e., the chain

$$(z) = (d)(\underline{\varepsilon}_X - \underline{a}\varepsilon(1)) \subset (d)\mu \subset \mu^2 = (z)$$

of inclusion of ideals forces a principal and a non-principal ideal to be equal. Thus, $\underline{b} \neq 0$. □

Assume now that $\underline{\varepsilon}_X \neq 0$. Then $\varepsilon(X) \neq 0$ and

$$d = (1 - c\varepsilon(1))\varepsilon(X)^{-1}, \quad c\varepsilon(X)^2 + (1 - c\varepsilon(1))(a\varepsilon(X) + b\varepsilon(1)) = 0,$$

by (2.25) for the former and for the latter, substitute in $\underline{t}$ in (2.40) into (2.42), and then substitute for c in (2.42) using (2.41). The latter equation simplifies as

$$c\left(\varepsilon(X)^2 - a\varepsilon(X)\varepsilon(1) - b\varepsilon(1)^2\right) = -\left(a\varepsilon(X) + b\varepsilon(1)\right). \tag{3.5}$$

where the left hand side and the right hand side can be rewritten as

$$c\left(\varepsilon(X)^2 - \varepsilon(1)\varepsilon(X^2)\right) = -\varepsilon(X^2). \tag{3.6}$$

Inserting this expression for c into the formula for d, we obtain

$$d\left(\varepsilon(X)^2 - \varepsilon(1)\varepsilon(X^2)\right) = \varepsilon(X). \tag{3.7}$$

Using this formula and (2.43) gives

$$d'\left(\varepsilon(X)^2 - \varepsilon(1)\varepsilon(X^2)\right) = -z^{-1}\varepsilon(1). \tag{3.8}$$

We can write the above three equations as

$$c\widetilde{\Delta} = \varepsilon(X^2), \quad d\widetilde{\Delta} = -\varepsilon(X), \quad d'\widetilde{\Delta} = z^{-1}\varepsilon(1), \tag{3.9}$$

cf. (2.17). What are the integrality properties of $\varepsilon(X)^2 - \varepsilon(1)\varepsilon(X^2) = -\widetilde{\Delta}$? We have

$$\varepsilon(X) \in z^{-1}\mu, \; \varepsilon(1) \in \mathcal{O}, \quad \varepsilon(X^2) = t \in z^{-1}\mathcal{O}.$$

Consequently, $\widetilde{\Delta} \in z^{-1}\mathcal{O}$.

Let us further specialize to particular values of $\varepsilon(X)$.

TABLE 3.1

Integrality conditions for $\varepsilon(X) = 1$		
$\mathcal{O}$	μ	$\mathcal{O}^\times$
$\underline{b}, \varepsilon(1)$	$\underline{a}$	$\underline{d}$

3.2.1. *Case* $\varepsilon(X) = 1$. We have $\varepsilon(X) \in z^{-1}\mu$. Let us try $\varepsilon(X) = 1$ so that $\underline{\varepsilon}_X = z$. Then $\underline{t} = \underline{a} + \underline{b}\varepsilon(1)$ and the relations (2.41)–(2.43) and (2.45) simplify to

$$d = 1 - c\varepsilon(1), \tag{3.10}$$

$$cz = -d\underline{t}, \tag{3.11}$$

$$d'z = -d\,\varepsilon(1), \tag{3.12}$$

$$d'\,\underline{t} = c\varepsilon(1). \tag{3.13}$$

The last relation follows from the previous two. Let us further assume that μ is prime, which we can do if $\mathcal{O}$ is a ring of integers of a number field. (Recall that the Chebotarev Density Theorem implies that any element in $\mathsf{Cl}(K)$ can be represented by a prime ideal $\mu \in \mathcal{O}_K$, in any number field K, see [**ME05**, Exercise 11.2.8] and [**Aph21, Sac16, Guy24**]). Relation (3.12) shows that $d\varepsilon(1) \in (z)$, with $d \in \mu$, so there are two possilities: (I) $\varepsilon(1) \in \mu$ or (II) $d \in (z)$. Case (I) implies, via (3.10), that $1 \in \mu$, a contradiction. Consequently, $d \in (z)$ and we can write $d = z\underline{d}$, with $\underline{d} \in \mathcal{O}$.

Let us rewrite (3.10)–(3.12) and the definition of $\underline{t}$ with the parameter $\underline{d}$:

$$z\underline{d} = 1 - c\varepsilon(1), \tag{3.14}$$

$$c = -\underline{d}\,\underline{t}, \tag{3.15}$$

$$d' = -\underline{d}\varepsilon(1), \tag{3.16}$$

$$\underline{t} = \underline{a} + \underline{b}\varepsilon(1). \tag{3.17}$$

Equations (3.15) and (3.16) allow to get rid of parameters c and d', reducing the relations to

$$1 = \underline{d}(z - \underline{t}\varepsilon(1)), \tag{3.18}$$

$$\underline{t} = \underline{a} + \underline{b}\varepsilon(1). \tag{3.19}$$

Relation (3.18) implies that $\underline{d}$ and $z - \underline{t}\varepsilon(1)$ are in $\mathcal{O}^\times$. Using the expression for $\underline{t}$ equation (3.18) changes to

$$\underline{b}\varepsilon(1)^2 + \underline{a}\varepsilon(1) - z = -\underline{d}^{-1}. \tag{3.20}$$

This is a single equation left, and our integrality conditions reduce to Table 3.2.1. Any solution to (3.20) with the above integrality conditions gives us an example of a Frobenius algebra over $\mathcal{O}$ with desired properties. Condition that μ be prime is not necessary (but we used it to derive the above equation).

Equation (3.20) can also be written as

$$(z - \underline{t}\varepsilon(1))\underline{d} = 1. \tag{3.21}$$

For more concrete examples, let us further assume that $\varepsilon(1) \in \mathcal{O}^\times$. Then we can solve (3.20) for $\underline{b}$:

$$\underline{b} = \varepsilon(1)^{-2}\left(z - \underline{a}\varepsilon(1) - \underline{d}^{-1}\right). \tag{3.22}$$

This gives a family of solutions, parametrized by $\underline{a} \in \mu$ and $\varepsilon(1), \underline{d} \in \mathcal{O}^\times$:

- z is already given, with $\mu^2 = (z)$,
- $X^2 = z^{-1}\underline{a}X + z^{-1}\underline{b}$, with $\underline{b}$ given by (3.22),
- $\varepsilon(1) \in \mathcal{O}^\times$ is one of the parameters, $\varepsilon(X) = 1$,
- $c = -\underline{d}\,\underline{t}$, $\underline{t} = \underline{a} + \underline{b}\varepsilon(1)$ and $d = z\underline{d} = z(z - \underline{t}\varepsilon(1))^{-1}$ satisfy $\widetilde{\varepsilon}(c + dX) = 1^*$,
- $c' = z^{-1}d$ and $d' = -d\varepsilon(1)$ satisfy $\widetilde{\varepsilon}(c' + d'X) = z^{-1}X^*$.

3.2.2. *Case $\varepsilon(X)$ is invertible in $\mathcal{O}^\times$.* Condition $\varepsilon(X) = 1$ in Section 3.2.1 can be rescaled to cover the case when $\varepsilon(X) \in \mathcal{O}^\times$. If $\varepsilon(X) = \lambda \in \mathcal{O}^\times$, define the new trace $\varepsilon_{\mathsf{new}} := \lambda^{-1}\varepsilon$. Likewise, rescale comultiplication to $\Delta_{\mathsf{new}} = \lambda\Delta$, but keep the multiplication in A the same. This modifies the Frobenius algebra structure of A and makes it satisfy $\varepsilon_{\mathsf{new}}(X) = 1$, without changing the $\mathcal{O}$-module decomposition $A = \mathcal{O}1 \oplus \mu X$.

Alternatively, one can rescale the generator of A and define $X_{\mathsf{new}} := \lambda^{-1}X$ without changing ε and Δ. This changes multiplication in A to

$$u_1 X_{\mathsf{new}}\, u_2 X_{\mathsf{new}} = \frac{u_1 u_2}{z}\left(\lambda^{-1}\underline{a}X_{\mathsf{new}} + \lambda^{-2}\underline{b}\right), \quad u_1, u_2 \in \mu, \tag{3.23}$$

cf. (2.47).

3.3. Other values of $\varepsilon(X)$. The integrality condition says that $\varepsilon(X) \in z^{-1}\mu$. So far we have found examples with $\varepsilon(X) = 0$ and $\varepsilon(X) \in \mathcal{O}^\times$. Can we find examples where $\varepsilon(X) \neq 0$ and $\varepsilon(X)$ is not an invertible element of $\mathcal{O}$?

Consider one of the simplest examples of the ring of integers $\mathcal{O} = \mathbb{Z}[\sqrt{-5}]$, with $K = \mathbb{Q}(\sqrt{-5})$ and a nontrival ideal class group $\mathsf{Cl}(\mathcal{O}) \cong \mathbb{Z}/2$, see [**Con19**]. The ideal $\mu = (2, 1 + \sqrt{-5})$ in $\mathcal{O}$ is non-principal, and $\mu^2 = (2)$, so we set $z = 2$. The quotient $\mathcal{O}/\mu \cong \mathbf{F}_2$, so μ is prime in $\mathcal{O}$. The only invertible elements in $\mathcal{O}$ are ± 1. Let us try

$$\varepsilon(X) = \frac{1 + \sqrt{-5}}{2} \in 2^{-1}\mu$$

so that $\underline{\varepsilon_X} = 1 + \sqrt{-5}$. We need to find a system of parameters so that relations (2.41)–(2.43) and (2.45) hold, subject to the integrality conditions in the table preceeding equation (2.40). Looking at equations (2.42), (2.43), specialized to our case:

$$c\,(1 + \sqrt{-5}) = -d\,\underline{t}, \quad d'(1 + \sqrt{-5}) = -d\,\varepsilon(1),$$

we can set $d = s(1 + \sqrt{-5})$ for $s \in \{\pm 1\}$ and reduce the two equations to

$$c = -s\underline{t}, \quad d' = -s\varepsilon(1),$$

allowing us to replace c and d' in the other two equations. Equation (2.45) becomes trivial, and Equations (2.40) and (2.44) are reduced to a formula and an equation

$$\underline{t} = 2^{-1}\underline{a}(1 + \sqrt{-5}) + \underline{b}\varepsilon(1), \qquad 1 = -s\underline{t}\varepsilon(1) + s(\sqrt{-5} - 2),$$

respectively. Inserting $\underline{t}$ into the 2nd equation above, one is left with

$$(2^{-1}\underline{a}(1 + \sqrt{-5}) + \underline{b}\varepsilon(1))\varepsilon(1) = \sqrt{-5} - 2 - s.$$

Specialize to $\underline{a} = 1 - \sqrt{-5} \in \mu$ to get

$$(3 + \underline{b}\varepsilon(1))\varepsilon(1) = \sqrt{-5} - s - 2.$$

Further specializing to $\varepsilon(1) \in \{\pm 1\} \subset \mathcal{O}$ results in

$$\underline{b} = \varepsilon(1)\Big((\sqrt{-5} - s - 2)\varepsilon(1) - 3\Big) \in \mathcal{O}.$$

One can check that all integrality conditions are satisfied by this solution, giving us an example of A as above with $\varepsilon(X)$ neither 0 nor in $\mathcal{O}^\times$ (and not in $\mathcal{O}$). We were quite loose with our choices, and there should be many examples with $\varepsilon(X)$ neither 0 nor an invertible element of $\mathcal{O}$.

One possible family of such examples would be first to set $d = s\underline{\varepsilon}_X$ for $s \in \mathcal{O}^\times$, simplifying equations (2.42) and (2.43) to formulas for c and d' and making (2.45) an identity, substitute in $\underline{t}$ in (2.40) and then look for various solutions of the remaining single equation

$$-s(z^{-1}\underline{a}\,\underline{\varepsilon}_X + \underline{b}\varepsilon(1))\,\varepsilon(1) + z^{-1}s\underline{\varepsilon}_X^2 = 1, \tag{3.24}$$

subject to the integrality conditions.

4. Towards link homology

Consider the tensor product $A^{\otimes 2} = A \otimes_{\mathcal{O}} A$ and denote by ℓ_x the operator of multiplication by $x \in A$ on the first term,

$$\ell_x(y_1 \otimes y_2) = xy_1 \otimes y_2. \tag{4.1}$$

Multiplication by elements of A on the first term turns $A^{\otimes 2}$ into a projective A-module isomorphic to

$$A \otimes A = (A \otimes 1) \oplus (A \otimes \mu X) \cong A \oplus (A \otimes_{\mathcal{O}} \mu).$$

Indeed, tensoring with μ over $\mathcal{O}$ is an involution on the category of A-modules, taking projectives to projectives. It would be interesting to understand under what conditions there is an isomorphism of A-modules $A \cong A \otimes_{\mathcal{O}} \mu$.

Consider the multiplication map $m : A \otimes A \longrightarrow A$ and recall the formula (2.47) for it:

$$u_1 X\, u_2 X = \frac{u_1 u_2}{z}(\underline{a}X + \underline{b}), \quad u_1, u_2 \in \mu. \tag{4.2}$$

There are isomorphisms $\mu \otimes \mu \cong (z) \cong \mathcal{O}$ of $\mathcal{O}$-modules. Pick $k \geq 2$ and elements $u_1, \dots, u_k \in \mu$ that generate the ideal μ. Any ideal in a Dedekind domain can be generated by two elements, so one can always choose $k = 2$ above. Pick elements $u'_1, \dots, u'_k \in \mu$ so that

$$\sum_{j=1}^{k} u_j u'_j = z. \tag{4.3}$$

We use shorthand $u_j \otimes u'_j$ to denote the sum $\sum_{j=1}^k u_j \otimes u'_j$ and likewise for $u_j u'_j = z$.

Introduce the following elements

$$\begin{aligned} X_u &:= uX \otimes 1 - 1 \otimes uX \in A^{\otimes 2}, \quad u \in \mu, \\ \widehat{X} &:= u_j X \otimes u'_j X - (\underline{a}X \otimes 1 + \underline{b}1 \otimes 1) \in A^{\otimes 2}, \end{aligned} \tag{4.4}$$

where in the second formula in (4.4), the sum sign over the index j is suppressed. Note that

$$u_j X u'_j X = (u_j u'_j)X^2 = z(z^{-1}\underline{a}X + z^{-1}\underline{b}1) = \underline{a}X + \underline{b}1, \tag{4.5}$$

where the second equality in (4.5) holds due to (2.2) and the first two rescaling of the parameters in (2.39). Then

$$X_{u_1} + X_{u_2} = X_{u_1+u_2}, \quad sX_u = X_{su}, \ s \in \mathcal{O},$$

and

$$\widehat{X} = u_j X \otimes u'_j X - u_j X u'_j X \otimes 1, \tag{4.6}$$

in view of (4.5). The set $X_\mu := \{X_u\}_{u\in\mu}$ is an $\mathcal{O}$-submodule of $A^{\otimes 2}$.

PROPOSITION 4.1. *We have*

$$\ker(m) = X_\mu \oplus \mathcal{O}\widehat{X}. \tag{4.7}$$

PROOF. Formulas (4.5) and (4.6) imply that $\widehat{X} \in \ker(m)$, and it is clear that $X_\mu \oplus \mathcal{O}\widehat{X} \subset \ker(m)$. Vice versa, given an element $y \in A^{\otimes 2}$, by adding the term X_u for some u, we can assume that y does not contain a multiple of $1 \otimes X$. Due to (4.6) and subtracting an $\mathcal{O}$-multiple of $\widehat{X}$, we can assume that y does not contain a multiple of $X \otimes X$ and reduce it to the form

$$y = y_1 X \otimes 1 + y_2 1 \otimes 1,$$

for some $y_1 \in \mu$ and $y_2 \in \mathcal{O}$. The condition $y \in \ker(m)$ forces $y_1 = y_2 = 0$. □

Recall (4.1). Using (4.3), we have

$$\ell_{uX}(\widehat{X}) = -X_{\underline{b}u} + \frac{u\underline{a}}{z}\widehat{X}, \quad \ell_{uX}(X_{u'}) = -\frac{uu'}{z}\widehat{X}. \tag{4.8}$$

Consequently, there is an isomorphism of $\mathcal{O}$-modules $A \otimes_{\mathcal{O}} \mu \cong \ker(m)$ taking $1 \otimes u$ to X_u.

To have an isomorphism of A-modules $\ker(m) \cong A$ requires $\ker(m)$ to have a single generator as an A-module. Such a generator must have the form

$$\widetilde{X} = \gamma\widehat{X} - X_u, \tag{4.9}$$

for some $\gamma \in \mathcal{O}$ and $u \in \mu$. Using (4.8) we have

$$\ell_{u'X}(\widetilde{X}) = \frac{uu'}{z}\widehat{X} + \gamma\left(-X_{\underline{b}u'} + \frac{u'\underline{a}}{z}\widehat{X}\right) = \frac{u'}{z}(\gamma\underline{a} + u)\widehat{X} - X_{\gamma\underline{b}u'}. \tag{4.10}$$

A necessary condition for $\widetilde{X}$ to generate $\ker(m)$ is that $\mu = \mathcal{O}u + \gamma\underline{b}\mu$ and that $\mathcal{O}$ is generated by γ and elements $\frac{u'}{z}(\gamma\underline{a} + u)$ over $u' \in \mu$.

Let us specialize to $\gamma = 1$ (this also covers the case $\gamma \in \mathcal{O}^\times$). Then

$$\widetilde{X} = \widehat{X} - X_u, \quad \ell_{u'X}(\widetilde{X}) = \frac{u'}{z}(\underline{a} + u)\widehat{X} - X_{\underline{b}u'}, \tag{4.11}$$

and

$$\ell_{u'X}(\widetilde{X}) - \frac{u'}{z}(\underline{a} + u)\widetilde{X} = X_{u'(-\underline{b}+\frac{u}{z}(\underline{a}+u))}. \tag{4.12}$$

Consequently, a sufficient condition for $\ker(m) \cong A$ is that for some $u \in \mu$ the element

$$-\underline{b} + \frac{u(\underline{a} + u)}{z} \tag{4.13}$$

is invertible in $\mathcal{O}$. Taking $u = 0$, we see that $\widehat{X}$ is a generator of $\ker(m)$ if and only if $\underline{b}$ is invertible. The Frobenius algebras with $\varepsilon(X) = 0$ described in Section 3.1

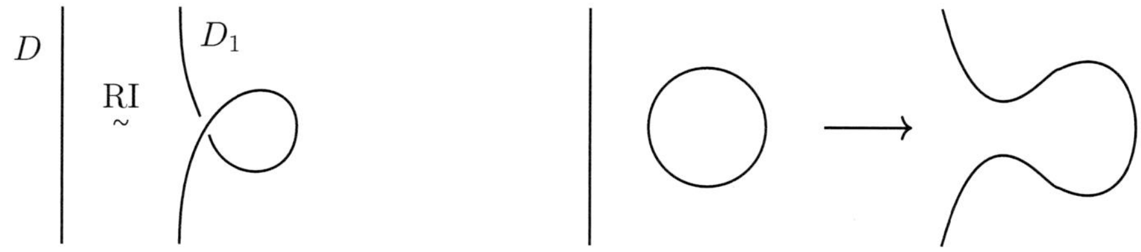

FIGURE 4.1. Left: One of the two flavors of the Reidemeister I move of link diagrams. Middle: two resolutions of the curl diagram D_1 on the right hand side of the RI move. Right: Depicting the tensor product $A \otimes \mu$ by a dot on a strand; two adjacent dots can be removed due to the isomorphism $\mu \otimes_{\mathcal{O}} \mu \cong \mathcal{O}$.

have this property. This at least gives us some examples when $\ker(m) \cong A$ as an A-module.

We do not know a classification or a convenient description of Frobenius algebras as above for which there is an isomorphism $\ker(m) \cong A$.

A Frobenius algebra $A \cong R[X]/(X^2 - aX - b)$ over the ground ring R, of rank two with a basis $\{1, X\}$ and a fixed Frobenius structure gives a link homology theory [**Kho06**, **BN05**, **KR22**]. Starting with a planar diagram D of a link L, one can repeat the construction of [**Kho00**]. One resolves D into 2^k planar diagrams, where k is the number of crossings of D. To a diagram with n circles, one associates $A^{\otimes n}$. These R-modules fit into a commutative cube via maps given by the multiplication m and comultiplication Δ in A on proper terms in the tensor product. The commutative cube is collapsed into the complex $C(D)$, with an appropriate grading shift. Homology $H(D) = H(C(D))$ of the complex can be shown to be invariant under the Reidemeister moves I, II, III, resulting in a link homology theory $H(L)$ for oriented links L.

The key property of the Frobenius algebra A is that it is of rank 2 over R. This is necessary for the invariance under the Reidemeister I move, shown in Figure 4.1 on the left. In that move, one adds a curl to an interval of the diagram D resulting in new diagram D_1. For one of the two possible types of the curl, the complex assigned to D_1 at the crossing is

$$0 \longrightarrow A \otimes A \xrightarrow{\ m\ } A \longrightarrow 0. \tag{4.14}$$

It breaks down into the sum of a contractible complex $0 \longrightarrow A \xrightarrow{\ 1\ } A \longrightarrow 0$ and the complex $0 \longrightarrow A \longrightarrow 0$. The latter complex is also what the interval contributes to $C(D)$. Removing the contractible summands leads to a homotopy equivalence $C(D_1) \cong C(D)$ and an isomorphism of homology groups $H(D_1) \cong H(D)$. Reidemeister I move for the opposite type of crossing is dealt with by passing to the dual of the complex (4.14), via the isomorphism $\widetilde{\varepsilon} : A \cong A^*$, and simplifying it in the same manner.

If R and A are additionally $\mathbb{Z}$-graded, and Frobenius structure respects the grading, then $H(L)$ is bigraded. Otherwise $H(L)$ has a single grading.

Using a cyclic Frobenius algebra A of higher rank over R, with a higher degree relation $X^n = f(X)$, for $n > 2$ and $\deg(f) \leq n-1$, does not work, since removing the contractible summand leaves one with the complex $0 \longrightarrow (A/R\cdot 1) \otimes_R A \longrightarrow 0$ isomorphic to $n-1$ copies of $0 \longrightarrow A \longrightarrow 0$, since $\mathsf{rk}_R(A/R \cdot 1) = n-1$. The higher

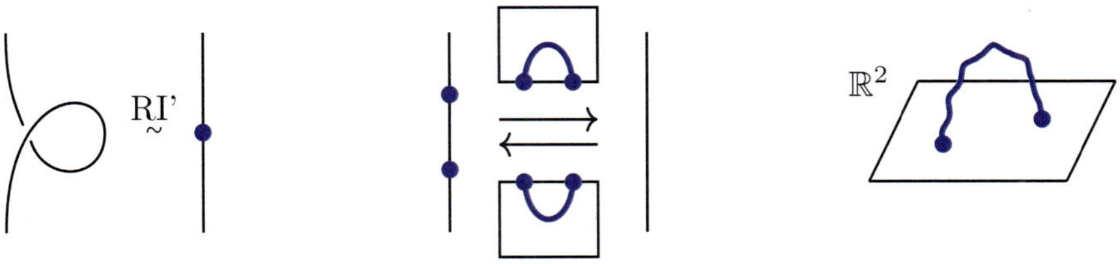

FIGURE 4.2. Left: RI' move, a variation on the Reidemeister I move. The tensor product $A \otimes \mu$ is depicted by a dot on a strand. Middle: two adjacent dots on a strand can be canceled due to an isomorphism $\mu \otimes_{\mathcal{O}} \mu \cong \mathcal{O}$. Right: dots can be moved out of the strands to the real plane, with the canceling cobordism on a pair of dots shown.

rank cyclic extensions do give rise to link homology, but that requires working with matrix factorizations or foam evaluation or bringing in suitable categories to make it work. Some of these approaches can be found in papers [**KR08**, **RW20**, **CK08**, **Man07**, **Web17**, **RW16**].

Consider a rank 2 Frobenius extension $(A, \mathcal{O})$ from the present paper where A is a projective but not a free $\mathcal{O}$-module. Given a planar diagram D of a link L, one can form the complex $C(D)$ as above, using the tensor powers of A over $\mathcal{O}$. The complex is singly-graded (no q-grading) and has homology groups $H(D)$.

Tensoring the complex with the field of fractions K results in the complex $C(D) \otimes_{\mathcal{O}} K$ of K-vector spaces whose homology groups $H(D) \otimes K$ are invariant under the Reidemeister moves and give a version of the Eun Soo Lee link homology theory [**Lee05**], due to $X^2 = aX + b$ having two distinct roots in some quadratic extension of K.

It is natural to ask whether, for some of these Frobenius extensions, groups $H(D)$ themselves are link invariants. The easiest way for this to happen is to have homotopy equivalences $C(D) \cong C(D_1)$ for diagrams D, D_1 related by a Reidemeister move.

Let us examine Reidemeister move I, already discussed above. The complex (4.14) breaks down into the sum of a contractible complex and the complex $0 \longrightarrow \ker(m) \longrightarrow 0$. The latter is isomorphic to $A \otimes \mu$. We would like it to be isomorphic to A, see the earlier discussion in this section. In particular, it is isomorphic to A if $\varepsilon(X) = 0$, but that is a rather special condition.

It would be useful to further develop this construction, and

- understand under what conditions there is an isomorphism $\ker(m) \cong A$,
- assuming $\ker(m) \cong A$, check what additional conditions are required for the invariance of the complex $C(D)$ under Reidemeister moves II, III and look for examples of such Frobenius $\mathcal{O}$-algebras A.

REMARK 4.2. Since $\ker(m) \cong A \otimes \mu$, instead of looking for an isomorphism $A \otimes \mu \cong A$, another possibility is to depict the tensor product $A \otimes \mu$ by a dot on a strand and replace the Reidemeister I move by the move RI' shown in Figure 4.2. Two dots on a strand encode the tensor product with $\mu \otimes \mu \cong \mathcal{O}$ and could be erased. More precisely, one sets up cobordisms between a strand with no dots and a strand with two dots that are mutually-inverse module isomorphisms $\mu \otimes \mu \leftrightarrow \mathcal{O}$.

To a collection of k planar circles, one attaches $A^{\otimes k}$. Whichever copy of A in this tensor product is tensored with μ does not make a difference for the tensor product. Consequently, the dot representing μ does not have to be on any particular edge or circle and can be placed anywhere in the plane. A pair of dots in the plane can be canceled, which can be represented by an arc in $\mathbb{R}^2 \times [0,\infty)$ connecting them. This arc can intersect facets of a cobordism between 1-manifolds in $\mathbb{R}^2$ without interacting with the facets.

In this setup, however, there is a problem with the Reidemeister II move, where, in the resolution of the 2-crossing diagram, two of the resolutions D_{00}, D_{11} are isotopic and the third resolution, say D_{10}, is isotopic to each of them with an additional circle. This circle contributes A to the state space of diagram D_{10}. The summand $\mathcal{O}1$ can be canceled with the state space of D_{11}, but for the remaining term μX to cancel with D_{00} via a contractible summand one still expects an isomorphism $A \otimes \mu \cong A$.

REMARK 4.3. Our examples of rank two Frobenius $\mathcal{O}$-algebras that are not free over $\mathcal{O}$ are not graded. It is easy to produce examples of rank $N > 2$ commutative graded Frobenius $\mathcal{O}$-algebras. For this, assume that μ is a non-principal ideal in $\mathcal{O}$ of order $N-1$ in $\mathsf{Cl}(\mathcal{O})$. Let $\mu^{N-1} = (z)$, so that an isomorphism $\mu^{\otimes N-1} \cong (z) \cong \mathcal{O}$ is fixed.

Start with the Frobenius algebra $A'_N = \mathcal{O}[X]/(X^N)$ with the trace $\varepsilon(X^{N-1}) = z^{-1}$ and $\varepsilon(X^m) = 0$ for $m < N-1$. The comultiplication is given by

$$\Delta(1) = z(X^{N-1} \otimes 1 + X^{N-2} \otimes X + \ldots + 1 \otimes X^{N-1}).$$

Tensoring with K produces a commutative Frobenius K-algebra $A'_N \otimes K$. Set $\deg X = 2$ to make A'_N and $A'_N \otimes K$ into graded Frobenius algebras.

Consider $\mathcal{O}$-subalgebra of $A'_N \otimes K$ given by

$$A_N := \mathcal{O}1 \oplus \mu X \oplus \mu^2 X^2 \oplus \ldots \oplus \mu^{N-1} X^{N-1}.$$

Due to the isomorphism $\mu^{N-1} \cong \mathcal{O}$, the trace ε restricts to an isomorphism $\mu^{N-1} X^{N-1} \cong \mathcal{O}$. It is straightforward to see that A_N is a Frobenius $\mathcal{O}$-algebra.

The algebras $A'_N \otimes K$, when N is odd and $\sqrt{N} \in K$, give examples of Turaev–Turner unoriented 2D TQFTs [**TT06**]. Algebra A'_3 is used in the construction of $\mathfrak{sl}(3)$-link homology [**Kho04**]. It is possible that some twists of this algebra can give rise to modifications of $\mathfrak{sl}(3)$-homology, and likewise for the $\mathfrak{sl}(N)$-link homology.

REMARK 4.4. An isomorphism of $\mathcal{O}$-modules $A \cong A^*$, see Section 1, constraints us to elements P of order two in $\mathsf{Cl}(\mathcal{O})$ as a summand of A. This isomorphism comes from a diffeomorphism between an oriented circle $\mathbb{S}^1$ and the same circle with the opposite orientation. Going one dimension up, if looking for an oriented 3D TQFT over $\mathcal{O}$, one would likewise be forced into elements of order 2 in the ideal class group as summands of $\mathcal{F}(M)$, for an oriented surface M, due to diffeomorphisms $M \cong (-M)$.

Returning to dimension two, one way to avoid this restriction is to add additional decorations to 1-manifolds and 2-cobordisms between them. For example, one can pick a group G and consider 2-cobordisms S equipped with a principal G-bundle over them. A oriented circle $\mathbb{S}^1$ is then replaced by a pair $(\mathbb{S}^1, c)$, where c is a conjugacy class in G. Reversing the orientation results in the pair $(\mathbb{S}^1, c^{-1})$, which is diffeomorphic to $(\mathbb{S}^1, c)$ only if the conjugacy classes c and c^{-1} coincide.

It would be interesting to investigate such decorated 2D TQFTs over Dedekind domains $\mathcal{O}$, including the special case $G = \mathsf{Cl}(\mathcal{O})$.

Acknowledgment

This paper resulted from a reading and research course given to one of the authors by the other two in the fall of 2024 at Johns Hopkins University.

References

[Aph21] Aphelli, *Is every element in the class group is represented by a prime ideal?*, Mathematics Stack Exchange, 2021, `https://math.stackexchange.com/q/4122212` (version: 2021-04-30).

[BN05] D. Bar-Natan, *Khovanov's homology for tangles and cobordisms*, Geom. Topol. **9** (2005), 1443–1499, DOI 10.2140/gt.2005.9.1443. MR2174270

[CK08] S. Cautis and J. Kamnitzer, *Knot homology via derived categories of coherent sheaves. II. $\mathfrak{sl}_m$ case*, Invent. Math. **174** (2008), no. 1, 165–232, DOI 10.1007/s00222-008-0138-6. MR2430980

[CL05] Q. Chen and T. Le, *Almost integral TQFTs from simple Lie algebras*, Algebr. Geom. Topol. **5** (2005), 1291–1314, DOI 10.2140/agt.2005.5.1291. MR2171810

[Con19] Keith Conrad, *Ideal classes and the Kronecker bound*, Online lecture notes for Algebraic Number Theory (preprint), `https://kconrad.math.uconn.edu/blurbs/gradnumthy/classgroupKronecker.pdf` (2019), 1–16.

[Fre14] D. S. Freed, *Anomalies and invertible field theories*, String-Math 2013, Proc. Sympos. Pure Math., vol. 88, Amer. Math. Soc., Providence, RI, 2014, pp. 25–45, DOI 10.1090/pspum/088/01462. MR3330283

[GIK+23] P. Gustafson, M. S. Im, R. Kaldawy, M. Khovanov, and Z. Lihn, *Automata and one-dimensional TQFTs with defects*, Lett. Math. Phys. **113** (2023), no. 5, Paper No. 93, 38, DOI 10.1007/s11005-023-01701-y. MR4636793

[GPV17] S. Gukov, P. Putrov, and C. Vafa, *Fivebranes and 3-manifold homology*, J. High Energy Phys. **7** (2017), 071, front matter+80, DOI 10.1007/JHEP07(2017)071. MR3686727

[GSWZ24] Stavros Garoufalidis, Peter Scholze, Campbell Wheeler, and Don Zagier, *The Habiro ring of a number field*, Preprint `arXiv:2412.04241` (2024), 1–71.

[Guy24] Luc Guyot, *Finding prime ideals for ideal classes in arbitrary dedekind domains*, MathOverflow, 2024, `https://mathoverflow.net/q/482280` (version: 2024-11-17).

[Hab02] K. Habiro, *On the quantum* sl_2 *invariants of knots and integral homology spheres*, Invariants of knots and 3-manifolds (Kyoto, 2001), Geom. Topol. Monogr., vol. 4, Geom. Topol. Publ., Coventry, 2002, pp. 55–68, DOI 10.2140/gtm.2002.4.55. MR2002603

[Hab04] K. Habiro, *Cyclotomic completions of polynomial rings*, Publ. Res. Inst. Math. Sci. **40** (2004), no. 4, 1127–1146. MR2105705

[Kas90] M. Kashiwara, *Crystalizing the q-analogue of universal enveloping algebras*, Comm. Math. Phys. **133** (1990), no. 2, 249–260. MR1090425

[Kas91] M. Kashiwara, *On crystal bases of the Q-analogue of universal enveloping algebras*, Duke Math. J. **63** (1991), no. 2, 465–516, DOI 10.1215/S0012-7094-91-06321-0. MR1115118

[Kho00] M. Khovanov, *A categorification of the Jones polynomial*, Duke Math. J. **101** (2000), no. 3, 359–426, DOI 10.1215/S0012-7094-00-10131-7. MR1740682

[Kho04] M. Khovanov, *sl(3) link homology*, Algebr. Geom. Topol. **4** (2004), 1045–1081, DOI 10.2140/agt.2004.4.1045. MR2100691

[Kho06] M. Khovanov, *Link homology and Frobenius extensions*, Fund. Math. **190** (2006), 179–190, DOI 10.4064/fm190-0-6. MR2232858

[KR08] M. Khovanov and L. Rozansky, *Matrix factorizations and link homology*, Fund. Math. **199** (2008), no. 1, 1–91, DOI 10.4064/fm199-1-1. MR2391017

[KR22] M. Khovanov and L.-H. Robert, *Link homology and Frobenius extensions II*, Fund. Math. **256** (2022), no. 1, 1–46, DOI 10.4064/fm912-6-2021. MR4361584

[Lee05] E. S. Lee, *An endomorphism of the Khovanov invariant*, Adv. Math. **197** (2005), no. 2, 554–586, DOI 10.1016/j.aim.2004.10.015. MR2173845

[Lus90] G. Lusztig, *Canonical bases arising from quantized enveloping algebras*, J. Amer. Math. Soc. **3** (1990), no. 2, 447–498, DOI 10.2307/1990961. MR1035415

[Lus10] G. Lusztig, *Introduction to quantum groups*, Modern Birkhäuser Classics, Birkhäuser/Springer, New York, 2010. Reprint of the 1994 edition, DOI 10.1007/978-0-8176-4717-9. MR2759715

[Man07] C. Manolescu, *Link homology theories from symplectic geometry*, Adv. Math. **211** (2007), no. 1, 363–416, DOI 10.1016/j.aim.2006.09.007. MR2313538

[ME05] M. R. Murty and J. Esmonde, *Problems in algebraic number theory*, 2nd ed., Graduate Texts in Mathematics, vol. 190, Springer-Verlag, New York, 2005. MR2090972

[RT91] N. Reshetikhin and V. G. Turaev, *Invariants of 3-manifolds via link polynomials and quantum groups*, Invent. Math. **103** (1991), no. 3, 547–597, DOI 10.1007/BF01239527. MR1091619

[RW16] D. E. V. Rose and P. Wedrich, *Deformations of colored $\mathfrak{sl}_N$ link homologies via foams*, Geom. Topol. **20** (2016), no. 6, 3431–3517, DOI 10.2140/gt.2016.20.3431. MR3590355

[RW20] L.-H. Robert and E. Wagner, *A closed formula for the evaluation of foams*, Quantum Topol. **11** (2020), no. 3, 411–487, DOI 10.4171/qt/139. MR4164001

[Sac16] Hans Sachs, *Is every ideal class represented by a prime?*, Mathematics Stack Exchange, 2016, `https://math.stackexchange.com/q/1883790` (version: 2016-08-05).

[SP24] C. Schommer-Pries, *Invertible topological field theories*, J. Topol. **17** (2024), no. 2, Paper No. e12335, 64, DOI 10.1112/topo.12335. MR4821355

[TT06] V. Turaev and P. Turner, *Unoriented topological quantum field theory and link homology*, Algebr. Geom. Topol. **6** (2006), 1069–1093, DOI 10.2140/agt.2006.6.1069. MR2253441

[Web17] B. Webster, *Knot invariants and higher representation theory*, Mem. Amer. Math. Soc. **250** (2017), no. 1191, v+141, DOI 10.1090/memo/1191. MR3709726

[Wit89] E. Witten, *Quantum field theory and the Jones polynomial*, Comm. Math. Phys. **121** (1989), no. 3, 351–399. MR990772

Department of Mathematics, Johns Hopkins University, Baltimore, Maryland 21218
Email address: `fespino5@jhu.edu`

Department of Mathematics, Johns Hopkins University, Baltimore, Maryland 21218
Email address: `meeseong@jhu.edu`

Department of Mathematics, Johns Hopkins University, Baltimore, Maryland 21218
Email address: `khovanov@jhu.edu`

Contemporary Mathematics
Volume **834**, 2026
https://doi.org/10.1090/conm/834/16683

Towards the affine and geometric invariant theory quotients of the Borel moment map

Mee Seong Im and Meral Tosun

ABSTRACT. We study the Borel moment map $\mu_B : T^*(\mathfrak{b} \times \mathbb{C}^n) \to \mathfrak{b}^*$, given by $(r, s, i, j) \mapsto [r, s] + ij$, and describe an algorithm to explicitly construct the geometric invariant theory (GIT) quotients $\mu_B^{-1}(0)/\!/_{\det} B$ and $\mu_B^{-1}(0)/\!/_{\det^{-1}} B$, and the affine quotient $\mu_B^{-1}(0)/\!/B$. We also provide an insight of the singular locus of 2^n irreducible components of μ_B. Finally, analogous to the Hilbert–Chow morphism, we discuss that the GIT quotient for the Borel setting is a resolution of singularities.

1. Introduction

Parabolic equivariant geometry frequently appears in algebraic geometry, representation theory, and mathematical physics. They generalize equivariant morphisms for reductive groups, with multitude of connections to quiver Hecke algebras (cf. [**KL09**, **KL11**, **Rou12**, **Rou08**]), isospectral Hilbert schemes (cf. [**Hai01**, **Im18**]), flag Hilbert schemes (cf. [**GNR21**]), and partial and complete (Grothendieck–)Springer resolutions (cf. [**CG10**, **Nev11**, **Im18**, **Im14**]). Due to the extensive deep connections it provides, the study of parabolic equivariant geometry is worthwhile on its own.

We will work over the set of complex numbers. Let $G = \mathsf{GL}_n(\mathbb{C})$, the set of $n \times n$ invertible matrices over $\mathbb{C}$. Let B be the set of invertible upper triangular matrices in G. Let $\mathfrak{g} = \text{Lie}(G)$, and $\mathfrak{b} = \text{Lie}(B)$.

Consider the matrix variety $\mathfrak{b} \times \mathbb{C}^n$, and its cotangent bundle $T^*(\mathfrak{b} \times \mathbb{C}^n)$, where we make the following identification:

$$T^*(\mathfrak{b} \times \mathbb{C}^n) := \mathfrak{b} \times \mathfrak{b}^* \times \mathbb{C}^n \times (\mathbb{C}^n)^*,$$

where $\mathfrak{b}^* \overset{\text{tr}}{\cong} \mathfrak{g}/\mathfrak{u}^+$ and $\mathfrak{u}^+$ is the nilpotent radical of $\mathfrak{b}$ and $(\mathbb{C}^n)^*$ is the dual of the vector space $\mathbb{C}^n$. There is a natural B-action on $\mathfrak{b} \times \mathbb{C}^n$ via $b.(r, i) = (brb^{-1}, bi)$.

2020 *Mathematics Subject Classification.* Primary 14M10, 53D20, 17B08, 14L30; Secondary 14L24, 13A50, 20G20.

Key words and phrases. Grothendieck–Springer resolution, moment map, complete intersection, parabolic subgroup, Borel moment map, parabolic moment map, geometric invariant theory, Hilbert scheme, singular locus, resolution of singularities.

Taking the derivative of this action gives us the map

$$a : \mathfrak{b} \to \Gamma(T_{\mathfrak{b}\times\mathbb{C}^n}) \subseteq \mathbb{C}[T^*(\mathfrak{b}\times\mathbb{C}^n)],$$

$$\text{given by} \quad a(v)(r,i) = \frac{d}{dt}\left(g_t.(r,i)\right)\Big|_{t=0} = ([v,r], vi),$$

where $g_t = \exp(tv)$. We dualize a to obtain the moment map

$$\mu_B = a^* : T^*(\mathfrak{b}\times\mathbb{C}^n) \to \mathfrak{b}^*, \quad \text{where} \quad (r,s,i,j) \mapsto [r,s] + ij. \tag{1.1}$$

Also note that the B-action is induced onto the cotangent bundle, giving us

$$B \circlearrowright T^*(\mathfrak{b}\times\mathbb{C}^n) \quad \text{via} \quad b.(r,s,i,j) = (brb^{-1}, bsb^{-1}, bi, jb^{-1}).$$

The quotient $[(\mathfrak{b}\times\mathbb{C}^n)/B]$ is an Artin stack since it has an infinite stabilizer group over some points in the semistable locus of $\mu_B^{-1}(0)$, but we will not go into a discussion of (global, non-smooth) quotient stacks in this paper.

In [**Im18**], the author restricts to the regular semisimple locus $\mu_B^{-1}(0)^{\text{rss}}$ of the components of μ_B, i.e., r has pairwise distinct eigenvalues, and studies its affine quotient $\mu_B^{-1}(0)^{\text{rss}} /\!\!/ B$. The author shows that the affine quotient is isomorphic to $\mathbb{C}^{2n}\backslash\Delta$, where $\Delta = \{(a_1,\dots,a_n,0,\dots,0) : a_\iota = a_\gamma \text{ for some } \iota \neq \gamma\}$ (cf. [**Im18**, Theorem 1.6]). So this locus $\mu_B^{-1}(0)^{\text{rss}} /\!\!/ B$ is contained in the smooth locus of the affine quotient, which we will denote by X^{sm}. For this regular semisimple locus, B-invariant polynomials are explicitly given for the affine quotient.

We now recall a description of the irreducible components of μ_B (see [**Nev11**, Prop. 4.2]), i.e., in symplectic algebraic geometry, this means the irreducible components of $\mu_B^{-1}(0)$. But first, we recall some results in [**Nev11**].

LEMMA 1.1 (Nevins, Lemma 4.1). *Suppose r is a diagonal $n\times n$ matrix with pairwise distinct eigenvalues. Let $i \in \mathbb{C}^n$ and $j \in (\mathbb{C}^n)^*$. Then given diagonal entries s_{aa} for an $n\times n$ matrix s, there is a unique s satisfying $[r,s]+ij = 0$. In particular, if $\overline{s} \in \mathfrak{g}/\mathfrak{u}^+$ and $\mu_B(r,\overline{s},i,j) = 0$, then there is a unique lift of $\overline{s}$ to $s \in \mathfrak{g}$ such that $\mu_G(r,s,i,j) = 0$, where $\mu_G : T^*(\mathfrak{g}\times\mathbb{C}^n) \to \mathfrak{g}^* \overset{\text{tr}}{\cong} \mathfrak{g}$.*

PROPOSITION 1.2 (Nevins, Proposition 4.2). *The subset C_ℓ consists of the orbits of quadruples $(r,\overline{s},i,j)$, where r is diagonal and has pairwise distinct eigenvalues, $i_k = \ell(k)$, $j_k = 1-\ell(k)$, and $\overline{s}$ is the image of a matrix $s \in \mathfrak{g}$ in $\mathfrak{g}/\mathfrak{u}^+$ that has arbitrary diagonal entries and off-diagonal entries*

$$s_{ab} = -\frac{(ij)_{ab}}{r_{aa} - r_{bb}}$$

for $a \neq b$. For $n \leqslant 5$, the irreducible components of the locus $\mu_B^{-1}(0)$ are the closures of subsets C_ℓ, where $\ell : \{1,\dots,n\} \to \{0,1\}$ is a function.

Thus the closures $\overline{C}_\ell$, as discussed in Proposition 1.2, are precisely the 2^n irreducible components of the preimage $\mu_B^{-1}(0)$. We will write $\overline{C}_0$ to be the closure of the irreducible component enumerated by $\ell(k) = 0$ for all $1 \leqslant k \leqslant n$, i.e., $i = 0 \in \mathbb{C}^n$ while $j = (1,1,\dots,1) \in (\mathbb{C}^n)^*$.

The first author has verified using `Macaulay2` (cf. [**GS02**]) and computational algebraic geometry techniques that μ_B is a complete intersection for up to $n = 5$.

Let P be a parabolic group of G consisting of invertible block upper triangular matrices. Note that $P \supseteq B$. Let $\mathfrak{p} = \text{Lie}(P)$. Using the fact from the previous paragraph, Im–Scrimshaw in [**IS21**] prove that given a parabolic subalgebra $\mathfrak{p}$ with at most 5 Jordan blocks, the components of μ_P form a complete intersection, where $\mu_P : T^*(\mathfrak{p} \times \mathbb{C}^n) \to \mathfrak{p}^*$, which is a generalization of the Borel moment map μ_B given in (1.1), but in this case, we restrict to at most 5 Jordan blocks. The irreducible components of the fiber $\mu_P^{-1}(0)$ are enumerated, where the vector i and the covector j are $\{0, 1\}$-vectors (cf. [**IS21**, Theorem 1.1]), which are similar to Proposition 1.2. Furthermore, they are equidimensional (cf. [**IS21**, Theorem 1.2]).

In this manuscript, we specialize when $P = B$ and provide a strategy to explicitly describe the entire affine quotient $\mu_B^{-1}(0) /\!\!/ B$ (Section 3), and the geometric invariant theory (GIT) quotients $\mu_B^{-1}(0) /\!\!/_{\det} B$ and $\mu_B^{-1}(0) /\!\!/_{\det^{-1}} B$ (Section 4). We initiate this construction by explicitly giving det-semi-invariant polynomials in Theorem 4.2.

THEOREM 1.3. *The restriction of functions via the embedding ε_{i_0} in* (4.1) *induces a surjection $\varepsilon_{i_0}^* : \mathbb{C}[\mu_B^{-1}(0)]^{B,\det^k} \twoheadrightarrow D^k$ of vector spaces for each $k \geqslant 1$, where*

$$D := \mathbb{C}[\ell^J,\ L^{[n]\setminus\{\iota\}}r,\ \ell^{[n]\setminus\{n\}}s\ell^J,\ \ell^J s\ell^{[n]\setminus\{1\}}]^{\epsilon},$$

a subspace of S_n-alternating polynomials.

We furthermore explicitly describe $\det^{-1}$-semi-invariant polynomials in Theorem 4.4:

THEOREM 1.4. *The restriction of functions via the embedding ε_{j_0} in* (4.2) *induces a surjection $\varepsilon_{j_0}^* : \mathbb{C}[\mu_B^{-1}(0)]^{B,\det^k} \twoheadrightarrow D^{-k}$ of vector spaces for each $k \leqslant -1$, where D is the S_n-alternating submodule given in* (4.5).

See Section 2.2 for the definition of affine and GIT quotients. Although we have explicitly constructed B-invariant polynomials and B-semi-invariant polynomials as generators of these quotients, it is difficult to show that they generate the B-invariant subalgebra and the semi-invariant module, respectively. In particular, the surjections in Theorems 1.3 and 1.4 are only maps of vector spaces as it is computationally heavy to study their ring structures. We leave this as an open problem to investigate their ring properties.

Moreover, although the regular semisimple results for the Borel subalgebra in [**Im18**] hold for all n, the parabolic subalgebra results in [**IS21**] hold only for those cases with less than or equal to 5 Jordan blocks since there does not appear to be appropriate methods to show that $\mu_P^{-1}(0)$ is a complete intersection when P has more than 5 Jordan blocks (cf. proof of [**IS21**, Prop. 4.2]), at least to our knowledge. One of the key assumptions in [**Im18**] is that the author restricts to the locus where $r \in \mathfrak{b}$ has pairwise distinct eigenvalues, which makes studying parabolic equivariant geometry (for any n) straightforward since r is diagonalizable. However, for the entire Borel subalgebra r, the equivariant geometry is no longer clear. We conclude by describing the singular locus of the irreducible components of μ_B for $n \leqslant 5$ (Theorem 5.1).

THEOREM 1.5. *Each irreducible component $\overline{\mathcal{C}}_\nu$, for $1 \leqslant \nu \leqslant 2^n$ and $n \leqslant 5$, is singular, whose singular locus has codimension 1 in $\overline{\mathcal{C}}_\nu$. Thus, the singular locus $\mu_B^{-1}(0)^{\text{sing}}$ of the components of μ_B has codimension 1 in $\mu_B^{-1}(0)$. Furthermore, the irreducible components do not intersect transversely.*

We further describe the regular semisimple locus and the singular locus for the det-twisted locus (Section 6), followed by conjectures on their connections to the Hilbert scheme $(\mathbb{C})^{[n]}$ of n points on a complex plane (Section 7).

2. Parabolic equivariant geometry

2.1. Parabolic invariant functions. We begin with a preliminary background. We work over $\mathbb{C}$. Let $B \subseteq \mathsf{GL}_n$ be the set of invertible upper triangular matrices, $\mathfrak{b} = \mathrm{Lie}(B)$, and $\mathfrak{b}^* = \mathfrak{gl}_n/\mathfrak{u}^+$, where $\mathfrak{u}^+$ is the nilpotent radical of $\mathfrak{b}$. So one can identify $\mathfrak{b}^*$ as the set of lower triangular matrices. Let B act on $\mathfrak{b}$ (and $\mathfrak{b}^*$) via the conjugation action, i.e., $b.r = brb^{-1}$ (and $b.s = bsb^{-1}$), where $b \in B$, $r \in \mathfrak{b}$, and $s \in \mathfrak{b}^*$. Write $\mathfrak{b} = \{(r_{\iota\gamma}) : 1 \leqslant \iota \leqslant \gamma \leqslant n\} \subseteq \mathfrak{gl}_n$. Let $\mathbb{C}[\mathfrak{b}] = \mathbb{C}[\{r_{\iota\gamma}\}_{1\leqslant\iota\leqslant\gamma\leqslant n}]$, a polynomial ring, and let

$$\mathbb{C}[\mathfrak{b}]^B = \{f(r_{\iota\gamma}) \in \mathbb{C}[\mathfrak{b}] : b.f(r_{\iota\gamma}) = f(r_{\iota\gamma}) \text{ for any } b \in B\}, \tag{2.1}$$

a subring of B-invariant polynomials.

Write $\mathrm{diag}(r) = (r_{11}, \ldots, r_{nn})$ to be an $n \times n$ diagonal matrix whose (ordered) coordinates along the diagonal are $r_{11}, \ldots, r_{nn}$. We can also identify $\mathrm{diag}(r)$ with the n-tuple $(r_{11}, \ldots, r_{nn}) \in \mathbb{C}^n$.

In Lemmas 2.1 and 2.2, we give the building blocks to B-invariant functions for the adjoint action on the Lie algebra of the Borel and its dual. These computations are foundational, before computing generalized B-invariant polynomials.

LEMMA 2.1. *We have* $\mathbb{C}[\mathfrak{b}]^B \cong \mathbb{C}[r_{11}, \ldots, r_{nn}]$.

PROOF. For $1 \leqslant \iota \leqslant n$, define a map $f_\iota : \mathfrak{b} \to \mathbb{C}$, where $f_\iota(r) = r_{\iota\iota}$. For $b \in B$, we have $b.f_\iota(r) = f_\iota(b^{-1}rb) = r_{\iota\iota}$. So $\mathbb{C}[f_1, \ldots, f_n] \subseteq \mathbb{C}[\mathfrak{b}]^B$. Now, let $\lambda\colon \mathbb{C}^* \to B$ be a 1-parameter subgroup defined by

$$\lambda(t)_{\iota\gamma} = \begin{cases} t^{n-a} & \text{if } \iota = \gamma = a, \\ \delta_{\iota\gamma} & \text{otherwise,} \end{cases} \tag{2.2}$$

where $\delta_{\iota\gamma}$ is the Kronecker delta. Then

$$(\lambda(t).r)_{\iota\gamma} = (\lambda(t)r\lambda(t)^{-1})_{\iota\gamma} = \begin{cases} r_{\iota\iota} & \text{if } \iota = \gamma, \\ r_{\iota\gamma}t^{\gamma-\iota} & \text{if } \iota < \gamma, \\ 0 & \text{if } \iota > \gamma. \end{cases}$$

So

$$\lim_{t\to 0}(\lambda(t).r)_{\iota\gamma} = \begin{cases} r_{\iota\iota} & \text{if } \iota = \gamma, \\ 0 & \text{if } \iota \neq \gamma. \end{cases}$$

Since the off-diagonal entries of r are zero, a B-invariant polynomial is independent of the off-diagonal coordinate functions. Thus $\mathbb{C}[\mathfrak{b}]^B \cong \mathbb{C}[\mathrm{diag}(r)]$. □

LEMMA 2.2. *We have* $\mathbb{C}[\mathfrak{b}^*]^B \cong \mathbb{C}[\mathrm{tr}(s)]$.

PROOF. Let F be a polynomial in $\mathbb{C}[\mathrm{tr}(s)]$. Since

$$F(\mathrm{tr}(bsb^{-1})) = F(\mathrm{tr}(sb^{-1}b)) = F(\mathrm{tr}(s))$$

for any $b \in B$, F is in the B-invariant subring $\mathbb{C}[\mathfrak{b}^*]^B$.

Now suppose $F \in \mathbb{C}[\mathfrak{b}^*]^B$ and let $s \in \mathfrak{b}^*$. Then for a 1-parameter subgroup $\lambda_1(t)$ with coordinates

$$\lambda_1(t)_{\iota\gamma} = \begin{cases} t^{\iota-1} & \text{if } \iota = \gamma, \\ 0 & \text{if } \iota \neq \gamma, \end{cases}$$

we have

$$(\lambda_1(t).s)_{\iota\gamma} = (\lambda_1(t)s(\lambda_1(t))^{-1})_{\iota\gamma} = \begin{cases} * & \text{if } \iota < \gamma, \\ s_{\iota\iota} & \text{if } \iota = \gamma, \\ t^{\iota-\gamma}s_{\iota\gamma} & \text{if } \iota > \gamma. \end{cases}$$

Taking the limit as $t \to 0$, we have

$$\lim_{t\to 0}(\lambda_1(t).s)_{\iota\gamma} = \begin{cases} * & \text{if } \iota < \gamma, \\ s_{\iota\iota} & \text{if } \iota = \gamma, \\ 0 & \text{if } \iota > \gamma. \end{cases}$$

Since off-diagonal entries of s are zero, our B-invariant polynomial F is independent of the coordinates $\{s_{\iota\gamma}\}_{\iota>\gamma}$. Now consider another 1-parameter subgroup $\lambda_2(t)$, where

$$\lambda_2(t)_{\iota\gamma} = \begin{cases} t^{\iota-1} & \text{if } \iota \leqslant \gamma, \\ 0 & \text{if } \iota > \gamma. \end{cases}$$

Then

$$\begin{aligned}(\lambda_2(t).s)_{\iota\gamma} &= \begin{cases} * & \text{if } \iota < \gamma, \\ t^{\iota-\gamma}\left(\sum_{k=\iota}^n s_{k\gamma} - \sum_{k=\iota}^n s_{k,\gamma-1}\right) & \text{if } \iota \geqslant \gamma, \end{cases} \\ &= \begin{cases} * & \text{if } \iota < \gamma, \\ s_{\iota\iota} - \sum_{k=\iota}^n s_{k,\iota-1} + \sum_{k=\iota+1}^n s_{k\iota} & \text{if } \iota = \gamma, \\ t^{\iota-\gamma}\left(\sum_{k=\iota}^n s_{k\gamma} - \sum_{k=\iota}^n s_{k,\gamma-1}\right) & \text{if } \iota > \gamma. \end{cases}\end{aligned} \tag{2.3}$$

Since $F(s) = F(s')$ for any $s' \in \overline{B.s}$ (the polynomial F must take the same value on any orbit closure), the equality $\lim_{t\to 0} F(\lambda_1(t).s) = \lim_{t\to 0} F(\lambda_2(t).s)$ must hold for any values of $\{s_{\alpha\beta}\}_{\alpha>\beta}$. So for each $1 \leqslant \iota < n$ (starting with $\iota = 1$ in ascending order), choose $\{s_{k\iota}\}_{k>\iota}$ in $\sum_{\iota<k\leqslant n} s_{k\iota}$ such that

$$-\sum_{k=\iota+1}^{n} s_{k\iota} = s_{\iota\iota} - \sum_{k=\iota}^{n} s_{k,\iota-1}. \tag{2.4}$$

Move the sum in (2.4) to the left-hand side so that

$$\sum_{\iota-1<k\leqslant n} s_{k,\iota-1} - \sum_{\iota<k\leqslant n} s_{k,\iota} = s_{\iota\iota} \qquad \text{for each } 1 \leqslant \iota < n.$$

This implies the sum of all such sum as ι varies from 1 to $n-1$ is

$$\begin{aligned}&\sum_{\iota=1}^{n-1}\left(\sum_{k=\iota}^{n} s_{k,\iota-1} - \sum_{k=\iota+1}^{n} s_{k\iota}\right) \\ &= \sum_{k=1}^{n} \cancel{s}_{k0} + \left(\sum_{\iota=2}^{n-1}\sum_{k=\iota}^{n} s_{k,\iota-1} - \sum_{\iota=1}^{n-2}\sum_{k=\iota+1}^{n} s_{k\iota}\right) - \sum_{k=n}^{n} s_{k,n-1} \\ &= -\sum_{k=n}^{n} s_{k,n-1} = \sum_{\iota=1}^{n-1} s_{\iota\iota}.\end{aligned}$$

By (2.3) and by choosing appropriate choices for $\{s_{\alpha\beta}\}_{\alpha>\beta}$ in (2.4), we have $(\lambda_2(t).s)_{\iota\iota} = 0$ for each $1 \leqslant \iota < n$ while

$$(\lambda_2(t).s)_{nn} = s_{nn} - \sum_{n-1<k\leqslant n} s_{k,n-1} = \mathrm{tr}(s).$$

This means all coordinate entries are zero except the (n,n)-entry, which is $\mathrm{tr}(s)$. Thus for F in $\mathbb{C}[\mathfrak{b}^*]^B$, $F(s)$ must be of the form $F(s')$, where all coordinates of s' are zero except the entry s'_{nn}, which equals $\mathrm{tr}(s)$. So F is a polynomial in $\mathrm{tr}(s)$. □

We will now generalize [**Im18**, Definition 2.2]. Let $[n] := \{1,\ldots,n\}$.

DEFINITION 2.3. Let $J \subseteq [n]$. Define

$$\ell^J := \ell^J(r) = \prod_{k\in J}(r - r_{kk}\,\mathbf{I}), \tag{2.5}$$

where $\mathbf{I}$ is the $n\times n$ identity matrix. Let $\mathrm{tr}^J := \mathrm{tr}(\ell^J)$. Then

$$L^J := L^J(r) = (\mathrm{tr}^J)^{-1}\ell^J.$$

In the case when $J = \{k\}$, we may write $\ell^k := \ell^k(r) = r - r_{kk}\,\mathbf{I}$, and when $J = \varnothing$, then $\ell^J := \mathbf{I}$. Furthermore, we will write the coordinates of ℓ^J as $\ell^J_{\gamma\mu}$.

LEMMA 2.4. *For any $b \in B$, $\ell^J(\mathrm{Ad}_b(r)) = \mathrm{Ad}_b(\ell^J(r))$, where the adjoint action is by conjugation.*

PROOF. For any $b \in B$,

$$\ell^J(\mathrm{Ad}_b(r)) = \prod_{k\in J} brb^{-1} - r_{kk}\,\mathbf{I} = b\left(\prod_{k\in J}(r - r_{kk}\,\mathbf{I})\right)b^{-1} = \mathrm{Ad}_b(\ell^J(r)).$$ □

COROLLARY 2.5. *For any $b \in B$, $L^J(\mathrm{Ad}_b(r)) = \mathrm{Ad}_b(L^J(r))$, where conjugation is the adjoint action.*

PROOF. This follows from Lemma 2.4. □

2.2. Geometric invariant theory. We refer to [**MFK94, New09**] for an extensive background in geometric invariant theory.

Let X be a variety (or a scheme) with an action by an algebraic group G. Then the G-invariant polynomial ring is defined as

$$\mathbb{C}[X]^G := \{f \in \mathbb{C}[X] : f(g.x) = f(x) \text{ for all } x \in X, g \in G\}. \tag{2.6}$$

Equation (2.6) is a generalization of Equation (2.1).

Let $\chi : G \to \mathbb{C}^*$ be a character of G, i.e., a group homomorphism. Then χ-semi-invariant module is defined to be

$$\mathbb{C}[X]^{G,\chi} := \{f \in \mathbb{C}[X] : f(g.x) = \chi(g)f(x) \text{ for all } x \in X, g \in G\}. \tag{2.7}$$

The affine quotient of X by G is defined to be

$$X /\!\!/ G := \mathrm{Spec}(\mathbb{C}[X]^G), \tag{2.8}$$

while the GIT quotient (twisted by χ) is defined as

$$X /\!\!/_{\chi} G := \mathrm{Proj}\left(\bigoplus_{i\geqslant 0}\mathbb{C}[X]^{G,\chi^i}\right). \tag{2.9}$$

The affine quotient of X under the G-action is an affine variety, a scheme which is usually highly singular, while the GIT quotient twisted by χ is a projective variety.

Invariant and semi-invariant functions play fundamental roles in the geometric construction of quotient spaces in algebraic geometry.

3. Affine quotient of the Borel moment map

Recall Definition 2.3. Let $\iota \in [n]$ and let $J \subseteq [n]$. Let

$$
\begin{aligned}
f_J(r,s,i,j) &= \operatorname{tr}(j\ell^J i),\\
g_J(r,s,i,j) &= \operatorname{tr}(\ell^J s),\\
h_\iota(r,s,i,j) &= \operatorname{tr}(L^{[n]\setminus\{\iota\}} r),\\
k_{n,J}(r,s,i,j) &= \operatorname{tr}(j\ell^{[n]\setminus\{n\}} s\ell^J i),\\
l_{J,1}(r,s,i,j) &= \operatorname{tr}(j\ell^J s\ell^{[n]\setminus\{1\}} i).
\end{aligned}
\tag{3.1}
$$

REMARK 3.1. One can easily see that $k_{n,[n]\setminus\{1\}} = l_{[n]\setminus\{n\},1}$.

REMARK 3.2. Note that $k_{n,J}$ and $l_{J,1}$ are carefully chosen so that they are well-defined polynomials. That is, since $s \in \mathfrak{g}/\mathfrak{u}^+$, a representative of an equivalence class, we need to be careful that polynomials involving s do not depend on the coordinates in $\mathfrak{u}^+$.

We have given in [**Im18**, §6.2] rational versions of the polynomials in (3.1), but in this paper, we have more B-invariant polynomials than the ones in [**Im18**].

LEMMA 3.3. *The polynomials in* (3.1) *are B-invariant.*

PROOF. First, f_J is a B-invariant polynomial since for $b \in B$,

$$
\begin{aligned}
f_J(b.(r,s,i,j)) &= f_J(brb^{-1}, bsb^{-1}, bi, jb^{-1})\\
&= \operatorname{tr}(jb^{-1}\ell^J(\mathrm{Ad}_b(r))bi)\\
&= \operatorname{tr}(jb^{-1}(b\ell^J b^{-1})bi) \text{ by Lemma 2.4}\\
&= \operatorname{tr}(j\ell^J i) = f_J(r,s,i,j).
\end{aligned}
$$

Similarly, g_J is B-invariant since for $b \in B$,

$$
\begin{aligned}
g_J(b.(r,s,i,j)) &= \operatorname{tr}(\ell^J(\mathrm{Ad}_b(r))bsb^{-1})\\
&= \operatorname{tr}(\mathrm{Ad}_b(\ell^J(r))bsb^{-1})\\
&= \operatorname{tr}((b\ell^J(r)b^{-1})bsb^{-1}) \text{ by Lemma 2.4}\\
&= \operatorname{tr}(\ell^J(r)s) = g_J(r,s,i,j).
\end{aligned}
$$

Next, for $b \in B$, we have

$$
\begin{aligned}
h_\iota(b.(r,s,i,j)) &= \operatorname{tr}(L^{[n]\setminus\{\iota\}}(\mathrm{Ad}_b(r))brb^{-1})\\
&= \operatorname{tr}(\mathrm{Ad}_b(L^{[n]\setminus\{\iota\}}(r))brb^{-1}) \text{ by Corollary 2.5}\\
&= \operatorname{tr}((bL^{[n]\setminus\{\iota\}}(r)b^{-1})brb^{-1})\\
&= \operatorname{tr}(L^{[n]\setminus\{\iota\}}(r)r) = h_\iota(r,s,i,j).
\end{aligned}
$$

Now, for $b \in B$, we have

$$\begin{aligned} k_{n,J}(b.(r,s,i,j)) &= \operatorname{tr}(jb^{-1}\ell^{[n]\setminus\{n\}}(\operatorname{Ad}_b(r))bsb^{-1}\ell^J(\operatorname{Ad}_b(r))bi) \\ &= \operatorname{tr}(jb^{-1}\operatorname{Ad}_b(\ell^{[n]\setminus\{n\}}(r))bsb^{-1}\operatorname{Ad}_b(\ell^J(r))bi) \text{ by Lemma 2.4} \\ &= \operatorname{tr}(jb^{-1}(b\ell^{[n]\setminus\{n\}}(r)b^{-1})bsb^{-1}(b\ell^J(r)b^{-1})bi) \\ &= \operatorname{tr}(j\ell^{[n]\setminus\{n\}}(r)s\ell^J(r)i) = k_{n,J}(r,s,i,j). \end{aligned}$$

Since the proof for $l_{J,1}$ is similar to the one for $k_{n,J}$, we omit the proof. □

PROPOSITION 3.4. *The B-invariant ring $\mathbb{C}[T^*(\mathfrak{b} \times \mathbb{C}^n)]^B$ contains the subalgebra generated by the polynomials in* (3.1).

PROOF. The theorem holds by Lemma 3.3. Thus, we have an inclusion

$$\mathbb{C}[f_J, g_J, h_\iota, k_{n,J}, l_{J,1}] \hookrightarrow \mathbb{C}[T^*(\mathfrak{b} \times \mathbb{C}^n)]^B$$

of algebras. □

CONJECTURE 3.5. *Let $\iota \in [n]$ and $J \subseteq [n]$. The polynomials $f_J, g_J, h_\iota, k_{n,J}, l_{J,1}$ in* (3.1) *generate the B-invariant subalgebra $\mathbb{C}[T^*(\mathfrak{b} \times \mathbb{C}^n)]^B$.*

REMARK 3.6. Although μ_B is a complete intersection for $n \leqslant 5$, the reverse inclusion

$$\mathbb{C}[T^*(\mathfrak{b} \times \mathbb{C}^n)]^B \subseteq \mathbb{C}[f_J, g_J, h_\iota, k_{n,J}, l_{J,1}]$$

is much more difficult to prove for the nonreductive group setting. We thus leave this as an open problem.

REMARK 3.7. One of the reasons why B-invariant polynomials are important is because they provide an alternative way to study the affine quotient $\mu^{-1}(0) /\!\!/ B$ via its ring of functions, i.e., since $\mathbb{C}[\mu^{-1}(0) /\!\!/ B] \cong \mathbb{C}[T^*(\mathfrak{b} \times \mathbb{C}^n)]/\langle([r,s]+ij)_{\gamma\nu}\rangle$,

$$\begin{aligned} \mathbb{C}[\mu^{-1}(0) /\!\!/ B] &= \mathbb{C}[\mu^{-1}(0)]^B \\ &= \frac{\mathbb{C}[T^*(\mathfrak{b} \times \mathbb{C}^n)]^B}{\langle([r,s]+ij)_{\gamma\nu}\rangle^B} \\ &\supseteq \frac{\mathbb{C}[f_J, g_J, h_\iota, k_{n,J}, l_{J,1}]}{\langle\text{syzygies}\rangle + \langle([r,s]+ij)_{\gamma\nu}\rangle^B}, \end{aligned}$$

where the syzygies can explicitly be computed using `Macaulay2` (cf. [**GS02**]). The syzygies, i.e., relations among generators, for our setting are inhomogeneous and appear to be difficult to generalize for higher n.

4. GIT quotients of the Borel moment map

Let $\chi : B \to \mathbb{C}^*$ be a character, i.e., a group homomorphism. In this section, we will consider when $\chi = \det$ and $\det^{-1}$. We refer to [**GG06**, §8] for the construction of the semi-invariants for the classical setting $\mathsf{GL}_n(\mathbb{C})$.

Consider the closed embedding

$$(4.1) \qquad \begin{aligned} \varepsilon_{i_0} :& \mathfrak{h} \times \mathfrak{h} \hookrightarrow \mathfrak{b} \times \mathfrak{b}^* \times \mathbb{C}^n \times (\mathbb{C}^n)^* \\ & (a_1, \ldots, a_n, b_1, \ldots, b_n) \mapsto (\operatorname{diag}(a_1, \ldots, a_n), \operatorname{diag}(b_1, \ldots, b_n), i_0, 0) \end{aligned}$$

and

$$(4.2) \qquad \begin{aligned} \varepsilon_{j_0} :& \mathfrak{h} \times \mathfrak{h} \hookrightarrow \mathfrak{b} \times \mathfrak{b}^* \times \mathbb{C}^n \times (\mathbb{C}^n)^* \\ & (a_1, \ldots, a_n, b_1, \ldots, b_n) \mapsto (\operatorname{diag}(a_1, \ldots, a_n), \operatorname{diag}(b_1, \ldots, b_n), 0, j_0), \end{aligned}$$

where $i_0 = (1, \ldots, 1) \in \mathbb{C}^n$ and $j_0 = (1, \ldots, 1) \in (\mathbb{C}^n)^*$.

4.1. Twisted by det. Let $J \subseteq [n]$, and let $v^* \in \bigwedge^n(\mathbb{C}^n)^*$ be a nonzero volume form. Let

$$f = (f_1, \ldots, f_n), \quad \text{where } f_i \in A := \mathbb{C}\left\langle \ell^J,\ L^{[n]\backslash\{\iota\}}r,\ \ell^{[n]\backslash\{n\}}s\ell^J,\ \ell^J s\ell^{[n]\backslash\{1\}} \right\rangle \tag{4.3}$$

and f is an n-tuple of noncommutative polynomials. Consider polynomial functions of the form

$$\psi_f = \langle v^*, f_1 i \wedge \ldots \wedge f_n i\rangle \in \mathbb{C}[T^*(\mathfrak{b} \times \mathbb{C}^n)]. \tag{4.4}$$

LEMMA 4.1. *Polynomials ψ_f are* det*-semi-invariant.*

PROOF. Consider ψ_f in (4.4). First, it is clear that ℓ^J and $L^{[n]\backslash\{\iota\}}r$ are well-defined since they are elements in $\mathfrak{b}$. Next, we will check that the product of matrices

$$\ell^{[n]\backslash\{n\}}s\ell^J \quad \text{and} \quad \ell^J s\ell^{[n]\backslash\{1\}}$$

are well-defined. First, consider $\ell^{[n]\backslash\{n\}}s\ell^J$. Since

$$\ell^{[n]\backslash\{n\}}_{\gamma\mu} = \begin{cases} * & \text{if } \mu = n, \\ 0 & \text{otherwise,} \end{cases}$$

where $*$ represents a nonzero coordinate entry, $\ell^{[n]\backslash\{n\}}$ in $\ell^{[n]\backslash\{n\}}s$ will kill all the coordinates of s in $\mathfrak{u}^+$, i.e., $\ell^{[n]\backslash\{n\}}s$ does not depend on $\mathfrak{u}^+$. Thus, $\ell^{[n]\backslash\{n\}}s\ell^J$ is well-defined.

Similarly, for $\ell^J s\ell^{[n]\backslash\{1\}}$,

$$\ell^{[n]\backslash\{1\}}_{\gamma\mu} = \begin{cases} * & \text{if } \gamma = 1, \\ 0 & \text{otherwise,} \end{cases}$$

where $*$ represents a nonzero coordinate entry. So $s\ell^{[n]\backslash\{1\}}$ only involves the first column of s. Thus, $s\ell^{[n]\backslash\{1\}}$ doesn't depend on $\mathfrak{u}^+$, and $\ell^J s\ell^{[n]\backslash\{1\}}$ is well-defined.

Next, since we have

$$\ell^J(\mathrm{Ad}_b(r)) = \mathrm{Ad}_b(\ell^J(r)) \text{ by Lemma 2.4,}$$
$$L^{[n]\backslash\{\iota\}}(\mathrm{Ad}_b(r))\,\mathrm{Ad}_b(r) = \mathrm{Ad}_b(L^{[n]\backslash\{\iota\}}(r))\,\mathrm{Ad}_b(r) = \mathrm{Ad}_b(L^{[n]\backslash\{\iota\}}(r)r) \text{ by Cor. 2.5,}$$
$$\ell^{[n]\backslash\{n\}}(\mathrm{Ad}_b(r))\,\mathrm{Ad}_b(s)\ell^J(\mathrm{Ad}_b(r)) = \mathrm{Ad}_b(\ell^{[n]\backslash\{n\}}(r)s\ell^J(r)) \text{ by Lemma 2.4, and}$$
$$\ell^J(\mathrm{Ad}_b(r))\,\mathrm{Ad}_b(s)\ell^{[n]\backslash\{1\}}(\mathrm{Ad}_b(r)) = \mathrm{Ad}_b(\ell^J(r)s\ell^{[n]\backslash\{1\}}(r)) \text{ by Lemma 2.4,}$$

$\mathrm{Ad}_b(f_\nu(r,s)) = f_\nu(\mathrm{Ad}_b(r), \mathrm{Ad}_b(s))$ for $1 \leqslant \nu \leqslant n$. So

$$\mathrm{Ad}_b(f_\nu i) = \mathrm{Ad}_b(f_\nu)bi = bf_\nu b^{-1}bi = bf_\nu i \quad \text{for} \quad 1 \leqslant \nu \leqslant n.$$

Thus

$$b.\psi_f = \langle v^*, bf_1 i \wedge \ldots \wedge bf_n i\rangle = \det(b)\langle v^*, f_1 i \wedge \ldots \wedge f_n i\rangle = \det(b)\psi_f,$$

and this completes the proof. □

THEOREM 4.2. *For each $k \geqslant 1$, the restriction of functions via the embedding ε_{i_0} in (4.1) induces a surjection $\varepsilon_{i_0}^* : \mathbb{C}[\mu_B^{-1}(0)]^{B,\det^k} \twoheadrightarrow D^k$ of vector spaces, where*

$$D := \mathbb{C}[\ell^J,\ L^{[n]\backslash\{\iota\}}r,\ \ell^{[n]\backslash\{n\}}s\ell^J,\ \ell^J s\ell^{[n]\backslash\{1\}}]^\epsilon, \tag{4.5}$$

a subspace of S_n-alternating polynomials.

Proof. Let $k \geqslant 1$. The imbedding $\overline{\mathcal{C}}_0 \hookrightarrow \mu_B^{-1}(0)$ induces a bijection

$$\mathbb{C}[\mu_B^{-1}(0)]^{B,\det^k} \xrightarrow{\sim} \mathbb{C}[\overline{\mathcal{C}}_0]^{B,\det^k}.$$

It follows that $\mathbb{C}[\mu_B^{-1}(0)]^{B,\det^k}$ contains products $\psi_1 \cdots \psi_k$ as a $\mathbb{C}[\mu_B^{-1}(0)]^B$-module. Since $\varepsilon_{i_0}^* \psi_f \in D$ for any f, we see that $\varepsilon_{i_0}^*(\psi_1 \cdots \psi_k) \in D^k$. Hence

$$\varepsilon_{i_0}^*(\mathbb{C}[\mu_B^{-1}(0)]^{B,\det^k}) = \varepsilon_{i_0}^*(\mathbb{C}[\overline{\mathcal{C}}_0]^{B,\det^k}) \subseteq D^k.$$

Note that $\mathbb{C}[\mu_B^{-1}(0)]^{B,\det^k}$ contains the k-fold products $\psi_1 \cdots \psi_k$, as a $\mathbb{C}[\mu_B^{-1}(0)]^B$-module. Thus it suffices to prove surjectivity of the map $\varepsilon_{i_0}^*$ for $k = 1$. To prove this, we identify D with the n-th exterior power $\bigwedge^n E$ of the vector space of polynomials, where

$$E := \mathbb{C}[\ell^J,\ L^{[n]\setminus\{\iota\}}r,\ \ell^{[n]\setminus\{n\}}s\ell^J,\ \ell^J s\ell^{[n]\setminus\{1\}}].$$

With this identification, the space D is spanned by wedge products $f_1 \wedge \ldots \wedge f_n$, where $f_1, \ldots, f_n \in E$.

By the definition of the irreducible component $\overline{\mathcal{C}}_0$ for any $(r, s, i, j) \in \overline{\mathcal{C}}_0$, we have $[r, s] = [r, s] + ij = 0$. So for any $f \in E$, the expression f is a well-defined matrix. That is, for any lift of f to a noncommutative polynomial $\widehat{f} \in A$ (cf. (4.3)), i.e., for any $\widehat{f}$ in the preimage of f under the natural projection $A \twoheadrightarrow E$, we have $\widehat{f} = f$. Thus, given an n-tuple $f_1, \ldots, f_n$, we have a well-defined polynomial

$$\psi_f = \langle v^*, f_1 i \wedge \ldots \wedge f_n i\rangle \in \mathbb{C}[\mu_B^{-1}(0)]^{B,\det}. \qquad \square$$

4.2. Twisted by $\det^{-1}$**.** Let $v \in \bigwedge^n \mathbb{C}^n$ be a nonzero volume form. Similar as before, let

$$g = (g_1, \ldots, g_n), \qquad \text{where } g_i \in A,$$

A is defined in (4.3), and g is an n-tuple of noncommutative polynomials. Consider polynomials of the form

$$\phi_g = \langle jg_1 \wedge \ldots \wedge jg_n, v\rangle \in \mathbb{C}[T^*(\mathfrak{b} \times \mathbb{C}^n)]. \tag{4.6}$$

Lemma 4.3. *The polynomials ϕ_g are $\det^{-1}$-semi-invariant.*

Proof. The proof of Lemma 4.3 is analogous to the proof of Lemma 4.1, so we omit the details. $\square$

Similar to Theorem 4.2, we have the following:

Theorem 4.4. *For each $k \leqslant -1$, the restriction of functions via the embedding ε_{j_0} in (4.2) induces a surjection $\varepsilon_{j_0}^* : \mathbb{C}[\mu_B^{-1}(0)]^{B,\det^k} \twoheadrightarrow D^{-k}$ of vector spaces, where D is the S_n-alternating submodule given in (4.5).*

5. The singular locus of the components of μ_B

We investigate the singular locus using the computer software `Singular` (cf. [**DGPS19**]) for $n \leqslant 5$.

In algebraic geometry, components of μ_B are by definition the components of $\mu_B^{-1}(0)$. Recall from Proposition 1.2 that the irreducible components of μ_B are enumerated by $\overline{\mathcal{C}}_\nu$, where $\overline{\mathcal{C}}_\nu$ is the closure of $\mathcal{C}_\nu$.

THEOREM 5.1. *Each irreducible component $\overline{\mathcal{C}}_\nu$, for $1 \leqslant \nu \leqslant 2^n$ and $n \leqslant 5$, is singular, whose singular locus has codimension 1 in $\overline{\mathcal{C}}_\nu$. Thus, the singular locus $\mu_B^{-1}(0)^{\text{sing}}$ of the components of μ_B has codimension 1 in $\mu_B^{-1}(0)$. Furthermore, the irreducible components do not intersect transversely.*

It is a straight-forward calculation to check that μ_B is a complete intersection for $n \leqslant 5$ since we are working with explicit coordinate polynomials of $\mu_B^{-1}(0)$. One could also use `Macaulay2` (cf. [**GS**]) to check that μ_B is not only a local complete intersection, but a complete intersection. This gives us an upper bound that there are at most 2^n irreducible components, where $n \leqslant 5$.

On the other hand, we can restrict $\mu_B^{-1}(0)$ to the semisimple locus, i.e., $r \in \mathfrak{b}$ has pairwise distinct eigenvalues. Then there exists $b \in B$ such that brb^{-1} is diagonal and bsb^{-1} remains in $\mathfrak{b}^*$, but with the eigenvalues of s twisted by the coordinates of b. We now focus on $bi \in \mathbb{C}^n$ and $jb^{-1} \in (\mathbb{C}^n)^*$. It is clear that 2^n distinct irreducible components are determined by the vanishing and nonvanishing entries of the vector i, up to rescaling, i.e., by the values 0 and 1. This shows that μ_B has at least 2^n irreducible components. This latter argument of restricting to the semisimple locus holds for all n.

Combining the upper and lower bounds proves that μ_B has exactly 2^n irreducible components, where $n \leqslant 5$.

See [**Im18, Im14**] for a full description of the semisimple locus.

REMARK 5.2. To describe the singular locus explicitly, we found the Jacobian of the polynomials $\mu_B(r,s,i,j)_{\gamma\nu}$ for $n \geqslant \gamma \geqslant \nu \geqslant 1$ and then set $n(n+1)/2 \times n(n+1)/2$ minors equal to zero to describe the singular locus (cf. [**DGPS19, GS02, Res**]). We cannot generalize this locus for all n since it is still an open problem on whether or not $\mu_B^{-1}(0)$ is a complete intersection for $n > 5$.

EXAMPLE 5.3. Let $n = 2$. Then

$$\mathbb{C}[\mu_B^{-1}(0)] = \frac{\mathbb{C}[r_{11}, r_{12}, r_{22}, s_{11}, s_{21}, s_{22}, x_1, x_2, y_1, y_2]}{\langle r_{12}s_{21} + x_1y_1,\ -r_{12}s_{21} + x_2y_2,\ (r_{22} - r_{11})s_{21} + x_2y_1\rangle},$$

and its Jacobian matrix is

$$J = \begin{pmatrix} 0 & s_{21} & 0 & 0 & r_{12} & 0 & y_1 & 0 & x_1 & 0 \\ 0 & -s_{21} & 0 & 0 & -r_{12} & 0 & 0 & y_2 & 0 & x_2 \\ -s_{21} & 0 & s_{21} & 0 & r_{22} - r_{11} & 0 & 0 & y_1 & x_2 & 0 \end{pmatrix}.$$

Since $\dim \mu_B^{-1}(0) = 7$, the singular locus of $\mu_B^{-1}(0)$ consists of all points whose rank of the Jacobian J is less than 3. This is precisely the locus where all 3×3 minors of J vanish.

That is, the 3×3 minors are homogeneous cubic polynomials:

$$\begin{gathered} r_{12}s_{21}y_1, \quad r_{12}s_{21}y_2, \quad r_{12}s_{21}x_1, \quad r_{12}s_{21}x_2, \quad r_{12}x_2y_1, \\ s_{21}y_1y_2, \quad s_{21}x_2y_1, \quad s_{21}x_1y_2, \quad s_{21}x_1x_2, \quad s_{21}^2y_1, \\ s_{21}^2y_2, \quad s_{21}^2x_1, \quad s_{21}^2x_2, \quad s_{21}y_1^2, \quad s_{21}x_2^2, \\ (r_{11} - r_{22})s_{21}y_1, \quad (r_{11} - r_{22})s_{21}y_2, \quad (r_{11} - r_{22})s_{21}x_1, \quad (r_{11} - r_{22})s_{21}x_2, \\ s_{21}(x_1y_1 - x_2y_2), \quad (r_{22} - r_{11})y_1y_2 + r_{12}y_1^2, \quad (r_{11} - r_{22})x_2y_1, \\ (r_{11} - r_{22})x_1y_2 + r_{12}(x_2y_2 - x_1y_1), \quad (r_{22} - r_{11})x_1x_2 - r_{12}x_2^2, \\ x_2y_1y_2, \quad x_2y_1^2, \quad x_2^2y_1, \quad x_1x_2y_1, \end{gathered} \tag{5.1}$$

and the ideal of the singular locus $\dim \mu_B^{-1}(0)^{\mathrm{sing}}$ is generated by the 3 polynomials defining $\mu_B^{-1}(0)$ and all the 3×3 minors in (5.1).

One can easily check that this singular locus has dimension 6.

REMARK 5.4. Recall from [**IS21**, Theorem 1.2] that the irreducible components are equidimensional of dimension $\binom{n+1}{2} + 2n$. So the singular locus restricted to each irreducible component has dimension $\binom{n+1}{2} + 2n - 1$ for $n \leqslant 5$. There are no isolated singularities.

6. Restriction to the regular semisimple and the singular locus

When restricting GIT and affine quotients to the regular semisimple locus $\mu_B^{-1}(0)^{\mathrm{rss}}$, the isomorphism $\mu_B^{-1}(0)^{\mathrm{rss}} /\!\!/_{\det} B \cong \mu_B^{-1}(0)^{\mathrm{rss}} /\!\!/_{\det^{-1}} B$ is clear due to duality of $\mathfrak{b}$ and $\mathfrak{b}^*$, but furthermore, we expect an isomorphism

$$\mu_B^{-1}(0)^{\mathrm{rss}} /\!\!/_{\det} B \xrightarrow{\simeq} \mu_B^{-1}(0)^{\mathrm{rss}} /\!\!/ B$$

of varieties. Since the regular semisimple locus is precisely the nonsingular locus in both GIT and affine quotients, by [**Im18**, Theorem 1.6], we obtain the following:

PROPOSITION 6.1. *The singular locus $\mu_B^{-1}(0)^{\mathrm{sing}} /\!\!/ B$ in the B-affine quotient is the locus isomorphic to*

$$H = \{(r_1, \dots, r_n, 0, \dots, 0) : r_i = r_j \text{ for some } i \neq j\}.$$

Therefore,

$$\mu_B^{-1}(0)^{\mathrm{sing}} /\!\!/_{\det} B \overset{\dagger}{\cong} \mu_B^{-1}(0)^{\mathrm{sing}} /\!\!/_{\det^{-1}} B \twoheadrightarrow \mu_B^{-1}(0)^{\mathrm{sing}} /\!\!/ B$$

is expected to be a resolution of singularities, where $\dagger$ means that a certain notion of wall-crossing in GIT has occurred, and stability has been changed from det to $\det^{-1}$. We note that there are no isolated singularities in the affine quotient, which have also been verified for $n \leqslant 5$ using `Singular` (cf. [**DGPS19**]).

7. Connections to the Hilbert scheme of n points on a complex plane

Let $\mu_G : T^*(\mathfrak{gl}_n \times \mathbb{C}^n) \to \mathfrak{gl}_n^* \overset{\mathrm{tr}}{\cong} \mathfrak{gl}_n$, where $(r, s, i, j) \mapsto [r, s] + ij$. By [**Nak99**, Ch. 3], we have the well-known minimal resolution

$$\begin{array}{c} \mu_G^{-1}(0) /\!\!/_{\det} \mathsf{GL}_n(\mathbb{C}) \cong (\mathbb{C}^2)^{[n]} \\ \Big\downarrow \text{Hilbert–Chow} \\ \mu_G^{-1}(0) /\!\!/ \mathsf{GL}_n(\mathbb{C}) \cong S^n\mathbb{C}^2 = \mathbb{C}^2 \times \cdots \times \mathbb{C}^2 / S_n \end{array}$$

of symplectic singularities, where $(\mathbb{C}^2)^{[n]}$ is the Hilbert scheme of n points on a complex plane.

Analogously for our setting, we have:

CONJECTURE 7.1. *The following hold:*

(i) $\mu_B^{-1}(0) /\!\!/_{\det} B \cong \mu_B^{-1}(0) /\!\!/_{\det^{-1}} B$,

(ii) *there is a resolution* $\mu_B^{-1}(0) /\!\!/_{\det} B \twoheadrightarrow \mu_B^{-1}(0) /\!\!/ B$ *of singularities,*

(iii) $\mu_B^{-1}(0) /\!\!/_{\det} B$ *is isomorphic to the flag Hilbert scheme on a complex plane (line).*

Conjecture 7.1(i) is known as variational GIT, or wall-crossing.

CONJECTURE 7.2. *The following diagram*

$$\begin{array}{ccc} \mu_B^{-1}(0) /\!\!/_{\det} B & \dashrightarrow\!\!\!\!\twoheadrightarrow & \mu_G^{-1}(0) /\!\!/_{\det} \mathsf{GL}_n(\mathbb{C}) \\ \downarrow & \circlearrowleft & \downarrow {\scriptstyle \textit{Hilbert–Chow morphism}} \\ \mu_B^{-1}(0) /\!\!/ B & \dashrightarrow\!\!\!\!\twoheadrightarrow & \mu_G^{-1}(0) /\!\!/ \mathsf{GL}_n(\mathbb{C}) \end{array}$$

commutes, where the two horizontal maps are surjective morphisms.

REMARK 7.3. Although proving Conjectures 7.1 and 7.2 would be interesting, we have obtained the syzygies for our B-invariant and det-semi-invariant polynomials using `Macaulay2` [**GS02**]. Furthermore, although writing the syzygies in terms of our generators appear to be unsystematic and difficult to generalize for higher n, i.e., $n > 5$, constructing affine and GIT quotients with our current set of generators and relating them to other well-known schemes would be interesting in their own right.

Acknowledgment

The first author thanks the Department of Mathematics at Galatasaray University in Istanbul, Turkey for warm hospitality in February 2019 when this project was initiated, and January 2020 when this project was completed. The first author was partially funded by National Academy of Sciences in Washington, D.C.

References

[CG10] N. Chriss and V. Ginzburg, *Representation theory and complex geometry*, Modern Birkhäuser Classics, Birkhäuser Boston, Ltd., Boston, MA, 2010. Reprint of the 1997 edition, DOI 10.1007/978-0-8176-4938-8. MR2838836

[DGPS19] G.-M. Greuel, *Description of SINGULAR: a computer algebra system for singularity theory, algebraic geometry and commutative algebra*, Euromath Bull. **2** (1996), no. 1, 161–172. MR1413182

[GG06] W. L. Gan and V. Ginzburg, *Almost-commuting variety, $\mathcal{D}$-modules, and Cherednik algebras*, IMRP Int. Math. Res. Pap. (2006), 26439, 1–54. With an appendix by Ginzburg. MR2210660

[GNR21] E. Gorsky, A. Neguț, and J. Rasmussen, *Flag Hilbert schemes, colored projectors and Khovanov-Rozansky homology*, Adv. Math. **378** (2021), Paper No. 107542, 115, DOI 10.1016/j.aim.2020.107542. MR4192994

[GS] D. R. Grayson and M. E. Stillman, *Macaulay2, a software system for research in algebraic geometry*, Available at `http://www.math.uiuc.edu/Macaulay2/`.

[GS02] D. R. Grayson and M. E. Stillman, *Macaulay2, a software system for research in algebraic geometry*, Available at `https://faculty.math.illinois.edu/Macaulay2`, 1993-2002.

[Hai01] M. Haiman, *Hilbert schemes, polygraphs and the Macdonald positivity conjecture*, J. Amer. Math. Soc. **14** (2001), no. 4, 941–1006, DOI 10.1090/S0894-0347-01-00373-3. MR1839919

[Im14] M. S. Im, *On semi-invariants of filtered representations of quivers and the cotangent bundle of the enhanced Grothendieck-Springer resolution*, ProQuest LLC, Ann Arbor, MI, 2014. Thesis (Ph.D.)–University of Illinois at Urbana-Champaign. MR3312842

[Im18] M. S. Im, *The regular semisimple locus of the affine quotient of the cotangent bundle of the Grothendieck-Springer resolution*, J. Geom. Phys. **132** (2018), 84–98, DOI 10.1016/j.geomphys.2018.05.022. MR3836769

[IS21] M. S. Im and T. Scrimshaw, *The regularity of almost-commuting partial Grothendieck-Springer resolutions and parabolic analogs of Calogero-Moser varieties*, J. Lie Theory **31** (2021), no. 1, 127–148. MR4161536

[KL09] M. Khovanov and A. D. Lauda, *A diagrammatic approach to categorification of quantum groups. I*, Represent. Theory **13** (2009), 309–347, DOI 10.1090/S1088-4165-09-00346-X. MR2525917

[KL11] M. Khovanov and A. D. Lauda, *A diagrammatic approach to categorification of quantum groups II*, Trans. Amer. Math. Soc. **363** (2011), no. 5, 2685–2700, DOI 10.1090/S0002-9947-2010-05210-9. MR2763732

[MFK94] D. Mumford, J. Fogarty, and F. Kirwan, *Geometric invariant theory*, 3rd ed., Ergebnisse der Mathematik und ihrer Grenzgebiete (2) [Results in Mathematics and Related Areas (2)], vol. 34, Springer-Verlag, Berlin, 1994. MR1304906

[Nak99] H. Nakajima, *Lectures on Hilbert schemes of points on surfaces*, University Lecture Series, vol. 18, American Mathematical Society, Providence, RI, 1999, DOI 10.1090/ulect/018. MR1711344

[Nev11] T. Nevins, *Stability and Hamiltonian reduction for Grothendieck-Springer resolutions*, Available at `http://www.math.illinois.edu/~nevins/papers/b-hamiltonian-reduction-2011-0316.pdf`, 2011.

[New09] P. E. Newstead, *Geometric invariant theory*, Moduli spaces and vector bundles, London Math. Soc. Lecture Note Ser., vol. 359, Cambridge Univ. Press, Cambridge, 2009, pp. 99–127. MR2537067

[Res] Wolfram Research, *Mathematica, Version 11.3*, Champaign, IL, 2018.

[Rou08] R. Rouquier, *2-Kac-Moody algebras*, Available at `arXiv:0812.5023`, 2008.

[Rou12] R. Rouquier, *Quiver Hecke algebras and 2-Lie algebras*, Algebra Colloq. **19** (2012), no. 2, 359–410, DOI 10.1142/S1005386712000247. MR2908731

Department of Mathematical Sciences, United States Military Academy, West Point, New York 10996

Current address: Department of Mathematics, United States Naval Academy, Annapolis, Maryland 21402

Email address: `meeseongim@gmail.com`

Department of Mathematics, Galatasary University, Ortaköy 34357, Istanbul, Turkey

Email address: `mrltosun@gmail.com`

Contemporary Mathematics
Volume **834**, 2026
https://doi.org/10.1090/conm/834/16684

$ER = EPR$, entanglement topology and tensor networks

Louis H. Kauffman

Abstract. This paper discusses ER = EPR. Given a background space and a quantum tensor network, we describe how to construct a new topological space, that welds the network and the background space together. This construction embodies the principle that quantum entanglement and topological connectivity are intimately related.

1. Introduction

We discuss the relationship of space, spacetime and quantum entanglement in the context of the hypothesis of Susskind and Malcedena [**15–17**]. Their $ER = EPR$ hypothesis is based on the suggestion that connectivity in spacetime is equivalent to quantum entanglement. Susskind asserts that quantum entanglement of distant black holes is equivalent to the existence of an Einstein-Rosen bridge connecting them. If this hypothesis is true, then there is indeed a topological underpinning for quantum entanglement. Here we make foundational comments on the $ER = EPR$ hypothesis. In the discussion below we examine entanglement and teleportation in relation to the construction of a space that is augmented by quantum states. Since an entangled state such as $|\delta\rangle = \frac{1}{\sqrt{2}}(|01\rangle + |10\rangle)$ can be formulated without any background space, we point out that it is possible graphically to form a new space from the given space or spacetime S of the physics by attaching a corresponding quantum network to S. The new space S' has connectivity related to the entanglement. This construction can then be considered as a precursor to the spacetime with an Einstein-Rosen bridge connecting the sites of the entangled particles. This analogy is illustrated in Figure 1 where we show on the left the bare bones of a line space augmented by the tensor network for an entangled pair and on the right a schema for a wormhole connection of two entangled black holes. The event horizon of the wormhole plays the same topological role as the extra point E in the augmentation. Any neighborhood of E must contain neighborhoods of the ends of the network. Any neighborhood of the event horizon is a connection of the two black holes. We see that underlying the properties of an event horizon are the simplest possiblities for effecting a topological connection.

2020 *Mathematics Subject Classification.* Primary 81P40, 81P42.
Key words and phrases. Quantum entanglement,topological space, topological connectivity, linking, augmented space, tensor network, wormhole, Heyting algebra.

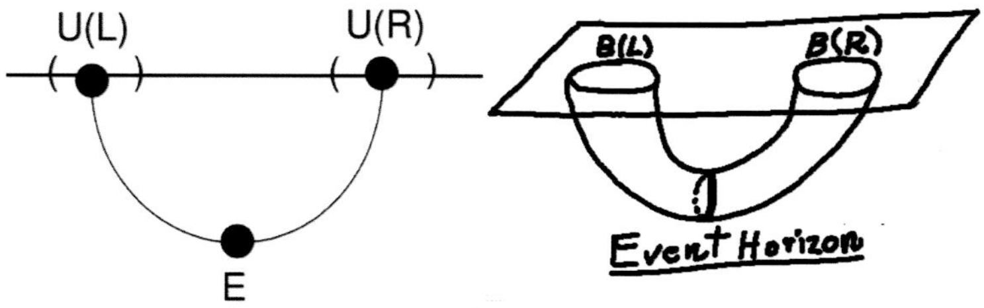

FIGURE 1. Augmented space and ER-bridge

In the paper we review the idea of tensor networks and their relationship with topology in Section 2. We consider network topologies and the augmentation referred to above in Section 3. We discuss topological entanglement and quantum entanglement in Section 4. We show how Heyting algebra structures are deeply related to topological connectivity and thus to $ER = EPR$ in Section 5. Section 6 is a summary of the ideas in the paper. Ideas in this paper are discussed from other viewpoints in our other papers [**10, 11**], and the present paper is intended as a source for further work.

2. Tensor networks

A tensor network is a graph G with tensors or matrices associated with each of its nodes and and index set I that can be used to label the edges of the graph. A contraction of the tensor net G is obtained by assigning fixed indices to all external edges of G and then summing over all possible assignments of indices to internal edges the products of the corresponding matrix entries for the nodes of the graph. See Figure 2 for illustrations of abstract tensor networks. It is often useful to choose a form for the nodes of the graph that is mnemonic for particular uses. For example, in Figure 3 will illustrate a tensor network that is associated with a knot diagram. The basic ingredients in this network are the cups, caps and crossings shown in the figure. Appropriate matrix choices including nodes that satisfy the Yang-Baxter equation, can be used to form the so-called quantum invariants of knots and links. Tensor networks of the type used here were originally devised by Roger Penrose [**6, 18**] to handle spin-networks for (quantum) angular momentum, pre-geometry and graph coloring problems.

A graph G without further structure is a pattern of the composition or connection of its nodes. Thus, even without the set theoretic concept of a topological space, a graph is a kind of pre-space through its indication of connection of given nodes with other nodes through the edges they share. When a graph is seen as a tensor network it becomes a computational structure relative to the assignment of matrices to its nodes. There is a direct relationship of tensor networks and the mathematical theory of categories. A category is a directed graph where extra assumptions are made about the existence of edges to ensure that if there is an edge from A to B and there is an edge from B to C, then there is a composition of these edges forming a new edge from A to C. Thus a category embodies certain compositionalities in its given structure. It is assumed that these compositions are

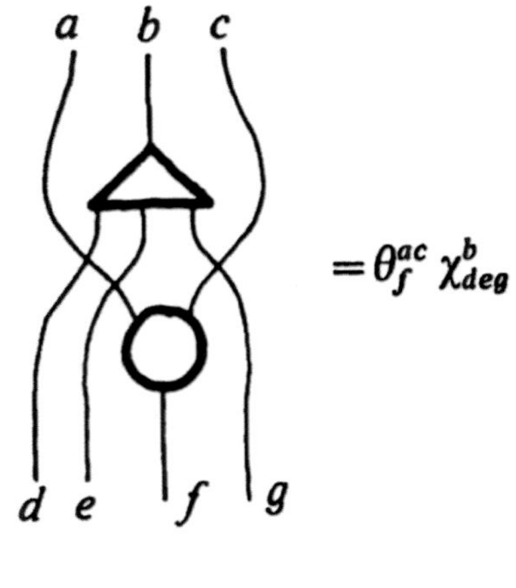

FIGURE 2. Tensor net

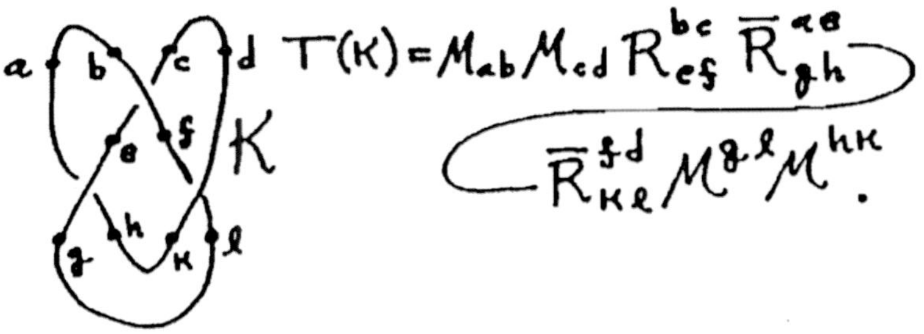

FIGURE 3. Topological tensor nets

associative and that every node has an edge from itself to itself that acts as an identity. A functor from a category to a category of linear transformations makes that category into a tensor network. Thus tensor networks and graphs are more general than categories, but categories can acquire tensor network structure quite naturally.

In this paper we will explain how to join tensor networks and topological spaces to form new spaces and sometimes new tensor networks. One can then start with a space that models a classical world and extend it so that it models a quantum world in its topology. Non-local connection, initally modeled by a bit of tensor network can be grafted to the space so that the tensor that embodies the entanglement becomes a part of the topological space. The tensor becomes an analog of a wormhole or Einstein-Rosen bridge that connects two points in the space superluminally.

3. Network topology

A *tensor network* is a graph G with tensors or matrices associated with each of its nodes and an index set I that can be used to label the edges of the graph. A *contraction* of the tensor net G is obtained by assigning fixed indices to all external edges of G and then summing over all possible index assignments to internal edges the products of the corresponding matrix entries for the nodes of the graph.

We now define a topological space $Top(G) = (V(G) \cup E(G), T(G))$ associated with any graph G. To do this we must give a collection of open sets $T(G)$ satisfying the axioms for a topology. It is sufficient to give a basis for the topology, consisting of a collection of sets that are designated as open and then to make the full collection of open sets $T(G)$ by taking those sets generated from the basis by finite intersection and arbitrary union. To this end, we take the set of points in the space to be the union of the vertex set of G, $V(G)$ and the edge set $E(G)$ of G. The basis consists in the whole space $V(G) \cup E(G)$, the empty set, and the neighborhoods of edges (edges of the graph are points in the space) defined by $N(e) = \{v | v$ is a vertex in G on the end of the edge $e\} \cup \{e\}$. Thus if e is an edge with endpoints v and v', then $N(e) = \{e, v, v'\}$. Note that this is not a Hausdorff topology. We have points in the topological space corresponding to edges in the graph, and the smallest open set containing a given vertex v can be the set consisting in v itself.

In the examples preceding this definition we have made a combination of graph topology and a given topological space such as a background geometrical space for measurement. Suppose that G is a graph with external edges and external nodes that we wish to attach to a given toological space X. Then we form the guotient topology between X and $Top(G)$ where the gluing is between end nodes of G and selected points in X. This formalizes the constructions we have previously indicated.

The usual topology on a graph is different from this topology. the usual topology is obtained by letting each edge have the topology of a unit interval and each then take the quotient topology obtained when joining all the edges together at the vertices to form the graph. Let $\tau(G)$ dente the usual topology on a graph G. We then can map

$$F : Top(G) \longrightarrow \tau(G)$$

by taking each vertex in G to the corresponding point in $\tau(G)$ and each edge e of G (seen as a point in $Top(G)$) to the midpoint of the interval that corresponds to e. This mapping is not continuous. The fact that F is not continuous in this way makes our model for the topology of a tensor net close to the interrelationship of the discrete and the continuous that is desired for the quantum model. Think of the half length of the continuous intervals assigned to edges of G as an analog of the Planck length. Then the geometry of the space $\tau(G)$ changes radically within that radius as measured by $Top(G)$. For sufficiently large neighborhoods of the vertices F will appear continuous.

The algebra of the net makes it possible for a single tensor net to be associated with many graphs since it is possible for a given matrix to factorize. Then a node in a given network will be replaced by a combination of nodes in a new network. This can result in the net making closer and closer approximations to a continuous space. The matrices in the nets we have considered represent entanglement and so one can regard the topology of $Top(G)$ as arising from quantum entanglement in these models.

An example of factorization is indicated in the Figure 4 where we show a net work whose nodes are projectors P such that $PP = P$. Then there is an infinite sequence of graphs and spaces with more and more nodes as we take more and more products of P in the form $P = PP = PPP = PPPP = \ldots$ This example

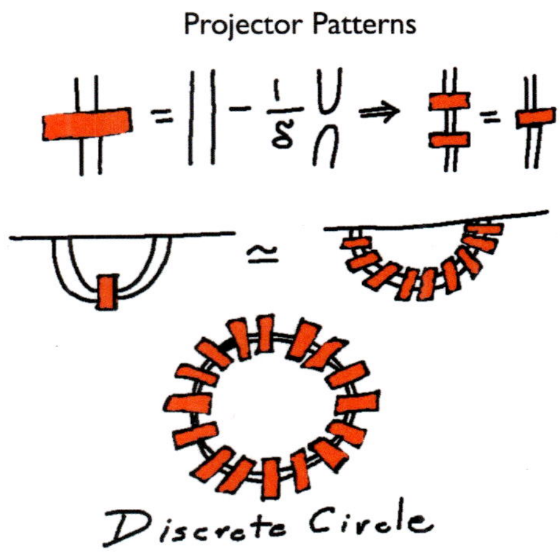

FIGURE 4. Projector networks

shows how a finite graph may fit into a family of infinitely many finite graphs whose appropriate limit is a continuum. In this case the corresponding continuum is a circle.

3.1. Space, time, quantum networks, and entanglement. Here is a summary for understanding quantum teleportation [**7**]. Take

$$|\delta\rangle = \frac{1}{\sqrt{2}}(|00\rangle + |11\rangle)$$

as a representative entangled state. Regard $|\delta\rangle$ as representing the state of two particles that we shall call L (left) and R (right) corresponding to δ's right and left tensor factors. Measuring $|\delta\rangle$ results either in $|00\rangle$ or $|11\rangle$. If an observer measures the left particle, and sees 0 then an observer who will measure the right particle must see 1 and vice versa. Nowhere in the quantum state $|\delta\rangle$ is there any information about the distance between the particles L and R or any information about the relative times for measurements to occur at the locales for these particles.

Note that the entangled state $|\delta\rangle$ is in the tensor product $V \otimes V$ where V is a qubit space spanned by $|0\rangle$ and $|1\rangle$. A general element in $V \otimes V$ has the form

$$|A\rangle = a_{00}|00\rangle + a_{01}|01\rangle + a_{10}|10\rangle + a_{11}|11\rangle,$$

and can presented as a 2×2 matrix

$$A = \begin{bmatrix} a_{00} & a_{01} \\ a_{10} & a_{11} \end{bmatrix}$$

Thus the matrix for $|\delta\rangle$ is the identity matrix

$$I = \begin{bmatrix} 1 & 0 \\ 0 & 1 \end{bmatrix}$$

By the same token, a successful measurement on two tensor lines can be represented in the dual basis spanned by elementary bras as

$$\langle M| = m_{00}\langle 00| + m_{01}\langle 01| + m_{10}\langle 10| + m_{11}\langle 11|,$$

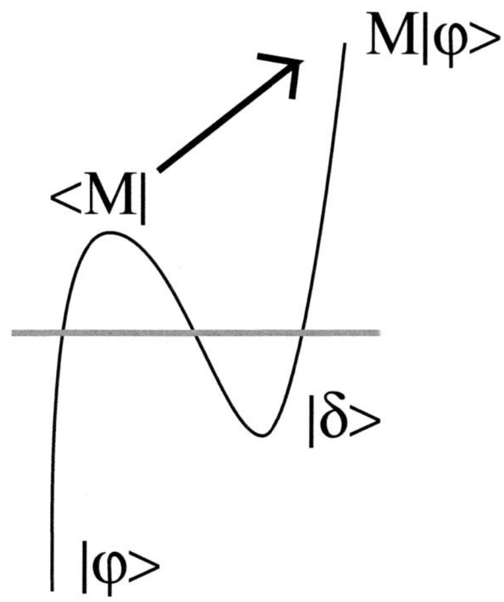

FIGURE 5. A teleportation scenario

with corresponding matrix

$$M = \begin{bmatrix} m_{00} & m_{01} \\ m_{10} & m_{11} \end{bmatrix}$$

Now consider Figure 5 where we have indicated an initial qubit state $|\phi\rangle$ tensored with the entangled state $|\delta\rangle$. A successful measurement has been made on the first two tensor lines. We assert that the state on the final tensor line is given by $M|\phi\rangle$ where this denotes the action of the matrix M of the measurement $\langle M|$ on the vector $|\phi\rangle$. This means that if Alice is at the site of the left particle and performs the measurement $\langle M|$, then she knows that Bob (at the site of the right particle) will have the quantum state $M|\phi\rangle$. If the matrix M is invertible and unitary, Alice can phone Bob and tell him to apply M^{-1} to the state that he has. The result will be that Bob will then have a perfect copy of the original state $|\phi\rangle$. This is the key to teleportation. There is a geometry in the tensor diagrams for the teleportation procedure. It is this geometry that we wish to pursue to understand the geometry and topology of entanglement.

In Figure 6 we illustrate the general case for a single qubit teleportation. The entangled state now has matrix E, not neccessarily the identity, an the measurement has matrix M. We then see from the figure that $|\psi'\rangle = EM|\psi\rangle$. The matrix E of an entangled state is neccessarily invertible, and so when E and M are unitary, our previous description of the teleportation procedure goes over mutatis mutandis. Using indices, the description of the state transformation is given by the equation

$$(\psi')^k = \psi^i M_{ij} E^{jk}.$$

The important point to note about this index version of the equation is that it is an exact translation of the structure of the tensor network given on the right part of the figure.

This tensor network is the detailed expression of the tensor diagram on the left part of Figure 6. This transformation from (ψ^i) to $(\psi')^k = \psi^i M_{ij} E^{jk}$ can be described by following the connectivity of the tensor network from Alice's locale to Bob's locale. The successful measurement $\langle M|$ completes the connection and transforms the quantum information at $|\psi\rangle$, located with Alice to $|\psi'\rangle = EM|\psi\rangle$,

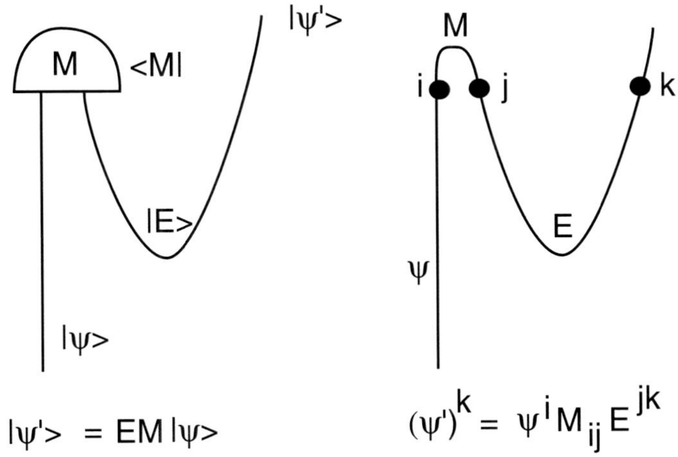

FIGURE 6. Teleportation tensors

located with Bob. It is a transfer of quantum information, a transfer of quantum states. To obtain observed information transfer one would need to control both measurement at Alice's end and corresponding measurement at Bob's end. Nevertheless, the tensor network for the entanglement can be viewed as a way to augment the simple space between Alice and Bob. This extra connectivity between Alice and Bob resides in the entangled state $|E\rangle$ that connects them.

3.2. Quantum tensor space. We formalize the idea that the tensor network for quantum entanglement can augment the original physical space to create a new connectivity. Let S be the given background space for the physical locations of particles. For each entangled state $|E\rangle$ with corresponding observers located at points L and R in the space S associate a new point E and a new open neighborhooda $N(E) = U(L) \cup \{E\} \cup U(R)$ for this new point E. Here $U(L)$ and $U(R)$ denote neighborhoods of L and R in the topology of S. Let S' be the new space with topology generated by these new neighborhoods of points corresponding to the entangled states. We call S' the *quantum tensor space* associated with S and its quantum network. See Figure 7 for an illustration of this concept.

Note that if the original space S is connected (so that it is not a union of disjoint open sets), then the augmented space S' is also connected, since in order to have an open cover containing the new point p there must be open sets $U(L)$ and (R) in the topology of S specified. Thus any open cover of S' will implicate an open cover of S. Such a cover cannot be a disjoint union of open sets in S, from which it follows that the cover of S' also cannot be a disjoint union of open sets.

On the other hand, suppose that there are disjoint open sets (each connected) $U(L)$ and $U(R)$ whose union is S. Then we see that the construction of S' produces a new space S' that is connected. Thus the augmentation construction provides a minimal way for disconnected L and R to achieve connection.

To construct the quantum tensor space, we introduce a least topological structure that can produce the special connection between L annd R. Note that the new space has non-Hausdorff points for each entangled state. The neighborhood

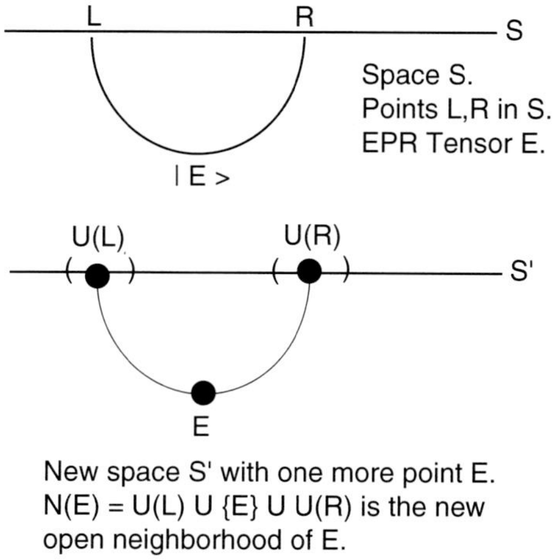

FIGURE 7. Augmented space

$N(E) = U(L) \cup \{E\} \cup U(R)$ is a combnatorial topological analogue of an Einstein-Rosen bridge connecting L and R. The analogy is important. Note that an observer in the space S', cannot move continuously from L to E without invoking an open neighborhood of E and the least such neighborhood contains R. Letting Alice be the observer at L and Bob the observer at R, we can say that Alice and Bob can meet together at the connecting point E in the analogue worm hole. The point E is the analogue of the event horizon of an Einstein-Rosen bridge between L and R. We will explore the analogies between connectivity in the the quantum tensor spaces and connectivity via Einstein-Rosen bridges in a later paper. It is possible that for larger networks and states with many particles these precursors to Einstein-Rosen Bridges will approximate the bridges in the continuum spacetime. For our purposes, we introduce this formalism to show how it is possible to weld a combinatorial quantum tensor network to a given background space.

Figure 8 illustrates the procedure known as *entanglement swapping*. Locations A and B are connected by a entangled state $|E\rangle$ and locations B and C are connected by an entangled state $|E'\rangle$. By performing a measurement $\langle M|$ at B we connect the two entangled states and make a new entangled state that connects A with C. In the process, the entanglement connection with B is lost. This example shows how the topology of the quantum tensor space will change under the act of measurement. Just as the quantum network undergoes graphical cut and rejoin operations under measurement, the corresponding quantum space, made by the prescription above, will change its connectivity properties. The actions on our simple spaces are easy to understand. In the case of the Susskind $ER = EPR$ hypothesis it will be very interesting to see what is the meaning of a procedure such as entanglement swapping.

3.3. Entangled states and wormholes. Here we describe a way to associate a possibly entangled state with a wormhole or Einstein-Rosen bridge, that is coherently related to the $ER = EPR$ hypothesis, and to our augmentation construction. Recall that a *cobordism* between two manifolds M and M' is a manifold W of one higher dimension such that the boundary of W is the union of M and M'.

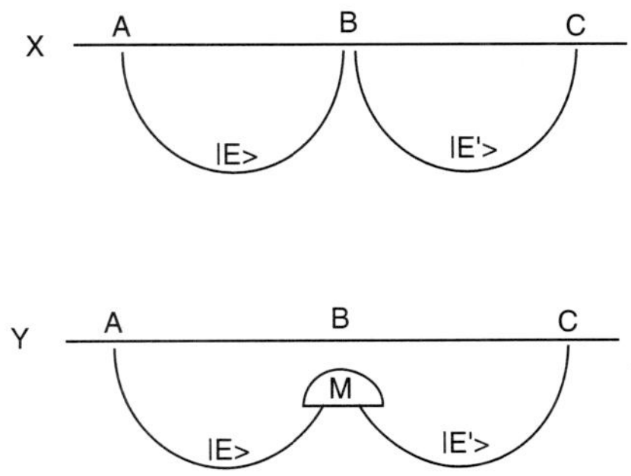

FIGURE 8. Entanglement swap

If M' is empty, then we say that W is a cobordism of M to the empty manifold. This simply means that the boundary of W is M. View Figure 9. A wormhole can be seen as a cobordism between an empty manifold and two spheres, drawn as circles in the figure. For a spacetime wormhole, the spheres would be two-dimensional (forming event horizons of the two black holes).

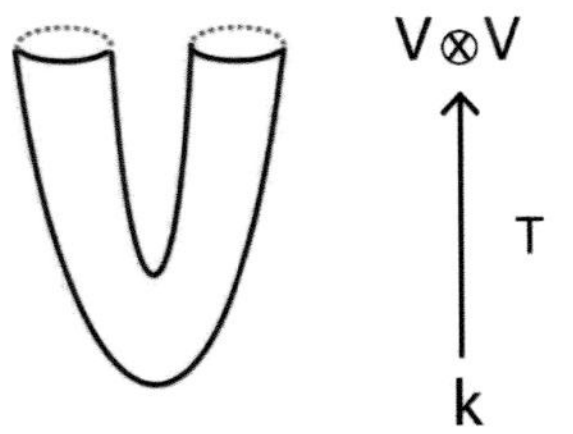

FIGURE 9. The quantum state of a wormhole

In topological quantum field theory [**3**, **4**] one considers functors from the category of manifolds (as objects) and cobordisms (as morphisms) to the category of vector spaces and linear transformations. In this point of view a wormhole as in Figure 9 would be sent by the functor to a linear mapping

$$T : k \longrightarrow V \otimes V$$

where the two-sphere S^2 (depicted as a circle in the figure) maps to V, the disjoint union of the two-spheres maps to $V \otimes V$, and the empty object maps to the ground field k.

Here T is the map corresponding to the wormhole itself. With this point of view, we can see how an entangled quantum state can be associated with a wormhole.

The possible state would occur using k, the complex numbers, and V a finite dimensional complex vector space associated with the two-sphere. $T(1)$ is a vector in the tensor product $V \otimes V$, and is a possibly entangled quantum state to be associated with the wormhole. The state $T(1)$ can be interpreted as an element of the tensor product of Hilbert spaces associated with each black hole (represented by their respective event horizons).

Note that the pattern of assignment of the state $T(1)$ to the wormhole is exactly in accord with our way of augmenting a space to include tensor nets that embody entangled states. The augmentation can always be done. The possible relationships of such augmentations with large scale modifications of spacetime such as Einstein-Rosen bridges or wormholes remains to be explored.

4. Topological entanglement and quantum entanglement

The purpose of this section of the paper is to show how topological connectivity in the form of spatially realized quantum networks can be used to make new points of view about quantum entanglement. These points of view shed new light on both the Aravind hypothesis and the Susskind $ER = EPR$ hypothesis. In our opinion the Susskind hypothesis is the deeper of the two and most likely to lead to new physics. Nevertheless, the connections between knot theory and the structure of quantum information are very strong and deserve further investigation.

Consider the Borommean Rings as illustrated in Figure 10 and Figure 11. The three rings are topologically linked. There is no way to separate them by continuous deformation of their configuration. If, however, any one of the three rings is removed, then the other two rings are disentangled. The linking of the three rings depends on the triplet. Now two of the rings are entangled without the presence of the third ring. Aravind [**1, 10**] has pointed out that the three-way entanglement of the Borommean rings is analogous to the three way entanglement of the GHZ state $GHZ = \frac{1}{\sqrt{2}}(|000\rangle + |111\rangle)$. The GHZ is an entangled state, but observation in any one of the three tensor factors renders it unentangled. No two of its tensor factors are entangled. This also led Aravind to suggest that an analog of measurement for topological links would be the removal of a component. Then other states such as $W = |100\rangle + |010\rangle + |001\rangle$ one could have entanglement properties corresponding to a different topological link. However, in the case of the state W we see that measurement in any given tensor factor can result (with equal probability) in a state that is either entangled or not entangled. Thus the relationship with the topology of links is, at best, more subtle than one could have suspected. See Figure 11 and the references [**8, 9**]. Nevertheless, as we have seen, it is natural to have a change in topology of the augmented network space correspond to quantum measurement.

In the reasoning about black holes and entanglement, Susskind [**15–17**] argues that the information limits on tripartite entanglement are decisive in creating the necessity of a wormhole connection between black holes that have entangled quantum states. The basis of the argument occurs in examining states $|p\rangle$ and $|p'\rangle$ wiithin the two black holes being entangled with each other and also entangled with a state $|q\rangle$ outside the first black hole. This leads to a tripartitie entanglement - where analysis shows that quantum monogamy would be violated if there were no wormhole connection.

Another point about tripartite entanglement and its geometric interpretations is discussed by Susskind as follows [**17**]. We begin with a tripartite entanglement from an observer of an entangled state $|\psi\rangle = |00\rangle + |11\rangle$ where it is assumed that this observer becomes entangled with the state $|\psi\rangle$. Letting the observer be denoted by the quantum state $|0\rangle$, we assume that on observing $|00\rangle$ from $|\psi\rangle$ the state is sent into the state $|000\rangle$, while upon observing $|11\rangle$ in $|\psi\rangle$ the state is sent into $|111\rangle$ (flipping the observing bit). Thus $(|00\rangle + |11\rangle) \otimes |0\rangle \longrightarrow |000\rangle + |111\rangle$ and the triply entangled state of observer and $|\psi\rangle$ becomes the state $GHZ = |000\rangle + |111\rangle$. One can then ask what a wormhole interpretation of the GHZ state would be like. See Figure 12. In this figure we illustrate the GHZ state and a schematic of that state as corresponding to a tripartite "black hole" in the form of a thrice punctured sphere. In the figure the spheres are circles. If we were working with 3+1 spacetime, then the spheres would be two dimensional wormhole horizons. For diagrammatic purposes it is useful to take the case of circles. Then the three circles can be seen as correspondent to the three components of the Borommean Rings illustrated in the same figure. The Rings are seen to be entangled topologically and in such a way that no two of them are topologically entangled, but there three rings taken together are entangled. This is the the link theoretic analog of the relationship of the Borommean rings with the GHZ state. The key point to notice in this analogy is that it is topological linking that has replaced topological connectivity.

What is the relationship between topological linking and topological connectivity? As we have seen in this paper and discussed in some detail, topological connectivity is a fundamental property that the most general topological spaces can have. Linking is, in topology, an apparently special property of submanifolds of a given manifold. Thus one dimensional curves can link one another in a three dimensional manifold such as Euclidean Three Space R^3 or in the Three Dimensional Sphere S^3. The GHZ state is a good exmple of a quantum state to consider in this regard. If there be connectivity at the base of this state it is a subtle one. Look again at Figure 12. We have told a story of the evolution of the state. The state $|\psi\rangle$ can itself represent the entanglement of an observer (Wigner, let us say) with the Schrodinger Cat. It is an entangled state. Then The new observer (Wigner's Friend, let us say) is entangled with this state, and that entanglement involves an interaction with Wigner's Friend that sends that resulting state into the GHZ. Once the state is in the GHZ, any observation of it by another observer (Einstein, let us say) will disentangle it. In the GHZ there is no longer any entanglement of pairs, only the tripartite entanglement remains. What sort of connectivity is this? Aravind has suggested that it is modeled by linking. Susskind has suggested that all entanglement is modeled by connectivity. What sort of connectivity will disentangle through the interaction with one tensor factor, with one black hole? There is no clear answer forthcoming in that topology. But the linking continues to suggest an answer.

Thus we are reminded of a topological problem. How does it come about that three rings can be linked, while any two of them are unlinked? Let us look at the geometry and topology of the situation by again examining Figure 12. In that Figure we have drawn the Borommean Rings in a topological form so that two of the rings are entirely disjoint from one another, and the third ring is seen to be effecting the connection between them. It is of course a matter of algebraic topology and knot theory to show that the rings really are linked, but this image shows how it is happening, and anyone who cares to make a model of the Rings from rope can experience how the third component connects the other two. Can there be a version of this kind of connectivity among three black holes?

5. Topological connectivity and Heyting algebras

Recall that a topological space X is said to be connected if and only if X is not the disjoint union of two disjoint, non-empty open sets U and V. That is, we say that X is disconnected if there are non-empty open sets U and V so that $X = U \cup V$ and $U \cap V = \phi$ where ϕ denotes the empty set.

FIGURE 10. Borommean rings

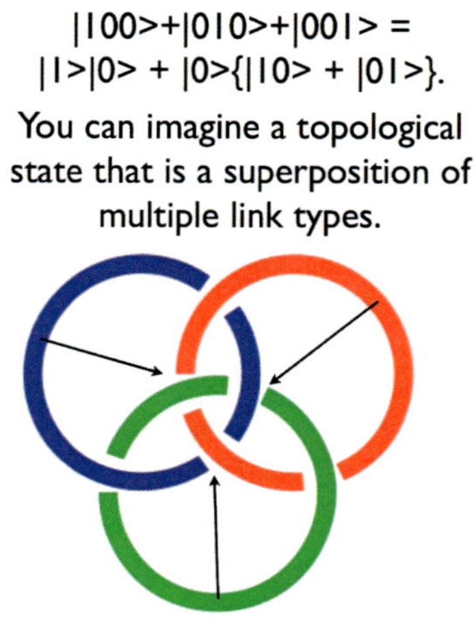

FIGURE 11. Borommean rings and Werner State

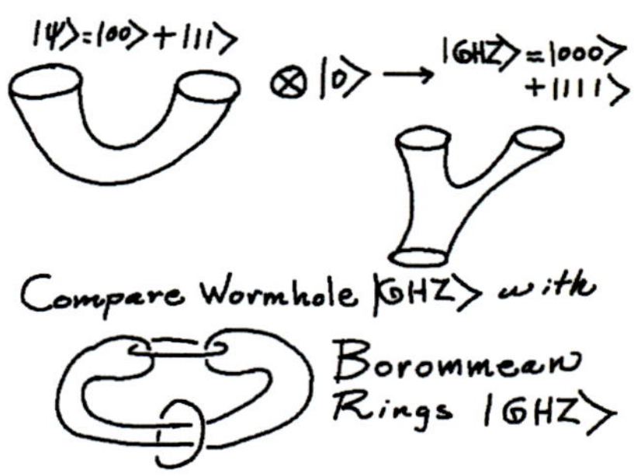

FIGURE 12. GHZ and Borommean rings

Given a topological space X, let $H(X)$ denote the collection of open sets in X. For $U \in H(X)$ define

$$\sim U = Int(U^c)$$

where Int denotes the operation of forming interior of a subset of X, and U^c denotes the set theoretic complement of U in X. The negation operation $\sim$ defines a Heyting algebra [**5**] structure on the collection of open sets $H(X)$. In particular, it can be the case that $(\sim U) \cup U$ is not equal to X. and it is generally the case that for any open set U in X, $\sim\sim\sim U = \sim U$. These are the characteristic properties of negation in a Heyting algebra. For our purposes, the Heyting algebra is a generalization of the Boolean algebra of subsets of a set where the (boolean) operations are set theoretic complementation, denoted here by U^c for U a subset of X, union and intersection denoted by $\cup$ and $\cap$ respectively. When $U^c = \sim U$ for all open sets U in X then the Heyting algebra $H(X)$ is Boolean. From this it follows that $H(X)$ is Boolean if and only if for every open set U in X we have that

$$Int(U^c) = U^c.$$

In other words, $H(X)$ is Boolean exactly when every open set is also closed. The simplest example of such a topology on a set X is the *indiscrete topology* where the open sets are just the set X and the empty set. Another example is the *discrete topology* on X where $\{x\}$ for every $x \in X$ is an open set. X with the discrete topology is totally disconnected and $H(X)$ is the Boolean algebra on the subsets of X under ordinary union, intersection and complementation.

For a non-trivial Heyting algebra consider $X = [0, 1]$ as the closed interval from 0 to 1 in the real line with the usual topology. Let $U = [0, 1/2)$ be the half open interval from 0 to 1/2 and let $V = (1/2, 1]$ be the half-open interval from 1/2 to 1. Then U and V are open sets in X with $U \cap V = \phi$ and $\sim U = V$ and $\sim V = U$. Thus $U \cup (\sim U)$ is not equal to the whole space X. Note also that if $W = U \cup V$ then $W^c = \{1/2\}$ and therefore $\sim W = \phi$. So we have $\sim\sim\sim W = \phi = \sim W$ but $\sim\sim W = X \neq W$.

More generally, suppose that X is a topological space and that U and V are disjoint open sets in X. Let $X' = X \cup \{p\}$ where p is a new point, not in the set X. Let $W = U \cup \{p\} \cup V$ be a neighborhood of p and take the topology generated by W and the open sets in X. Then we see that p acts to connect the disjoint opens U and V.

If a space is totally disconnected, then its Heyting algebra is Boolean. Each occurrence of connectivity in a topological space is accompanied by non-trivial Heyting structure. For example, suppose that X is the disjoint union of two open sets U and V. Add one one point p to form a space $\hat{X}$ with new open set $U \cup \{p\} \cup V$ the smallest open set containing p. In $H(\hat{X})$ we have $\sim U = V$ and $\sim V = U$ but the union of U and V is not equal to $\hat{X}$. The Heyting structure of $\hat{X}$ detects its connectivity. It is convenient to diagram the structure of $\hat{X}$ so that p is analogous to an event horizon and that the neighborhood $U \cup \{p\} \cup V$ acts like an Einstein-Rosen Bridge connecting the domains U and V. This analogy is shown in Figure 1 with the label E standing for the extra point p. When we regard $\hat{X}$ in this way, it is a precursor to the $ER = EPR$ scenario of Susskind and Maldacena.

REMARK. Here is another example for producing connectivity by adding one point. Suppose that X and Y are topological spaces, each connected, but disjoint from one another. Let $Z = X \cup Y$ with the topology generated by the open sets of X and the open sets of Y. Then the space Z is not connected since it is the union of the two disjoint open sets X and Y. We can create a new space $W = X \cup \{p\} \cup Y$ where p is a single point disjoint from both X and Y by adding neighborhoods for p defined by $N(p) = U \cup \{p\} \cup V$ where U and V are non-empty open sets from X and Y respectively. This makes p on the frontier between X and Y and the new space W is now connected. The interest in the example would be to think of p as representing a quantum entangled state $E = E(p)$ between X and Y. Then the left hand tensor end of E could be measured from anywhere in X and the the right hand end from anywhere in Y. It is not necessary to localize these measurements to specific points in the "distant" spaces.

DISCUSSION. Susskind suggests that connectivity and quantum entanglement are inseparable in quantum systems. In topology we can consider spaces that are disconnected and not necessarily associate them with physics. If X and Y are entirely disjoint topological spaces then we can form the topological space consisting in their union. In such a space we have no connection whatever between X and Y. Without any extra structure there can be no entanglement between a particle in X and another particle in Y. I mean this as a *mathematical* statement about X and Y. As far as the set theory and the topology is concerned, if two sets are disjoint, then there are no relationships given between them. If we speak of entanglement, it means that some extra structure beyond the two sets is given. The usual way of speaking in physics is to give this extra structure ideally quite independent of the spaces and the point locations for observations in those spaces. Thus I may write $|\psi\rangle = |00\rangle + |11\rangle$ to indicate an entangled state, but there is no information in this formalism about where in some space the observations are being made. When we give this information, we tie the left tensor factor to some neighborhood U in X and we tie the right tensor factor to some neighborhood V in Y. We say that observation at U is correlated with observation at V, and in this way we have succeeded in out language in *connecting* the disjoint spaces X and Y. It is our epistemological/topological experiment here to make this connection a minimal connectivity in an augmentation of the space $Z = X \cup Y$ to a new space that is topologically connected through a point-set topological wormhole going between X and Y. At the level of our construction this amounts to changing the language about the space so that it is a new space with the relationship between X and Y instantiated within its structure. The relationship was already in the physics. We have reified it into the spatial description.

The paradigm of entanglement is two particles that are produced at some point p and then each transported to X and to Y. This assumes that there is some connection between X and Y. The minimal such connection is a structure of the kind we construct by adding the single point p and taking the axiom that any neighborhood of p must contain a point from X and a point from Y. Then the entanglement can be represented by seeing $X \cup \{p\} \cup Y$ as the augmentation of X and Y by a tensor net that holds the entanglement for these particles. In this sense we affirm Susskind's position that topological connectivity and entanglement of quantum states are two sides of the same coin.

6. Summary

By studying the boundary between topological and quantum entanglement we can construct a correspondence between topological invariants and entangling R matrices that may have a significant impact on the study of quantum computing. In the final section of the paper we have discussed relations between the ideas of this paper and the entanglement hypotheses of Aravind and the ER = EPR hypothesis of Susskind and his collaborators. In the light of the latter hypothesis we have shown how to augment a space to a new space that contains a topological version of the tensor networks describing its quantum structure.

We have shown how to augment a space to a new space that contains a topological version of the tensor networks describing its quantum structure. This marks the beginning of a unification of properties of spatial connectivity and properties of quantum entanglement.

Lets go back to the matter of topologies. We have a simple notion of connectivity at the level of graphs (path connectivity corresponding to a sequence of edges from a node to another node). If we allow non-well-founded sets then every set corresponds to a graph [**2**], and the non well founded sets correspond to graphs with cycles in them. Thus we could have $S = \{A, B\}$ where A is a member of B and B is a member of A. The graph has a loop in it with A pointing to B and B pointing to A. We can regard this mutual pointing as a basic form of connectivity. This corresponds in the knot set theory [**12–14**] to a linking. The upshot is that connectivity and linking are intimately related at a foundational level of topology. We can associate a graph-topology to *any* set. It is then possible to reenter all the considerations of this paper and look at them again. We can take the graph of the set as its basic topology and also assign a set theoretic topology to the graph by our augmentation method. In this way it is possible to have knots and links in the very bottom of the foundational structure, and we can generalize the notion of topology and of graph associated with a set to include the intricacies of the combinatioral topology of knots and links. This is a project for the future.

References

[1] P. K. Aravind, *Borromean entanglement of the GHZ state*, Potentiality, entanglement and passion-at-a-distance (Boston, MA, 1994), Boston Stud. Philos. Sci., vol. 194, Kluwer Acad. Publ., Dordrecht, 1997, pp. 53–59, DOI 10.1007/978-94-017-2732-7_4. MR1739812

[2] P. Aczel, *Non-well-founded sets*, CSLI Lecture Notes, vol. 14, Stanford University, Center for the Study of Language and Information, Stanford, CA, 1988. With a foreword by Jon Barwise [K. Jon Barwise]. MR940014

[3] M. Atiyah, *The geometry and physics of knots*, Lezioni Lincee. [Lincei Lectures], Cambridge University Press, Cambridge, 1990, DOI 10.1017/CBO9780511623868. MR1078014

[4] J. C. Baez and J. Vicary, Wormholes and Entanglement. Classical and Quantum Gravity 31. DOI: 10.1088/0264-9381/31/21/214007 (2014).

[5] L. Esakia, *Heyting algebras*, Trends in Logic—Studia Logica Library, vol. 50, Springer, Cham, 2019. Duality theory; Edited by Guram Bezhanishvili and Wesley H. Holliday; Translated from the Russian edition [MR0847050] by Anton Evseev, DOI 10.1007/978-3-030-12096-2. MR3967703

[6] L. H. Kauffman, *Knots, abstract tensors and the Yang-Baxter equation*, Knots, topology and quantum field theories (Florence, 1989), World Sci. Publ., River Edge, NJ, 1989, pp. 179–334. MR1146944

[7] L. H. Kauffman, Teleportation Topology, quant-ph/0407224, (in the Proceedings of the 2004 Byelorus Conference on Quantum Optics), *Opt. Spectrosc.* 9, 2005, 227-232.

[8] L. H. Kauffman and S. J. Lomonaco Jr., Quantum knots, in *Quantum Information and Computation II, Proceedings of Spie, 12 -14 April 2004* (2004), ed. by Donkor Pirich and Brandt, pp. 268-284.

[9] S. J. Lomonaco and L. H. Kauffman, *Quantum knots and mosaics*, Quantum Inf. Process. **7** (2008), no. 2-3, 85–115, DOI 10.1007/s11128-008-0076-7. MR2420814

[10] L. H. Kauffman and E. Mehrotra, *Topological aspects of quantum entanglement*, Quantum Inf. Process. **18** (2019), no. 3, Paper No. 76, 36, DOI 10.1007/s11128-019-2191-z. MR3905533

[11] S. Antoniou, L. H. Kauffman, and S. Lambropoulou, *Topological surgery in cosmic phenomena*, Adv. Theor. Math. Phys. **23** (2019), no. 3, 701–765, DOI 10.4310/ATMP.2019.v23.n3.a3. MR4049074

[12] L. H. Kauffman, *Knots and physics*, Series on Knots and Everything, vol. 1, World Scientific Publishing Co., Inc., River Edge, NJ, 1991, DOI 10.1142/9789812796226. MR1141156
[13] L. H. Kauffman, *Knot logic*, Knots and applications, Ser. Knots Everything, vol. 6, World Sci. Publ., River Edge, NJ, 1995, pp. 1–110, DOI 10.1142/9789812796189_0001. MR1364188
[14] L. H. Kauffman, *Knot logic and topological quantum computing with Majorana fermions*, Logic and algebraic structures in quantum computing, Lect. Notes Log., vol. 45, Assoc. Symbol. Logic, La Jolla, CA, 2016, pp. 223–335. MR3524272
[15] J. Maldacena and L. Susskind, *Cool horizons for entangled black holes*, Fortschr. Phys. **61** (2013), no. 9, 781–811, DOI 10.1002/prop.201300020. MR3104458
[16] L. Susskind and Y. Zhao, *Teleportation through the wormhole*, Phys. Rev. D **98** (2018), no. 4, 046016, 18, DOI 10.1103/physrevd.98.046016. MR3933715
[17] L. Susskind, *Copenhagen vs Everett, teleportation, and* ER = EPR, Fortschr. Phys. **64** (2016), no. 6-7, 551–564, DOI 10.1002/prop.201600036. MR3513422
[18] R. Penrose, *Applications of negative dimensional tensors*, Combinatorial Mathematics and its Applications (Proc. Conf., Oxford, 1969), Academic Press, London-New York, 1971, pp. 221–244. MR281657

Department of Mathematics, Statistics and Computer Science, University of Illinois at Chicago, 851 South Morgan Street, Chicago, Illinois 60607-7045
Email address: kauffman@uic.edu

Contemporary Mathematics
Volume **834**, 2026
https://doi.org/10.1090/conm/834/16685

On the Khovanov homology of $T(2, n)$ via long exact sequence

Gabriel Montoya-Vega

ABSTRACT. Khovanov homology is a powerful link invariant: a categorification of the Jones polynomial that enjoys a rich and beautiful algebraic structure. This homology theory has been extensively studied and it has become an ubiquitous topic in contemporary knot theory research. In the same spirit, the Kauffman skein relation, which allows to define the Kauffman bracket polynomial up to normalization of the unknot, can be categorified by means of a long exact sequence. Following Oleg Viro, in an expository style we present how to build Khovanov homology from the Kauffman bracket polynomial and construct its long exact sequence. Furthermore, we present a deviceful and practical way in which this long exact sequence can be used for the computation of the Khovanov homology of torus links of the type $T(2, n)$.

1. Introduction

Khovanov homology (KH) offers a nontrivial generalization of the Jones polynomial and the Kauffman bracket polynomial of links in $\mathbb{R}^3$. A more powerful invariant than the Jones polynomial, it has been extensively studied over the last two decades. The essence of KH is that a bigraded chain complex is associated to a link, in such a way that the homology of the complex is a link invariant. Furthermore, the graded Euler characteristic of the chain complex is the Jones polynomial.

This article is organized as follows. In Section 2 we present how to build Khovanov homology from the Kauffman bracket polynomial and show how this approach connects with the original construction. In Section 3, the long exact sequence of KH is constructed and in Section 4 we use it to compute the KH of torus links $T(2, n)$. Section 5 consists of two examples illustrating the KH of two specific knots. Finally, in Section 6 we give some possibilities for further research paths in the context of different families of links.

2. Khovanov homology via Kauffman bracket polynomial

In this section we construct KH following Oleg Viro's approach [**Vir1**, **Vir2**]. Recall that the Kauffman bracket polynomial (KBP) gives a simple approach to the Jones polynomial [**Kau**]. Hence, it is natural to first categorify the KBP and then connect this theory to Khovanov's original [**Kho**]. The construction of KH that we present here is also referred to as the unoriented framed version of KH.

2020 *Mathematics Subject Classification.* Primary 57K10; Secondary 57K14, 57K18.
The author was support by the National Science Foundation through Grant DMS-2212736.

DEFINITION 2.1.

(i) The (unreduced) **Kauffman bracket polynomial** is a function from the set of unoriented link diagrams $\mathcal{D}$ to Laurent polynomials with integer coefficients in the variable A, $[\,]:\mathcal{D}\longrightarrow \mathbb{Z}[A^{\pm 1}]$. The polynomial is characterized by the rules $[\emptyset]=1$, $[\bigcirc]=(-A^2-A^{-2})$, and the skein relation:

(ii) Let D be an unoriented link diagram and let $cr(D)$ be its crossings set. A **Kauffman state** s, of D, is a function $s: cr(D)\longrightarrow \{A, B\}$. This function is understood as an assignment of a **marker** to each crossing according to the convention illustrated in Figure 2.1. Denote by KS the set of all Kauffman states. Every marker yields a natural **smoothing** of the crossing as shown in Figure 2.1.

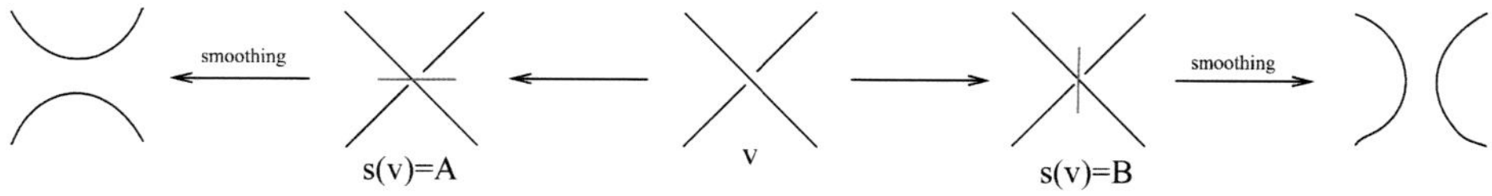

FIGURE 2.1. Markers at a crossing v of D and their corresponding smoothing.

The KBP of a link diagram D is given by the state sum formula:

$$[D]=\sum_{s\,\in\, KS} A^{|s^{-1}(A)|-|s^{-1}(B)|}(-A^2-A^{-2})^{|D_s|},$$

where D_s denotes the system of circles obtained after smoothing all crossings of D according to the markers of s, and $|D_s|$ denotes the number of circles in the system. By having the state sum formula from the unreduced version of the Kauffman bracket polynomial, it is possible to have an one-to-one correspondence between the circles in D_s and the factors $(-A^2-A^{-2})$. This leads to the idea of an enhanced Kauffman state, as defined below.

DEFINITION 2.2. An **enhanced Kauffman state** S of D is a Kauffman state s together with a function $\varepsilon: D_s\longrightarrow\{+,-\}$, assigning to each circle of D_s a positive or a negative sign.

Denote by EKS the set of all enhanced Kauffman states. Notice that for a Kauffman state s there are $2^{|D_s|}$ enhanced Kauffman states. Then the KBP can be written as a sum of monomial terms coming from the enhanced Kauffman states as follows:

$$[D]=\sum_{S\,\in\, EKS}(-1)^{|D_s|}A^{|s^{-1}(A)|-|s^{-1}(B)|}(A^2)^{|\varepsilon^{-1}(+)|-|\varepsilon^{-1}(-)|}.$$

Letting $\sigma(s)=|s^{-1}(A)|-|s^{-1}(B)|$ and $\tau(S)=|\varepsilon^{-1}(+)|-|\varepsilon^{-1}(-)|$, we obtain the following formula:

$$[D]=\sum_{S\,\in\, EKS}(-1)^{|D_s|}A^{\sigma(s)+2\tau(S)},$$

which is the enhanced Kauffman state sum formula for the unreduced KBP.

The enhanced Kauffman states form a basis for the chain groups of the Khovanov chain complex. We now define the bigrading on EKS, the chain groups, and the boundary maps.

DEFINITION 2.3.

(i) The **bidegree** on the enhanced Kauffman states is defined as the following set:

$$\mathcal{S}_{a,b}(D) = \mathcal{S}_{a,b} = \{S \in EKS \mid a = \sigma(s),\ b = \sigma(s) + 2\tau(S)\}.$$

(ii) The **chain groups** $C_{a,b}(D) = C_{a,b}$, are defined to be the free abelian groups with basis $\mathcal{S}_{a,b}(D) = \mathcal{S}_{a,b}$, i.e. $C_{a,b} = \mathbb{Z}\mathcal{S}_{a,b}$. Therefore, $C(D) = \bigoplus_{a,b \in \mathbb{Z}} C_{a,b}(D)$ is a bigraded free abelian group.

(iii) For a link diagram D we define the **chain complex** $\mathscr{C}(D) = \{(C_{a,b}, \partial_{a,b})\}$, where the **differential map** $\partial_{a,b} : C_{a,b} \longrightarrow C_{a-2,b}$ is defined by

$$\partial_{a,b}(S) = \sum_{S' \in \mathcal{S}_{a-2,b}} (-1)^{t(S,S')}(S,S')S'.$$

In the definition of the differential map above, $S \in \mathcal{S}_{a,b}$ and (S,S') is the incidence number of S and S'. Specifically, either $(S,S') = 0$ or $(S,S') = 1$, and it is equal to 1 if and only if the following conditions hold:

(1) S and S' are identical except at only one crossing, say v. Moreover $s(v) = A$ and $s'(v) = B$, where S and S' are enhancements of s and s', respectively.

(2) $\tau(S') = \tau(S)+1$ and every component of D_s not interacting with the crossing v keeps its sign for $D_{s'}$.

Condition (1) reflects the fact that the value of σ is decreasing by 2, while condition (2) indicates that either the number of negative signs decreases or the number of positive signs increases. Finally, $(-1)^{t(S,S')}$ requires an ordering of the crossings in the link diagram D. $t(S,S')$ is defined as the number of crossings of D with B markers in S bigger than v in the chosen ordering. This condition is sufficient for the differential to satisfy $\partial_{a-2,b} \circ \partial_{a,b} = 0$. It is important to remark that the homology does not depend on the ordering of crossings [**Kho**]. Figure 2.2 illustrates the cases when the enhanced states S and S' are incident.

DEFINITION 2.4. The **Khovanov homology** of the diagram D is defined to be the homology of the chain complex $\mathscr{C}(D)$:

$$H_{a,b}(D) = \frac{ker(\partial_{a,b})}{im(\partial_{a+2,b})}.$$

Certainly, the most important property of the framed version of Khovanov homology is its invariance under second and third Reidemeister moves. The following theorem summarizes the results. See [**Vir1, Vir2**] for a proof.

THEOREM 2.5. *Let D be a link diagram. The homology groups*

$$H_{a,b}(D) = \frac{ker(\partial_{a,b})}{im(\partial_{a+2,b})},$$

are invariant under Reidemeister moves of second and third type. Therefore, they are invariants of unoriented framed links. Moreover, the effect of the first Reidemeister move (positive or negative) R_1, is the shift in the homology, $H_{a,b}(R_{1+}(D)) =$

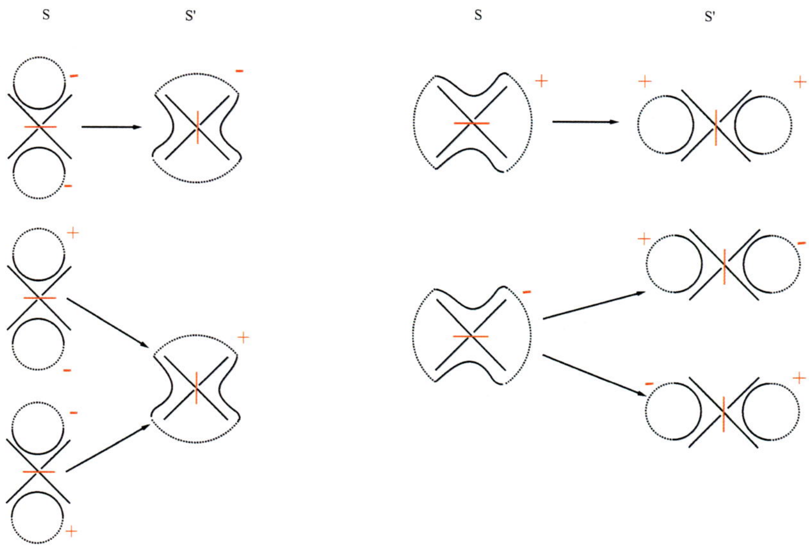

FIGURE 2.2. Signs of circles after fusion (left) and split (right) of components.

$H_{a+1,b+3}(D)$ and $H_{a,b}(R_{1-}(D)) = H_{a-1,b-3}(D)$. These groups categorify the unreduced Kauffman bracket polynomial and are called the framed Khovanov homology groups.

2.1. Classical Khovanov (co)homology. Khovanov originally associated a bigraded chain complex to an oriented link diagram whose homology is a link invariant. Let $\vec{D}$ be an oriented link diagram obtained after choosing an orientation on the unoriented link diagram D. Let $w(\vec{D}) = w$ be its writhe. Then, the classical Khovanov (co)homology, $\mathcal{H}^{i,j}(\vec{D})$, and the framed version of KH, $H_{a,b}(D)$, are related by the following equalities:

$$\mathcal{H}^{i,j}(\vec{D}) = H_{w-2i,3w-2j}(D) = H_{a,b}(D) = \mathcal{H}^{\frac{w-a}{2},\frac{3w-b}{2}}(\vec{D}).$$

3. Long exact sequence of Khovanov homology

In this section, we examine the fact that the skein relation used to define the KBP can be categorified by means of a long exact sequence [**Vir1**, **Vir2**]. Then this long exact sequence is used to calculate the KH of torus links of type $T(2, n)$.

3.1. Construction. Recall from Definition 2.3 that the set of enhanced Kauffman states $\mathcal{S}_{a,b}$ gives the basis for the free abelian groups $C_{a,b} = \mathbb{Z}\mathcal{S}_{a,b}$. Let v be a fixed crossing of the link diagram D. Consider the sets $\mathcal{S}^{A,v}_{a,b}$ and $\mathcal{S}^{B,v}_{a,b}$ defined as follows:

$$\mathcal{S}^{A,v}_{a,b} = \{S \in \mathcal{S}_{a,b} \mid s(v) = A\} \qquad \textit{and} \qquad \mathcal{S}^{B,v}_{a,b} = \{S \in \mathcal{S}_{a,b} \mid s(v) = B\},$$

this is, $\mathcal{S}^{A,v}_{a,b}$ consists of all enhanced states S with bigrading (a, b) having an A marker at the crossing v; analogously, $\mathcal{S}^{B,v}_{a,b}$ consists of all enhanced states S with bigrading (a, b) having a B marker at the crossing v. Then, it can be seen that

$$\mathcal{S}_{a,b} \quad = \quad \mathcal{S}^{A,v}_{a,b} \quad \sqcup \quad \mathcal{S}^{B,v}_{a,b}.$$

Continuing with the notation from the previous section, denote the free abelian groups generated by these sets as:

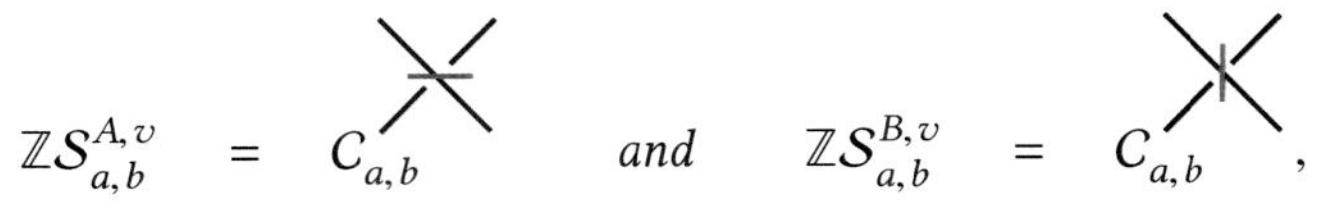

where $v =$ . Consequently, at the level of groups we have that:

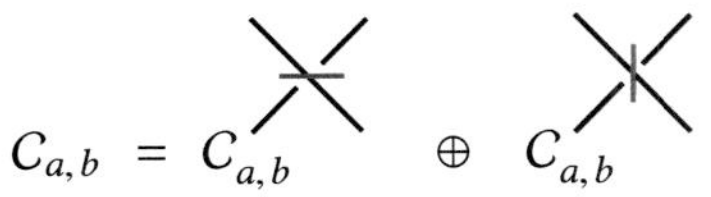 *or equivalently* $\mathbb{Z}S_{a,b} = \mathbb{Z}S^{A,v}_{a,b} \sqcup \mathbb{Z}S^{B,v}_{a,b}$.

Notice that the complex

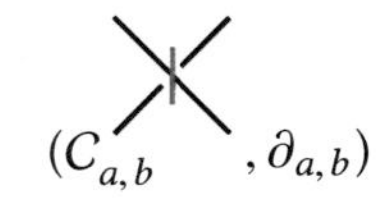

is a chain subcomplex of $(C_{a,b}, \partial_{a,b})$, this is

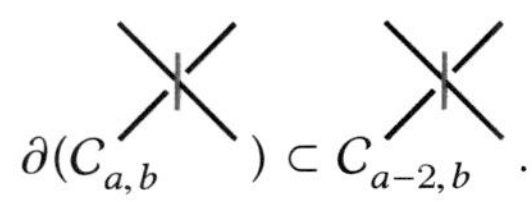

In contrast, this is not necessarily true for

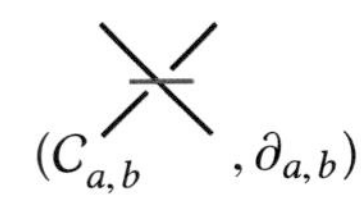

since $\partial_{a,b}$ may change the marker at the crossing from an A marker to a B marker, as discussed while constructing KH in Section 2. The following short exact sequence of chain complexes can be written:

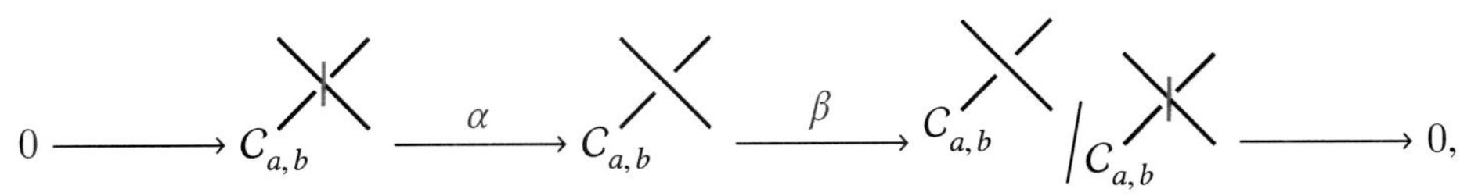

where the map α sends an enhanced Kauffman state of D having the crossing v with a B marker to the enhanced Kauffman state of D assigning a B label to v, while the other crossings keep the markers and the signs of the circles are preserved. Similarly, the map β sends an enhanced Kauffman state having a B label at v to zero, and sends each enhanced Kauffman state with A label at v to the enhanced Kauffman state of D with the crossing v given an A label, while the other markers of crossings and signs of circles are preserved.

Observe that as a group

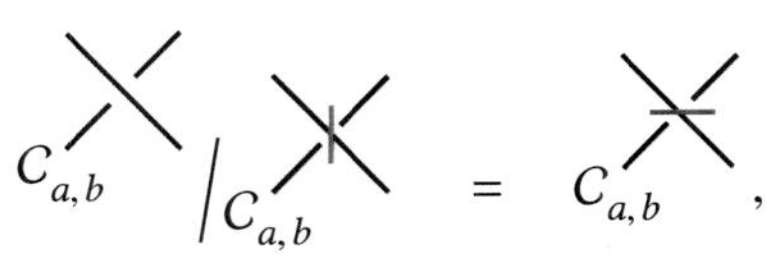

and there is a chain complex connecting map coming from the construction of KH:

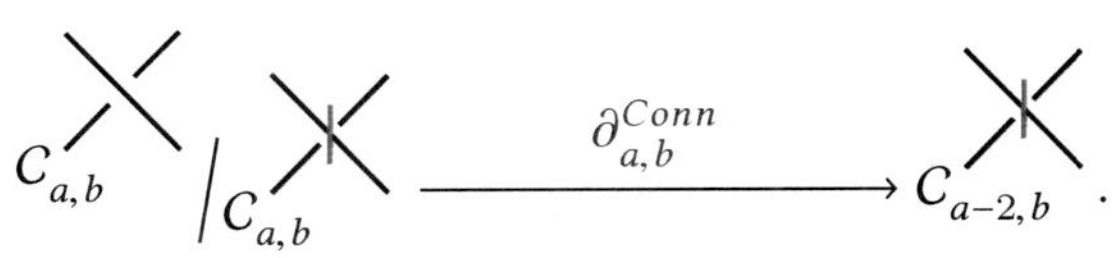

As it is standard, this leads to a long exact sequence of homology:

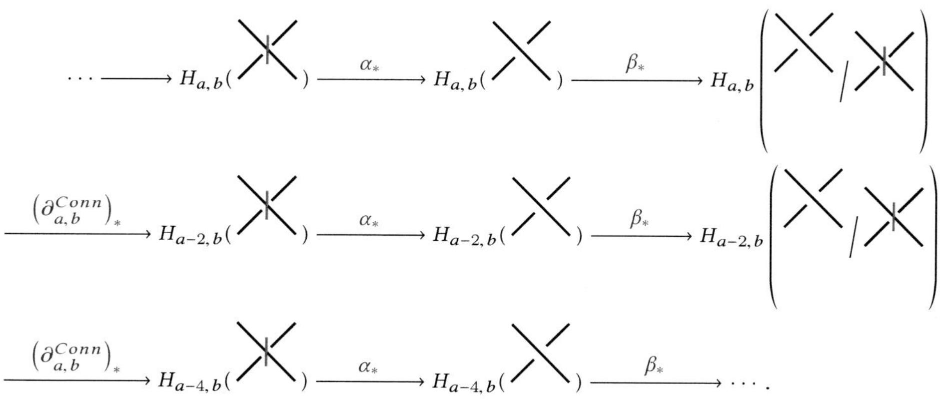

Observe that after smoothing the crossing, we have the following chain complex equalities:

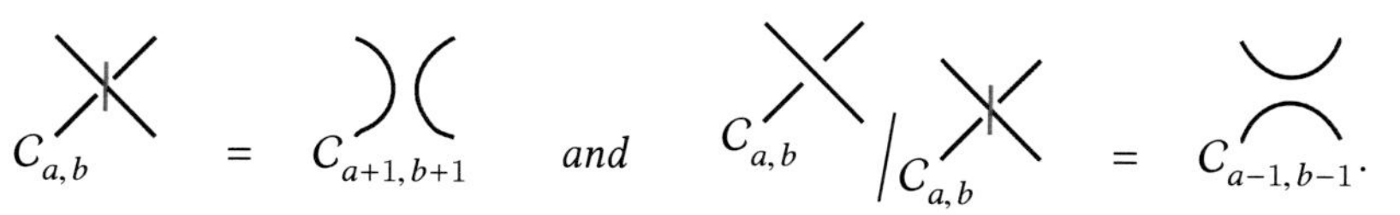

Hence, the following short exact sequence of chain complexes of diagrams is obtained:

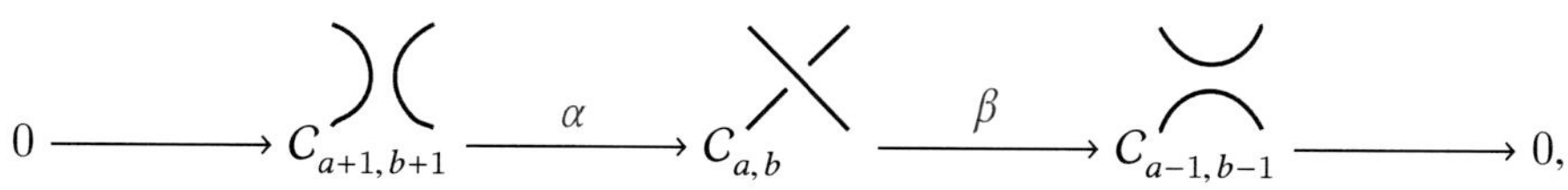

which in turn, yields the following long exact sequence, known as the Long Exact Sequence of Khovanov homology:

(3.1)

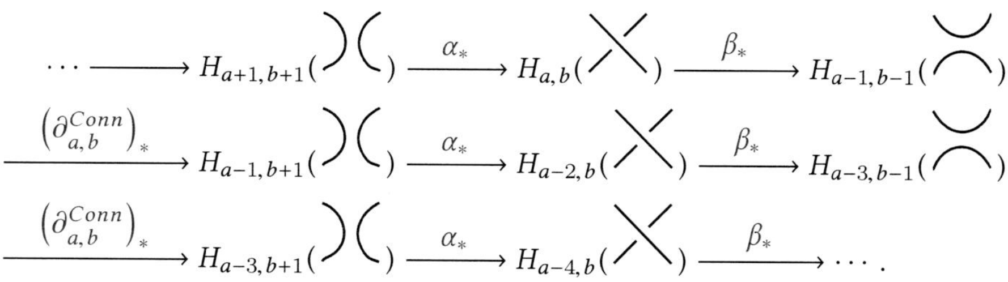

The following results follows directly from the previous construction.

THEOREM 3.1.

(1) *If* $H_{a+1,b+1}()()) = 0$ *then* $\beta_* : H_{a+1,b+1}(\diagup\!\!\!\!\diagdown) \to H_{a,b}(\asymp)$ *is a monomorphism.*

(2) *If* $H_{a-1,b+1}()()) = 0$ *then* $\beta_* : H_{a+1,b+1}(\diagup\!\!\!\!\diagdown) \to H_{a,b}(\asymp)$ *is an epimorphism.*

(3) *If* $H_{a+1,b+1}()()) = 0 = H_{a-1,b+1}()())$ *then* $\beta_* : H_{a+1,b+1}(\diagup\!\!\!\!\diagdown) \to H_{a,b}(\asymp)$ *is an isomorphism.*

4. Khovanov Homology of Torus links $T(2, n)$

Torus links of the type $T(2, n)$ have always enjoyed importance regarding research in knot theory. Indeed, the KH of this type of links was first computed by M. Khovanov [**Kho**]. Subsequently, J. H. Przytycki provided a different approach to this computation by observing the connection of Khovanov homology and Hochschild homology [**Prz**]. In this section, we present an alternative, deviceful method of calculating the homology of torus links $T(2, n)$ by explicitly using the long exact sequence of Khovanov homology [**MV1**].

DEFINITION 4.1. A **torus link** of type (p,q), also referred to as a (p, q)−torus link is a link ambient isotopic to a curve contained in a standard torus T^2. This curve wraps p times around the longitude and it wraps q times around the meridian. If p and q are relatively prime, then the torus link has only one component and therefore it is called a **torus knot**.

Let

$v =$ 

be a fixed crossing. Observe that a vertical smoothing at a crossing v of a torus link $T(2, n)$ produces the trivial knot with $1 - n$ twists. We now write the long exact sequence of KH. First, notice that maintaining the crossing (trivially) does not change the torus link $T(2, n)$, the horizontal smoothing of the crossing results in $T(2, n-1)$, and the vertical smoothing of the crossing yields the trivial knot with framing twisted $1 - n$ times. In Figure 4.1 this setting is illustrated for the torus knot $T(2, 3)$.

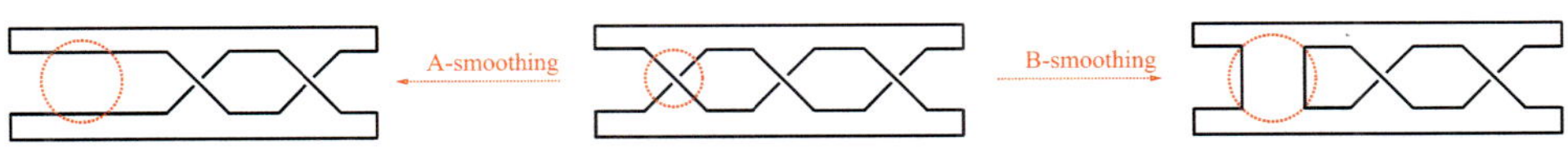

FIGURE 4.1. Torus knot $T(2, n) = T(2, 3)$ (center), Torus knot $T(2, n - 1) = T(2, 2)$ (left), and the trivial knot with framing $1 - n = -2$ (right).

The following theorem states the Khovanov homology of torus links $T(2, n)$ with $n > 0$. The long exact sequence of KH plays a key role in the proof.

THEOREM 4.2. *Let $T(2, n)$ be a torus link with $n > 0$. Its Khovanov homology $H_{a,b}(T(2, n))$ is given by:*

$$H_{a,b}(T(2,n)) = \begin{cases} \mathbb{Z} & \text{for } (a,b) = (n,n) \text{ or } (a,b) = (-n,-3n), \\ \mathbb{Z} & \text{for } a = n-2s, b = n-4s+4 \text{ where } s \text{ is even and } 0 \leq s \leq n, \\ \mathbb{Z} & \text{for } a = n-2s,\ b = n-4s \text{ where } s \text{ is odd and } 3 \leq s \leq n, \\ \mathbb{Z}_2 & \text{for } a = n-2s,\ b = n-4s+4 \text{ where } s \text{ odd and } 3 \leq s \leq n, \\ 0 & \text{otherwise.} \end{cases}$$

PROOF. The proof proceeds by induction on n.
For the trivial knot $T(2, 1)$, the theorem holds as only

$$H_{1,1}(T(2,1)) = H_{1,5}(T(2,1)) = \mathbb{Z}$$

and the other homology groups are trivial.

For the Hopf link $T(2,2)$ the theorem holds as can be verified with the Khovanov homology Table 1.[1]

TABLE 1. Khovanov homology table of the torus knot $T(2,2)$.

b | a	-2	0	2
6			$\mathbb{Z}$
2			$\mathbb{Z}$
-2	$\mathbb{Z}$		
-6	$\mathbb{Z}$		

As the inductive hypothesis, suppose the result holds for $n-1$ where $n > 2$. Furthermore, consider the case where the map $\beta_* : H_{a,b}(T(2,n)) \longrightarrow H_{a-1,b-1}(T(2,n-1))$ is not necessarily an isomorphism. Thus, the long exact sequence should be closely analyzed. First, recall that after solving the crossing v with a B-smoothing, the trivial knot with framing $1-n$ is obtained, i.e. $T_B(2,n) = \bigcirc^{1-n}$. Hence, its homology is given by:

$$H_{x,y}(\bigcirc^{1-n}) = \begin{cases} \mathbb{Z} & \text{when } (x,y) = (1-n, 3(1-n) \pm 2), \\ 0 & \text{otherwise.} \end{cases}$$

Therefore, the map β_* is not necessarily an isomorphism if $H_{x,y}(\bigcirc^{1-n}) \neq 0$. More precisely, $H_{x,y}(\bigcirc^{1-n}) \neq 0$ when $(x,y) = (1-n, 3(1-n)-2) = (1-n, 1-3n)$ or $(x,y) = (1-n, 3(1-n)+2)) = (1-n, 5-3n)$.

CASE (i). Let $(x,y) = (1-n, 1-3n)$. In the long exact sequence, the neighborhood of $H_{1-n,1-3n}(\bigcirc^{1-n})$ has the form:

$$\longrightarrow H_{1-n,-1-3n}(T(2,n-1)) \xrightarrow{\left(\partial^{Conn}\right)} H_{1-n,1-3n}(\bigcirc^{1-n}) \xrightarrow{\alpha_*}$$
$$H_{-n,-3n}(T(2,n)) \xrightarrow{\beta_*} H_{-n-1,-3n-1}(T(2,n-1)) \longrightarrow .$$

By the inductive hypothesis, $H_{1-n,-1-3n}(T(2,n-1))$ and $H_{-n-1,-3n-1}(T(2,n-1))$ are trivial. Hence, the previous long exact sequence becomes:

$$\longrightarrow 0 \xrightarrow{\left(\partial^{Conn}\right)} H_{1-n,1-3n}(\bigcirc^{1-n}) \xrightarrow{\alpha_*}$$
$$H_{-n,-3n}(T(2,n)) \xrightarrow{\beta_*} 0 \longrightarrow .$$

Then, $H_{1-n,1-3n}(\bigcirc^{1-n}) = H_{-n,-3n}(T(2,n)) = \mathbb{Z}$, and so the theorem holds as $(a,b) = (-n,-3n)$.

CASE (ii). Let $(x,y) = (1-n, 5-3n)$. In the long exact sequence, the neighborhood $H_{1-n,5-3n}(\bigcirc^{1-n})$ has the form:

$$0 \longrightarrow H_{2-n,4-3n}(T(2,n)) \xrightarrow{\beta_*} H_{1-n,3-3n}(T(2,n-1)) \xrightarrow{\left(\partial^{Conn}\right)}$$
$$H_{1-n,5-3n}(\bigcirc^{1-n}) \xrightarrow{\alpha_*} H_{-n,4-3n}(T(2,n)) \xrightarrow{\beta_*}$$
$$H_{-n-1,3-3n}(T(2,n-1)) \longrightarrow .$$

[1] A well-known computation. See, for instance, [**MV1**].

By the inductive hypothesis, $H_{-n-1,3-3n}(T(2,n-1)) = 0$ and $H_{1-n,3-3n}(T(2,n-1)) = H_{-(n-1),-3(n-1)} = \mathbb{Z}$. Thus, the previous long exact sequence becomes:

$$0 \longrightarrow H_{2-n,4-3n}(T(2,n)) \xrightarrow{\beta_*} \mathbb{Z} \xrightarrow{\left(\partial^{Conn}\right)}$$
$$\mathbb{Z} \xrightarrow{\alpha_*} H_{-n,4-3n}(T(2,n)) \xrightarrow{\beta_*} 0 \longrightarrow .$$

To determine the other entries of the long exact sequence we need to understand the map $\partial^{Conn}_{1-n,3-3n} : \mathbb{Z} \longrightarrow \mathbb{Z}$. There are two general possibilities: the map is the zero map or it is multiplication by $k > 0$. First suppose the map $\partial^{Conn}_{1-n,3-3n}$ is the zero map. In this case,

$$H_{2-n,4-3n}(T(2,n)) = H_{1-n,3-3n}(T(2,n-1)) = \mathbb{Z} \qquad \textit{and,}$$

$$H_{-n,4-3n}(T(2,n)) = H_{1-n,5-3n}(\bigcirc^{1-n}) = \mathbb{Z}.$$

Then, $H_{2-n,4-3n}(T(2,n)) = \mathbb{Z}$ and in this case we have that $(a,b) = (2-n, 4-3n) = (n-2n+2, n-4n+4) = (n-2(n-1), n-4(n-1)) = (n-2s, n-4s)$ for $s = n-1$. Similarly, $H_{-n,4-3n}(T(2,n)) = \mathbb{Z}$ where $(a,b) = (-n, 4-3n) = (n-2s, n-4s+4)$ for $s = n$ and thus, the theorem holds. Finally, suppose the map $\partial^{Conn}_{1-n,3-3n}$ is multiplication by $k > 0$. This case was addressed in 2006 by M. Pabiniak, J. H. Przytycki, and R. Sazdanović in [**PPS**] by studying the algebra of truncated polynomials $\mathcal{A}_m = \mathbb{Z}[x] / (x^m)$, which for $m = 2$ is closely related to Khovanov homology [**Kho, BN**]. More precisely, when n is even, the map $\partial^{Conn}_{1-n,3-3n}$ is the zero map and thus we have $H_{2-n,4-3n}(T(2,n)) = \mathbb{Z} = H_{-n,4-3n}(T(2,n))$ as noted in the previous part. On the other hand, when n is odd, the map $\partial^{Conn}_{1-n,3-3n}$ is multiplication by $k = 2$ and we have that $H_{2-n,4-3n}(T(2,n)) = 0$, $H_{-n,4-3n}(T(2,n)) = \mathbb{Z}_2$, and so the theorem holds. □

5. Example: KH tables for T(2,11) and T(2,12)

We present in this section the KH Table 2 and KH Table 3 for the torus links $T(2,11)$ and $T(2,12)$, respectively.

Observe that $H_{a,b}(T(2,n))$ has support on two diagonals containing (n,n) or $(n, n+4)$. In other words, $H_{a,b}(T(2,n))$ is nontrivial only for $(a,b) = (n-2s, n-4s)$ or $(a,b) = (n-2s, n-4s+4)$. Moreover, $H_{a,b}(T(2,n))$ has torsion only in groups for which $(a,b) = (n-2s, n-4s+4)$.

TABLE 2. Khovanov homology of the torus knot $T(2, 11)$.

b | a	-11	-9	-7	-5	-3	-1	1	3	5	7	9	11
15												$\mathbb{Z}$
11												$\mathbb{Z}$
7										$\mathbb{Z}$		
3									$\mathbb{Z}_2$			
-1								$\mathbb{Z}$	$\mathbb{Z}$			
-5							$\mathbb{Z}_2$					
-9						$\mathbb{Z}$	$\mathbb{Z}$					
-13					$\mathbb{Z}_2$							
-17				$\mathbb{Z}$	$\mathbb{Z}$							
-21			$\mathbb{Z}_2$									
-25		$\mathbb{Z}$	$\mathbb{Z}$									
-29	$\mathbb{Z}_2$											
-33	$\mathbb{Z}$											

TABLE 3. Khovanov homology of the torus link $T(2, 12)$.

b | a	-12	-10	-8	-6	-4	-2	0	2	4	6	8	10	12
16													$\mathbb{Z}$
12													$\mathbb{Z}$
8											$\mathbb{Z}$		
4										$\mathbb{Z}_2$			
0									$\mathbb{Z}$	$\mathbb{Z}$			
-4								$\mathbb{Z}_2$					
-8							$\mathbb{Z}$	$\mathbb{Z}$					
-12						$\mathbb{Z}_2$							
-16					$\mathbb{Z}$	$\mathbb{Z}$							
-20				$\mathbb{Z}_2$									
-24			$\mathbb{Z}$	$\mathbb{Z}$									
-28		$\mathbb{Z}_2$											
-32	$\mathbb{Z}$	$\mathbb{Z}$											
-36	$\mathbb{Z}$												

6. Future directions

Khovanov Homology has been computed for many links and the experimental evidence shows abundance of 2−torsion while other torsion groups appear (until recently) very seldom; see for instance [**Kho, Shu**]. In [**AP**] the authors studied the notion of adequate diagrams. They observed that for an n−crossing A−adequate link diagram D, the highest grading in KH is given by $H_{n,*}(D) = H_{n,n+2|s_A|}(D) = \mathbb{Z}$, where S_A denotes the Kauffman state with an A marker assigned at all its crossings. Moreover, they showed that the next nontrivial homology group $H_{n-2,n+2|s_A|-4}(D)$, has $\mathbb{Z}_2$−torsion as long as the graph associated to the state s_A, $G(D) = G_{s_A}(D)$, is not bipartite. Furthermore, the group $H_{n-2,n+2|s_A|-4}(D)$ was explicitly computed for

A–adequate links in [**PPS**]. In particular, for a connected A–adequate diagram D,

$$tor\ (H_{n-2,n+2|s_A|-4}(D)) = \begin{cases} \mathbb{Z}_2, & for\ \ G(D)\ \ having\ \ and\ \ odd\ \ cycle \\ 0, & for\ \ a\ \ bipartite\ \ graph. \end{cases}$$

They additionally showed that for a strongly A–adequate diagram D with the graph $G(D)$ containing an even cycle, this group contains $\mathbb{Z}_2$–torsion. An idea for further research might be to analyze whether the long exact sequence of KH can be used in full generality to compute the KH for this type of links and to detect torsion, extending the work in [**PS**], for instance.

Recently there have been significant developments in the study of torsion in Khovanov homology, see for instance [**PBIMW, Muk, MS**]. The method used in the calculation of the KH of torus links $T(2, n)$ can be used to construct links having torsion different from $\mathbb{Z}_2$ in its Khovanov homology. It is a reasonable and promising idea to test the full power of the long exact sequence of KH and expand on the results from [**MPSWY**]. For example, a natural research path is to adapt the long exact sequence to calculate the KH of other families of links. One can consider torus links of type $T(3, k)$ since a horizontal smoothing would produce a $T(2, n)$ torus link. Similarly, one could consider pretzel links with three columns, as an horizontal smoothing would produce a similar result.

Acknowledgments

The author thanks Hanna Makaruk and Robert Owczarek for organizing, in repeated occasions, amazing AMS special sessions. Furthermore, the author gives special thanks to his doctoral advisor, Józef H. Przytycki, whose guidance has been essential in exploring the beautiful world of knot theory.

*Hoping to popularize knot theory in Colombia and Latin-America in general, this paper serves as a partial translation to [**MV2**].

References

[AP] M. M. Asaeda and J. H. Przytycki, *Khovanov homology: torsion and thickness*, Advances in topological quantum field theory, NATO Sci. Ser. II Math. Phys. Chem., vol. 179, Kluwer Acad. Publ., Dordrecht, 2004, pp. 135–166, DOI 10.1007/978-1-4020-2772-7_6. MR2147419

[BN] D. Bar-Natan, *On Khovanov's categorification of the Jones polynomial*, Algebr. Geom. Topol. **2** (2002), 337–370, DOI 10.2140/agt.2002.2.337. MR1917056

[Kau] L. H. Kauffman, *An invariant of regular isotopy*, Trans. Amer. Math. Soc. **318** (1990), no. 2, 417–471, DOI 10.2307/2001315. MR958895

[Kho] M. Khovanov, *A categorification of the Jones polynomial*, Duke Math. J. **101** (2000), no. 3, 359–426, DOI 10.1215/S0012-7094-00-10131-7. MR1740682

[MV1] G. J. Montoya Vega, *A Historical Exploration of Knot Theory, Khovanov Homology, and Framing Changes of Links and Skein Modules*, ProQuest LLC, Ann Arbor, MI, 2022. Thesis (Ph.D.)–The George Washington University. MR4495185

[MV2] G. Montoya-Vega, *A first look at knot theory and Khovanov homology* (Spanish, with English and Spanish summaries), Rev. Integr. Temas Mat. **41** (2023), no. 2, 103–123. MR4699062

[Muk] S. Mukherjee, *On odd torsion in even Khovanov homology*, Exp. Math. **31** (2022), no. 3, 1039–1045, DOI 10.1080/10586458.2020.1743215. MR4477421

[MS] S. Mukherjee and D. Schütz, *Arbitrarily large torsion in Khovanov cohomology*, Quantum Topol. **12** (2021), no. 2, 243–264, DOI 10.4171/qt/149. MR4261658

[MPSWY] S. Mukherjee, J. H. Przytycki, M. Silvero, X. Wang, and S. Y. Yang, *Search for torsion in Khovanov homology*, Exp. Math. **27** (2018), no. 4, 488–497, DOI 10.1080/10586458.2017.1320242. MR3894728

[PPS] M. D. Pabiniak, J. H. Przytycki, and R. Sazdanović, *On the first group of the chromatic cohomology of graphs*, Geom. Dedicata **140** (2009), 19–48, DOI 10.1007/s10711-008-9307-4. MR2504733

[Prz] J. H. Przytycki, *When the theories meet: Khovanov homology as Hochschild homology of links*, Quantum Topol. **1** (2010), no. 2, 93–109, DOI 10.4171/QT/2. MR2657644

[PBIMW] J. H. Przytycki, R. P. Bakshi, D. Ibarra, G. Montoya-Vega, and D. Weeks, *Lectures in knot theory—an exploration of contemporary topics*, Universitext, Springer, Cham, [2024] ©2024, DOI 10.1007/978-3-031-40044-5. MR4738299

[PS] J. H. Przytycki and R. Sazdanović, *Torsion in Khovanov homology of semi-adequate links*, Fund. Math. **225** (2014), no. 1, 277–304, DOI 10.4064/fm225-1-13. MR3205574

[Shu] A. N. Shumakovitch, *Torsion of Khovanov homology*, Fund. Math. **225** (2014), no. 1, 343–364, DOI 10.4064/fm225-1-16. MR3205577

[Vir1] O. Viro, *Remarks on definition of Khovanov homology*, Preprint, `arXiv:arXiv:math/0202199 [math.GT]`, 2002.

[Vir2] O. Viro, *Khovanov homology, its definitions and ramifications*, Fund. Math. **184** (2004), 317–342, DOI 10.4064/fm184-0-18. MR2128056

Department of Mathematics, University of Puerto Rico at Río Piedras, San Juan 00925, Puerto Rico

Email address: `gabrielmontoyavega@gmail.com`

Contemporary Mathematics
Volume **834**, 2026
https://doi.org/10.1090/conm/834/16686

Exploring unimodality of the plucking polynomial with delay function

Dionne Ibarra, Alex Landry, Gabriel Montoya-Vega, and Józef H. Przytycki

ABSTRACT. The plucking polynomial is an invariant of rooted trees with connections to knot theory. The polynomial was constructed in 2014 as a tool to analyze lattice crossings after taking the quotient by the Kauffman bracket skein relations. In this paper we study the plucking polynomial and the plucking polynomial with delay function. We present a formula for the plucking polynomial of hedgehog rooted trees and explore the unimodality of this polynomial. In particular, we consider an anti-unimodal delay function and a delay function with a specific image set. Furthermore, we present a number of interesting examples and make some speculations on the unimodality of plucking polynomials with delay functions of hedgehog rooted trees.

1. Introduction

The plucking polynomial is an invariant of rooted trees that enjoys connections with the mathematical theory of knots. The fourth author introduced the plucking polynomial (originally called the q-polynomial of rooted trees) after completing a work with M. Dąbkowski and C. Li concerning the Kauffman bracket skein module of lattice crossings [**DLP**, **Prz**]. The plucking polynomial's connections to lattice crossings is explored in [**DP**]. A further study of the properties of the plucking polynomial can be found in [**CMPWY1**, **CMPWY2**, **CMPWY3**].

The plucking polynomial is a polynomial defined recursively from a plane rooted tree by "plucking" the leaves of the tree. This article is organized as follows. In this section we first define a rooted tree and introduce the standard version of the plucking polynomial together with a calculation example. Then we define the plucking polynomial with delay function. In Section 2 we define a hedgehog rooted tree and show a calculation of its plucking polynomial. There we present a formula for the calculation of the plucking polynomial with delay function and we study its unimodality under certain conditions. Section 3 consists of some examples and a generalization of the structure of the polynomial for a specific hedgehog rooted

2020 *Mathematics Subject Classification.* Primary 57K10; Secondary 57M15.
Key words and phrases. Rooted trees, plucking polynomial, unimodality, knot theory.
The first author acknowledges the support by the Australian Research Council through grant DP210103136. The third author acknowledges the support of the National Science Foundation through Grant DMS-2212736. The fourth author was partially supported by the Simons Collaboration Grant 637794.

tree. Finally, in Section 4 we present some questions and speculations that may guide future directions of related investigation.

DEFINITION 1.1. A **rooted tree** is a graph that is connected, does not contain any cycles (acyclic), and one vertex is designated as the root. A plane rooted tree is a rooted tree embedded in the plane. Figure 1 shows examples of plane rooted trees.

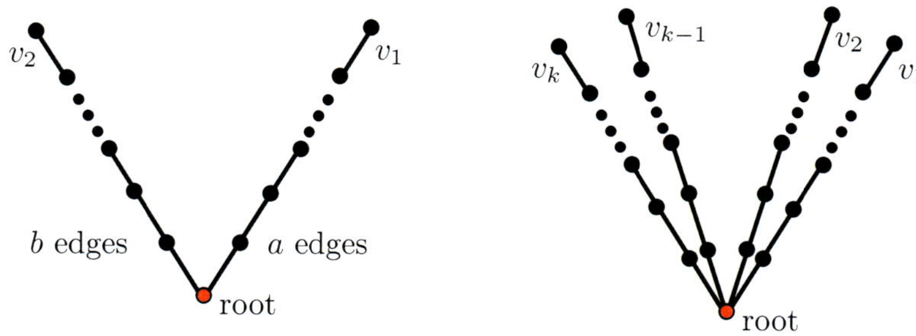

(a) $T_{b,a}$ (b) $T_{a_k,a_{k-1},\ldots,a_1}$

FIGURE 1. Example of plane rooted trees: (a) $T_{b,a}$ is a rooted tree with 2 branches of length b and a, and each branch has one leaf, v_2 and v_1, respectively. (b) $T_{a_k,a_{k-1},\ldots,a_1}$ is a rooted tree with k branches of length a_i for $1 \leq i \leq k$, and each branch has one leaf.

DEFINITION 1.2. Let T be a plane rooted tree and let $|E(T)|$ denote the number of edges of T. The **plucking polynomial** $Q(T, q)$ (or succinctly $Q(T)$) is the unique polynomial obtained from T by the initial condition and recursive formula shown below:

(1) If T is the one vertex tree, i.e. $|E(T)| = 0$, then $Q(T, q) = 1$.
(2) If $|E(T)| > 0$, then

$$Q(T, q) = \sum_{v \in L(T)} q^{r(T,v)} Q(T - v, q),$$

where the sum is taken over the set $L(T)$ of all leaves of T, that is, vertices of degree different from the root; $r(T, v)$ is the number of edges of T to the right of the unique path connecting v to the root; $T - v$ is a rooted tree obtained by removing the leaf v from T along with the edge directly connected to v.

EXAMPLE 1.3. Consider the following plane rooted tree T.

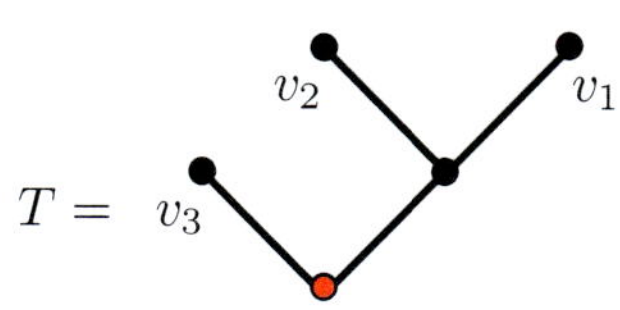

The unique path connecting each leaf v_i to the root for $i = 1, 2, 3$ is shown in Figure 2. We obtain $r(T, v_1) = 0$, $r(T, v_2) = 1$, and $r(T, v_3) = 3$.

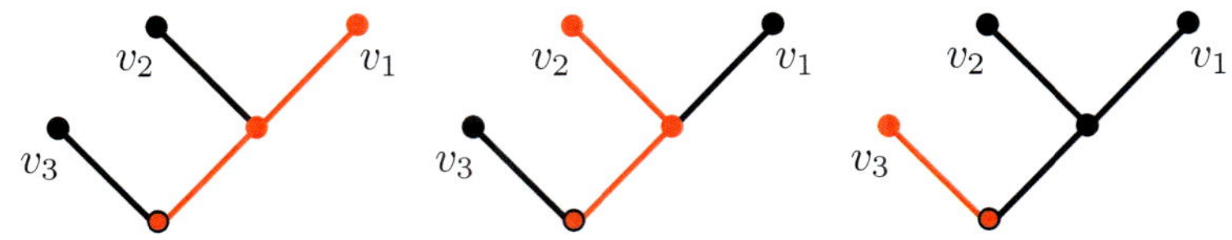

FIGURE 2. The red segments highlight the unique path from v_i to the root for $i = 1, 2, 3$.

Then $Q(T, q) = q^0 Q(T - v_1) + q^1 Q(T - v_2) + q^3 Q(T - v_3)$, where

$T - v_1 =$ 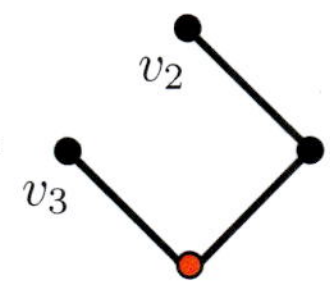 , $T - v_2 =$ 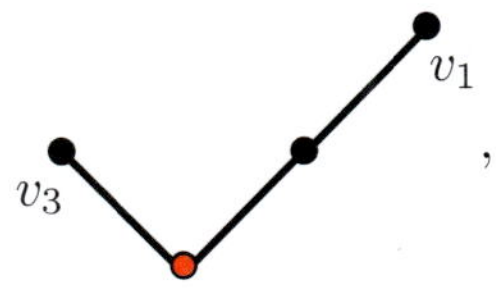 , $T - v_3 =$ 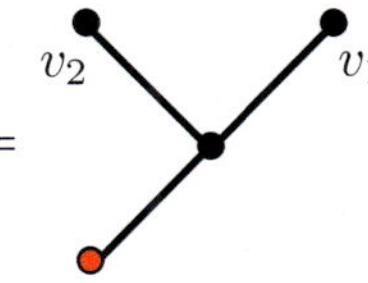 .

Observe that $Q(T - v_1) = Q(T - v_2)$, so we only need to compute $Q(T - v_1)$ and $Q(T - v_3)$.

$Q(T - v_1) = q^0 Q\Big($  $\Big) + q^2 Q\Big($ 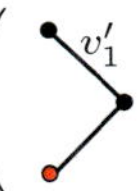 $\Big)$

$= \Big(q^0 Q\Big($  $\Big) + q^1 Q\Big($  $\Big)\Big) + q^2 \Big(q^0 Q\Big($  $\Big)\Big)$

$= Q(\bullet) + qQ(\bullet) + q^2 Q(\bullet)$

$= 1 + q + q^2,$

and

$Q(T - v_3) = q^0 Q\Big($ 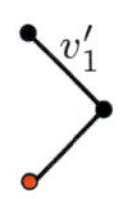 $\Big) + q^1 Q\Big($ 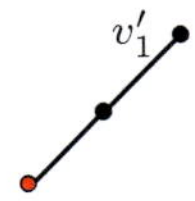 $\Big)$

$= (1 + q) Q\Big($ 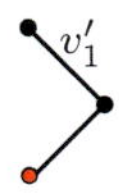 $\Big)$

$= (1 + q) q^0 Q\Big($  $\Big)$

$= (1 + q) q^0 Q(\bullet) = 1 + q.$

Therefore, $Q(T, q) = (1 + q)(1 + q + q^2) + q^3(1 + q) = (1 + q)(1 + q + q^2 + q^3)$.

THEOREM 1.4 ([**Prz**]). *The plucking polynomial $Q(T)$ of a rooted tree T does not depend on the plane embedding. $Q(T)$ is therefore an invariant of rooted trees.*

PROOF. See [**Prz**, Theorem 2.2 and Corollary 2.3]. □

Notice that in the previous example we calculated the plucking polynomial of a special rooted tree which was shown in Figure 1a. Definition 1.5 introduces the notions of quantum integers, q-factorials, and Gaussian polynomials. Thus we can write $Q(T - v_1) = Q(T_{1,2}) = [3]_q$ and $Q(T, q) = [2]_q [4]_q$.

DEFINITION 1.5. $[n]_q = 1 + q + \cdots + q^{n-1}$ is called a **quantum integer**, $[n]_q! = \prod_{k=1}^{n} [k]_q$ is called a ***q*-factorial**, and

$$\binom{n}{i, n-i}_q = \frac{[n]_q!}{[i]_q![n-i]_q!}$$

is called a **Gaussian polynomial**. By convention $[0]_q! = 1$, $\binom{n}{0}_q = 1 = \binom{n}{n}_q$, and $\binom{n}{-1} = 0$.

EXAMPLE 1.6. Consider the rooted tree with two long branches of length b and a respectively, shown in Figure 1a. One natural question that arises is how many different ways there are to "pluck" the tree $T_{b,a}$ one leaf at a time. In the case where the plane is commutative, the left branch should be plucked b times and the right branch a times, so the answer in this case is given by $\binom{a+b}{a}$. On the other hand, when the plane is q-commutative, i.e. $yx = qxy$, For a detailed discussion see [**Prz**, **PBIMW**].

The following definition introduces the notion of a delay function in the calculation of the plucking polynomial. This version of the polynomial is particularly relevant given its connection with the Kauffman bracket skein module via Catalan states of lattice crossings; see [**DP**, **Prz**] and Chapter 10 of [**PBIMW**].

DEFINITION 1.7. The plucking polynomial $Q(T, f)$ of a plane rooted tree T with a **delay function** $f : L(T) \to \mathbb{Z}^+$ is given by the initial conditions $Q(\bullet) = 1$, $Q(T, f) = 0$ if $f(w) \geq 2$ for any leaf w, and the recursive relation:

$$Q(T, f) = \sum_{v \,\in\, L_1(T)} q^{r(T,v)} Q(T - v, f_v),$$

where $L_1(T) = f^{-1}(1) \in L(T)$, and $f_v : L(T - v) \to \mathbb{Z}^+$ is given by

$$f_v(u) = \begin{cases} max\{1, f(u) - 1\} & \text{when } u \in L(T), \\ 1 & \text{when } u \notin L(T), u \in L(T - v). \end{cases}$$

EXAMPLE 1.8. Consider the following plane rooted tree T with delay function:

$(T, f) =$ 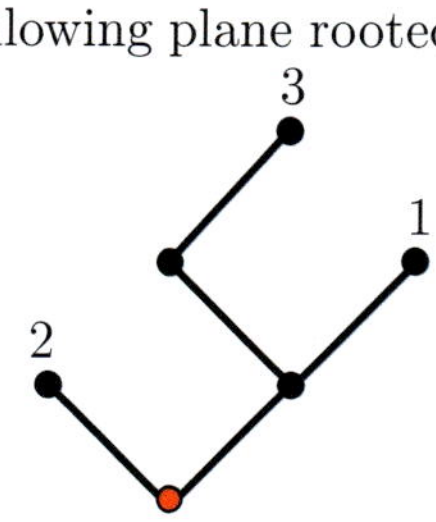

.

Then,

$Q(T, f) = Q$ 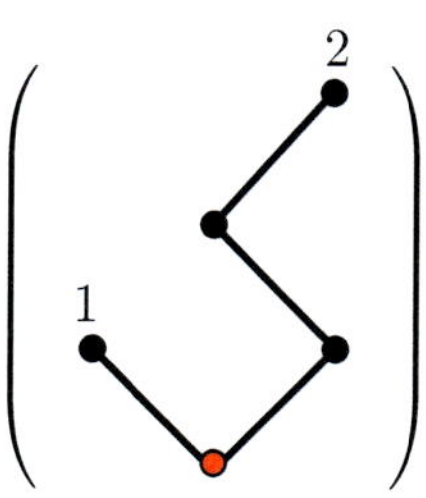

$= q^3 Q$ 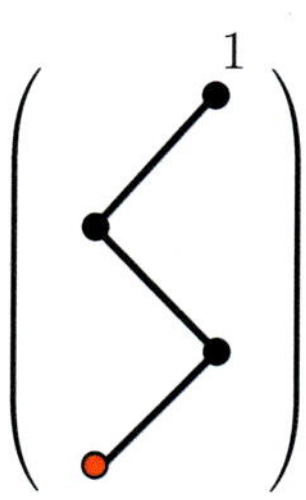

$= q^3.$

2. Plucking polynomial of the hedgehog rooted tree

A hedgehog graph is a rooted tree in which every edge is a leaf. A hedgehog graph can be written as $T_{1,1,\ldots,1}$, a root with n protruding edges; see Figure 3.

FIGURE 3. A hedgehog rooted tree.

EXAMPLE 2.1. Consider the hedgehog rooted tree $T_{1,1,\cdots,1}$ with n leaves as shown in Figure 3. Since $T_{1,1,\cdots,1} - v_1 = T_{1,1,\cdots,1} - v_2$ for all $v_1, v_2 \in L(T_{1,1,\cdots,1})$, then $Q(T_{1,1,\cdots,1}) = [n]_q Q(T_{1,1,\cdots,1} - v)$. By induction we have

$$Q(T_{1,1,\cdots,1}) = [n]_q!.$$

EXAMPLE 2.2. We use Definition 1.7 to calculate the plucking polynomial of the hedgehog rooted tree shown in Figure 4.

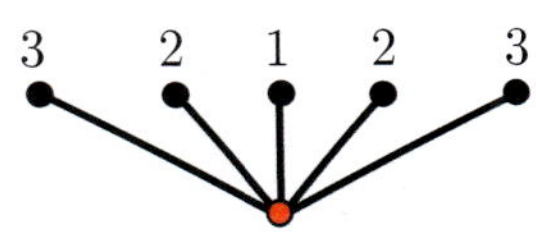

FIGURE 4. Hedgehog rooted tree with delay values 2 and 3.

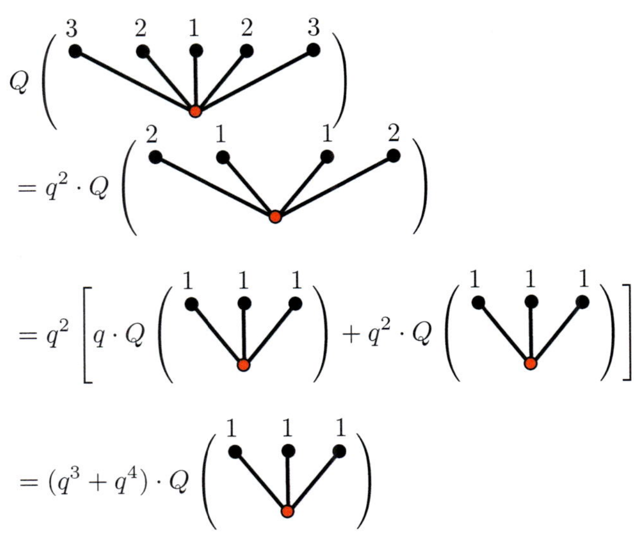

$$\begin{aligned}
&= [3]_q! \cdot (q^3 + q^4) \\
&= [3]_q \cdot [2]_q \cdot [1]_q \cdot (q^3 + q^4) \\
&= (1 + q + q^2) \cdot (1 + q) \cdot (q^3 + q^4) \\
&= q^7 + 3q^6 + 4q^5 + 3q^4 + q^3 \\
&= q^3 \, [3]_q \, [2]_q^2 .
\end{aligned}$$

In the following two subsections, we analyze special cases of the plucking polynomial of a hedgehog rooted tree with delay function. First we prove that the plucking polynomial is unimodal when the delay function f is anti-unimodal in Section 2.1. Then we consider the specific case of a delay function whose image is the set $\{1, 2\}$ in Section 2.2.

2.1. Unimodality. We say that that the sequence $a_1, a_2, \cdots a_n$ is unimodal if there exists $1 \leq j \leq n$ such that

$$a_1 \leq a_2 \leq \cdots \leq a_j \geq \cdots \geq a_n.$$

The polynomial P is called unimodal if its coefficients form a unimodal sequence. Consider a plane rooted tree T with delay function $f : L(T) \to \mathbb{Z}^+$. We say that this function is anti-unimodal if $-f$ is unimodal.

In this subsection we prove that in the case of f anti-unimodal, $Q(T_{1,\dots,1}, f)$ is the product of quantum integers therefore $Q(T_{1,\dots,1}, f)$ is unimodal.

Let T be the plane rooted tree $T_{1,1,\cdots,1}$, that is, a rooted tree embedded in the plane with a root and n other vertices which are all leaves. Denote the leaves (from left to right) by $v_1, \dots, v_n$. Let the delay function f be defined by $f(v_i) = a_i$. We use the notation $|f^{-1}(i)| = f_i$, and $M = max(a_1, a_n)$. Assume that the function f is anti-unimodal, that is, for some j we have that

$$a_1 \geq a_2 \geq \cdots \geq a_j \leq \cdots \leq a_n.$$

and that $f(v_j) = a_j = 1$; otherwise $Q(T, f)$ would be equal to zero. The last piece of notation we need is the following. For $i > 1$, let $f_i = f_i^- + f_i^+$, where f_i^+ is the number of vertices in $f^{-1}(i)$ which are to the right of v_j in the tree. The precise formulation is given in Theorem 2.3.

THEOREM 2.3. *Let $T = T_{1,1,\cdots,1}$ be a hedgehog rooted tree. Let f be an anti-unimodal delay function as described above. Then*

$$Q(T, f) = q^{\sum_{i=2}^{a_n}(i-1)f_i^+} \prod_{j=1}^{n} \left[(\sum_{i=1}^{j} f_i) - j + 1 \right]_q.$$

In particular, $Q(T, f)$ is unimodal as it is the product of quantum integers.

PROOF. The proof proceeds by induction on n - the number of leaves of T. Recall the recursive formula from the definition of the plucking polynomial:

$$Q(T, f) = \sum_{v \in L(T)} q^{r(T,v)} Q(T - v, f_v).$$

In this case, the rooted tree with delay function $(T - v, f_v)$ is independent of the vertex v. Therefore, it would be again a hedgehog with anti-unimodal function. Thus,

$$Q(T, f) = \left(\sum_{v \in L_1(T)} q^{r(T,v)} \right) \cdot Q(T - v, f_v),$$

for arbitrarily chosen v in $L_1(T)$. Let us analyze how we find $q^{r(T,v)}$ for any $v \in L_1$. Consider the vertex v' in L_1 and suppose it is the most to the right among all vertices. Then $r(T, v')$ is the number of vertices of L to the right of v'. In our notation:

$$r(T, v') = |f^{-1}(2)| + |f^{-1}(3)| + \cdots + |f^{-1}(a_n)|.$$

Then,

$$\sum_{v\in L_1(T)} q^{r(T,v)} = q^{r(T,v)} + qq^{r(T,v')} + q^2q^{r(T,v')} + \cdots + q^{|f^{-1}(1)-1|}q^{r(T,v')}$$
$$= (1+q+q^2+q^3+\cdots+q^{|f^{-1}(1)|-1})\cdot q^{r(T,v')}$$
$$= \left[f^{-1}(1)\right]_q \cdot q^{r(T,v')}.$$

Therefore,

$$Q(T,f) = \left(\sum_{v\in L_1(T)} q^{r(T,v)}\right)\cdot Q(T-v,f_v) = \left[f^{-1}(1)\right]_q \cdot q^{r(T,v)}\cdot Q(T-v',f_{v'}).$$

Observe that for $f\equiv 1$ we get $Q(T) = [n]_q!$ as needed. Moreover, notice that if $j > \sum_{i=1}^{j} f_i$ then the sum is equal to zero. □

EXAMPLE 2.4. Consider again the hedgehog rooted tree from Example 2.2 illustrated in Figure 4. Let us calculate its plucking polynomial using the formula from Theorem 2.3.

From the figure we can determine all the variables for the formula. First we have that $max(a_1,a_n) = max(3,3) = 3$. Then we see that $f(v_1) = a_1 = 3$, $f(v_2) = a_2 = 2$, $f(v_3) = a_3 = 1 = a_j$, $f(v_4) = a_4 = 2$, and $f(v_5) = a_5 = 3$. Recall that $f_i = f_i^- + f_i^+$ for $i>0$, where $f_i = |f^{-1}(i)|$. This means that we need to consider the cases for $i = 2,3,4,5$.

- For $i=2$, $f_2 = f_2^- + f_2^+ = |f^{-1}(2)| = 2$.
- For $i=3$, $f_3 = f_3^- + f_3^+ = 1+1 = 2$.
- For $i=4$, we have $f_4 = 0$.
- For $i=5$, we have $f_5 = 0$.

Moreover, observe that

$$\sum_{i=2}^{a_n}(i-1)f_i^+ = \sum_{i=2}^{3}(i-1)f_i^+ = (2-1)f_2^+ + (3-1)f_3^+ = (1)(1)+(2)(1) = 3,$$

which means that from here we get the factor q^3. Now observe that

$$\prod_{j=1}^{5}\left[\left(\sum_{i=1}^{j} f_i\right) - j + 1\right]_q$$
$$= \left[\left(\sum_{i=1}^{1} f_i\right) - 1 + 1\right]_q \left[\left(\sum_{i=1}^{2} f_i\right) - 1\right]_q \left[\left(\sum_{i=1}^{3} f_i\right) - 2\right]_q$$
$$\cdot \left[\left(\sum_{i=1}^{4} f_i\right) - 3\right]_q \left[\left(\sum_{i=1}^{5} f_i\right) - 4\right]_q$$
$$= [f_1]_q\,[f_1+f_2-1]_q\,[f_1+f_2+f_3-2]_q\,[f_1+f_2+f_3+f_4-3]_q$$
$$\cdot [f_1+f_2+f_3+f_4+f_5-4]_q$$
$$= [1]_q\,[2]_q\,[3]_q\,[2]_q\,[1]_q = ([2]_q)^2\,[3]_q\,.$$

Thus $Q(T,f) = q^3([2]_q)^2\,[3]_q$, which agrees with the result obtained in Example 2.2.

2.2. Special delay functions of the hedgehog rooted tree. Now we consider a delay function whose image is equal to the set $\{1, 2\}$. The precise formulation for computing the plucking polynomial in this case is given in Proposition 2.5.

PROPOSITION 2.5. *Let $T = T_{1,1,\cdots,1}$ be the Hedgehog (star) rooted tree with n rays and let f be a delay function with values 1 or 2. By ordering the leaves from right to left we let $\varepsilon_i = 1$ if f(i'th leaf) $= 1$ and $\varepsilon_i = 0$ if f(i'th leaf)$= 2$. Then*

$$Q(T, f) = p_n(q)[n-1]_q!,$$

where $p_n(q) = \sum_{i=0}^{n-1} \varepsilon_i q^i$.

PROOF. Observe that $(T - v, f_v)$ for any leaf v is the rooted tree $T_{1,1,\ldots,1,1}$ with $n-1$ rays and trivial delay function (that is $f_v \equiv 1$). Thus $Q(T - v, f_v) = [n-1]_q!$. □

It is still open whether the plucking polynomial of a rooted tree as in the previous proposition is always unimodal, though experimental calculations suggest this is the case. See Conjecture 4.1.

Proposition 2.5 suggests that it is important to analyze polynomials of the form $p_n(q)[n-1]_q!$, where $p_n(q) = \varepsilon_0 + \varepsilon_1 q + \cdots + \varepsilon_{n-1} q^{n-1}$ and $\varepsilon_i = 1$ or 0. This analysis is still open, however it is possible to show unimodality for the polynomial $p_{n+1}[n]_q$ where $p_{n+1}(q) = \varepsilon_0 + \varepsilon_1 q + \varepsilon_2 q^2 + \cdots + \varepsilon_n q^n$ and $\varepsilon_i = 0$ or 1.

PROPOSITION 2.6. *The polynomial $p_{n+1}(q)[n]_q$ is unimodal.*

PROOF. Observe that

$$\begin{aligned}
&(\varepsilon_0 + \varepsilon_1 q + \varepsilon_2 q^2 + \cdots + \varepsilon_n q^n)(1 + q + q^2 + \cdots + q^{n-1}) \\
&\quad = \varepsilon_0 + (\varepsilon_0 + \varepsilon_1)q + (\varepsilon_0 + \varepsilon_1 + \varepsilon_2)q^2 \\
&\qquad + \cdots + (\varepsilon_0 + \varepsilon_1 + \varepsilon_2 + \cdots + \varepsilon_{n-1})q^{n-1} + (\varepsilon_1 + \varepsilon_2 + \cdots + \varepsilon_n)q^n \\
&\qquad + (\varepsilon_2 + \cdots + \varepsilon_n)q^{n+1} + (\varepsilon_3 + \cdots + \varepsilon_n)q^{n+2} + \cdots + \varepsilon_n)q^{2n-1}.
\end{aligned}$$

Therefore, if $p_{n+1}(q)[n]_q = \sum_{i=0}^{2n-1} c_i q^i$ then

$$c_0 \leq c_1 \leq \cdots \leq c_{n-1} \text{ and } c_n \geq c_{n+1} \geq \cdots \geq c_{2n-1}.$$

From this we conclude that $p_{n+1}(q)[n]_q$ is unimodal. Moreover, notice that

$$c_n - c_{n-1} = \varepsilon_n - \varepsilon_{n-1} = \begin{cases} 1 & \text{when } \varepsilon_n = 1, \varepsilon_0 = 0, \\ 0 & \text{when } \varepsilon_n = \varepsilon_0, \\ -1 & \text{when } \varepsilon_n = 0, \varepsilon_0 = 1. \end{cases}$$ □

The proof of the previous theorem also leads to the result that $p_{n+1}[n]_q$ is strongly unimodal if and only if there are $i \leq j$ such that $\varepsilon_i = \varepsilon_{i+1} = \cdots = \varepsilon_j = 1$.

COROLLARY 2.7. *If $p_{n+1}(q)$ is symmetric then $p_{n+1}(q)[n]_q!$ is unimodal.*

PROOF. By Proposition 2.6 $p_{n+1}(q)[n]_q$ is unimodal. Moreover, it is also symmetric as a product of symmetric polynomials. Thus, by a classical result[1] that the product of symmetric unimodal polynomials is symmetric and unimodal, we conclude that $p_{n+1}(q)[n]_q!$ is unimodal. □

---

[1]See, for instance, [**Win, Sta**].

Observe that the same method we used in Proposition 2.6 can be applied to the polynomial $p_{n+2}(q)[n+1]_q[n]_q$ and succeed with some effort. However, our main focus is the polynomial $p_{n+1}(q)[n]_q!$. To this effect we have:

$$[n+1]_q[n]_q = 1+2q+3q^2+\cdots+nq^{n-1}+nq^n+(n-1)q^{n+1}+\cdots+2q^{2n-2}+q^{2n-1}.$$

After some laborious calculation for $p_{n+2}(q)[n+1]_q[n]_q$ we obtain:

PROPOSITION 2.8. *The polynomial* $p_{n+2}(q)[n+1]_q[n]_q$ *is unimodal. Here* $p_{n+2}(q)$ *is an arbitrary polynomial* $p_{n+2}(q) = \varepsilon_0+\varepsilon_1 q+..+\varepsilon_{n+1}q^{n+1}$, *where* $\varepsilon_i = 0$ *or* 1.

More precisely we can show that if $p_{n+2}(q)[n+1]_q[n]_q = \sum_{i=0}^{3n} c_i q^i$ *then*

$$c_0 \le c_1 \le \cdots \le c_n \le \cdots \le c_{n+j} \ge \cdots \ge c_{2n} \ge c_{2n-1} \ge \cdots \ge c_{3n-1} \ge c_{3n},$$

where j *is the smallest number,* $0 \le j \le n$ *such that*

$$c_{n+j+1} - c_{n+j} = -\varepsilon_0 - \varepsilon_1 - \varepsilon_j + \hat{\varepsilon}_{j+1} + \varepsilon_{j+2} + \cdots + \varepsilon_n + \varepsilon_{n+1} \le 0.$$

PROOF. we have

$$\begin{aligned}
p_{n+2}(q)[n+1]_q[n]_q &= (\varepsilon_0+\varepsilon_1 q+\cdots+\varepsilon_{n+1}q^{n+1}) \\
&\quad\cdot(1+2q+3q^2+\cdots+nq^{n-1}+nq^n+(n-1)q^{n+1} \\
&\quad+\cdots+2q^{2n-2}+q^{2n-1}) \\
&= \varepsilon_0+(2\varepsilon_0+\varepsilon_1)q+(3\varepsilon_0+2\varepsilon_1+\varepsilon_2)q^2 \\
&\quad+\cdots+(n\varepsilon_0+(n-1)\varepsilon_1+\cdots+\varepsilon_{n-1})q^{n-1} \\
&\quad+(n\varepsilon_0+n\varepsilon_1+(n-1)\varepsilon_2+\cdots+\varepsilon_n)q^n \\
&\quad+((n-1)\varepsilon_0+n\varepsilon_1+n\varepsilon_2+(n-1)\varepsilon_3+\cdots+\varepsilon_{n+1})q^{n+1} \\
&\quad+((n-2)\varepsilon_0+(n-1)\varepsilon_1+n\varepsilon_2+(n)\varepsilon_3+\cdots+2\varepsilon_{n+1})q^{n+2} \\
&\quad+\cdots+2\varepsilon_0+3\varepsilon_1+\cdots+n\varepsilon_{n-2}+n\varepsilon_{n-1} \\
&\quad+(n-1)\varepsilon_n+(n-2)\varepsilon_{n+1})q^{2n-2}+\varepsilon_0+2\varepsilon_1 \\
&\quad+\cdots+n\varepsilon_{n-1}+n\varepsilon_n+(n-1)\varepsilon_n+(n-1)\varepsilon_{n+1})q^{2n-1} \\
&\quad+\varepsilon_1+2\varepsilon_2+\cdots+(n-1)\varepsilon_{n-1}+n\varepsilon_n+n\varepsilon_{n+1})q^{2n-1} \\
&\quad+\varepsilon_2+2\varepsilon_3+\cdots+(n-1)\varepsilon_n+n\varepsilon_{n+1})q^{2n} \\
&\quad+\cdots+(2\varepsilon_{n+1}+\varepsilon_n)q^{3n-1}+\varepsilon_{n+1}q^{3n}.
\end{aligned}$$

From this we get in particular that:

$$c_{n+j+1} - c_{n+j} = -\varepsilon_0 - \varepsilon_1 - \varepsilon_j + \hat{\varepsilon}_{j+1} + \varepsilon_{j+2} + \cdots + \varepsilon_n + \varepsilon_{n+1} \le 0$$

and finally the unimodality of $p_{n+2}(q)[n+1]_q[n]_q$. □

3. Calculations and examples

EXAMPLE 3.1. Consider the hedgehog rooted tree $T = 114\cdots411$ (k-times 4) denoted by $1^2 4^k 1^2$. See Figure 5. We show that for $k = 2,3,4,5,6$ the plucking polynomial is not unimodal. Concrete calculations are given below.

$(T,f) =$ 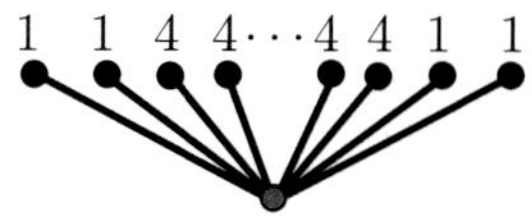

FIGURE 5. Hedgehog rooted tree $T = 114\cdots411$.

(a) $T = 1^2 4^2 1^2$:

$$\begin{aligned} Q(T,f) &= q^{13} + 5q^{12} + 12q^{11} + 18q^{10} + 19q^9 + 17q^8 \\ &\quad + 17q^7 + 19q^6 + 18q^5 + 12q^4 + 5q^3 + q^2 \\ &= q^2(1+q)^3(1+q+q^2)^2(1+q^4) \\ &= q^2[2]_q^2[3]_q[3]_q!(1+q^4). \end{aligned}$$

(b) $T = 1^2 4^3 1^2$:

$$\begin{aligned} Q(T,f) &= q^{18} + 6q^{17} + 18q^{16} + 36q^{15} + 53q^{14} + 61q^{13} + 59q^{12} + 54q^{11} \\ &\quad + 54q^{10} + 59q^9 + 61q^8 + 53q^7 + 36q^6 + 18q^5 + 6q^4 + q^3 \\ &= q^3(1+q)^5(1+q^2)(1+q+q^2)^2(1-q+q^2-q^3+q^4) \\ &= q^3[2]_q^2[3]_q[4]_q!(1+q^5). \end{aligned}$$

(c) $T = 1^2 4^4 1^2$:

$$\begin{aligned} Q(T,f) &= q^{24} + 7q^{23} + 25q^{22} + 61q^{21} + 114q^{20} + 173q^{19} + 221q^{18} \\ &\quad + 245q^{17} + 245q^{16} + 234q^{15} + 228q^{14} + 234q^{13} + 245q^{12} \\ &\quad + 245q^{11} + 221q^{10} + 173q^9 + 114q^8 + 61q^7 + 25q^6 + 7q^5 + q^4 \\ &= q^4(1+q)^4(1+q^2)^2(1+q+q^2)^2(1-q^2+q^4)(1+q+q^2+q^3+q^4) \\ &= q^4[2]_q^2[3]_q[5]_q!(1+q^6). \end{aligned}$$

(d) $T = 1^2 4^5 1^2$:

$$\begin{aligned} Q(T,f) &= q^5 + 8q^6 + 33q^7 + 94q^8 + 208q^9 + 381q^{10} + 600q^{11} + 832q^{12} \\ &\quad + 1034q^{13} + 1171q^{14} + 1232q^{15} + 1234q^{16} + 1212q^{17} + 1200q^{18} \\ &\quad + 1212q^{19} + 1234q^{20} + 1232q^{21} + 1171q^{22} + 1034q^{23} + 832q^{24} + 600q^{25} \\ &\quad + 381q^{26} + 208q^{27} + 94q^{28} + 33q^{29} + 8q^{30} + q^{31} \\ &= q^5[2]_q^2[3]_q[6]_q!(1+q^7). \end{aligned}$$

(e) $T = 1^2 4^6 1^2$:

$$\begin{aligned} Q(T,f) &= q^{39} + 9q^{38} + 42q^{37} + 136q^{36} + 344q^{35} + 725q^{34} + 1325q^{33} + 2155q^{32} \\ &\quad + 3174q^{31} + 4287q^{30} + 5364q^{29} + 6276q^{28} + 6934q^{27} + 7315q^{26} \\ &\quad + 7465q^{25} + 7477q^{24} + 7451q^{23} + 7451q^{22} + 7477q^{21} + 7465q^{20} \\ &\quad + 7315q^{19} + 6934q^{18} + 6276q^{17} + 5364q^{16} + 4287q^{15} + 3174q^{14} \\ &\quad + 2155q^{13} + 1325q^{12} + 725q^{11} + 344q^{10} + 136q^9 + 42q^8 + 9q^7 + q^6 \\ &= q^6[2]_q^2[3]_q[7]_q!(1+q^8). \end{aligned}$$

We have checked that the polynomial is unimodal for $k = 1, 7, 8, 9, 10$ and we conjecture that is unimodal for any $k > 6$. In particular for $k = 7$ we get:

$$\begin{aligned} Q(T,f) = q^{48} &+ 10q^{47} + 52q^{46} + 188q^{45} + 532q^{44} + 1257q^{43} + 2582q^{42} + 4737q^{41} \\ &+ 7909q^{40} + 12179q^{39} + 17468q^{38} + 23514q^{37} + 29896q^{36} + 36105q^{35} \\ &+ 41645q^{34} + 46137q^{33} + 49397q^{32} + 51464q^{31} + 52566q^{30} + 53031q^{29} \\ &+ 53170q^{28} + 53170q^{27} + 53031q^{26} + 52566q^{25} + 51464q^{24} + 49397q^{23} \\ &+ 46137q^{22} + 41645q^{21} + 36105q^{20} + 29896q^{19} + 23514q^{18} + 17468q^{17} \\ &+ 12179q^{16} + 7909q^{15} + 4737q^{14} + 2582q^{13} + 1257q^{12} + 532q^{11} + 188q^{10} \\ &+ 52q^{9} + 10q^{8} + q^{7} \\ = q^7[2]_q^2[3]_q[8]_q!(1+q^9). \end{aligned}$$

We can write in general the following proposition.

PROPOSITION 3.2. *Consider the hedgehog tree $T = 114\cdots 411$ (k-times 4). Its plucking polynomial is given by the formula:*

$$Q(T,f) = q^k[k+1]_q!(1+q)^2(1+q+q^2)(1+q^{k+2}).$$

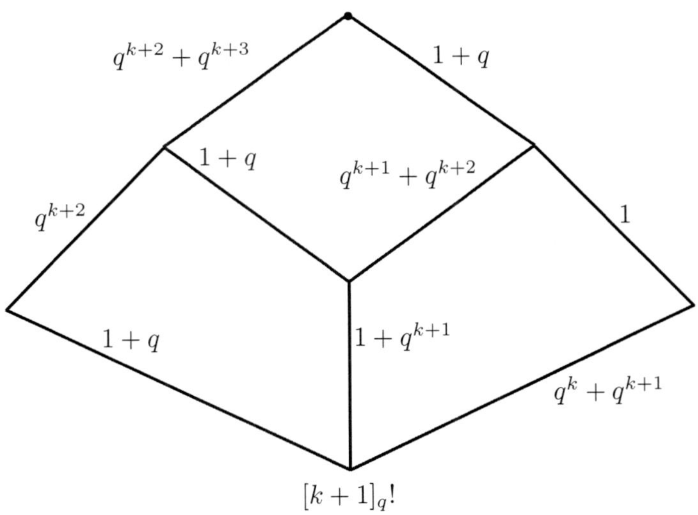

FIGURE 6. Computation of $Q(T, f)$ for the tree $T = 114\cdots 411$ (k-times 4).

PROOF.

$$\begin{aligned} Q(T,f) &= (q^{k+3}+q^{k+2})(q^{k+2}(q+1)[k+1]_q! + (1+q)(q^{k+1}+1)[k+1]_q!) \\ &\quad + (q+1)((q^{k+2}+q^{k+1})(q^{k+1}+1)[k+1]_q! + (q^{k+1}+q^k)[k+1]_q!) \\ &= [k+1]_q!(1+q)^2(q^k + q^{2k+4} + (1+q^{k+1})(q^{k+2}+q^{k+1})) \\ &= [k+1]_q!(1+q)^2(q^k)(1+q^{k+4}+q^2+q+q^{k+3}+q^{k+2}) \\ &= q^k[k+1]_q!(1+q)^2(1+q+q^2)(1+q^{k+2}). \end{aligned}$$

□

EXAMPLE 3.3. Consider the hedgehog rooted tree denoted by $1^a3^k1^b$, that is a-times 1, k-times 3, and b times 1. Then

$$Q(T_{1^a3^k1^b}, f) = [k+a+b-2]_q!\Big((q^{k+b-1}+q^{k+b})[a]_q[b]_q + q^{2k+2b}[a]_q[a-1]_q + [b]_q[b-1]_q\Big).$$

The proof of the formula for this example is explained in Figure 7.

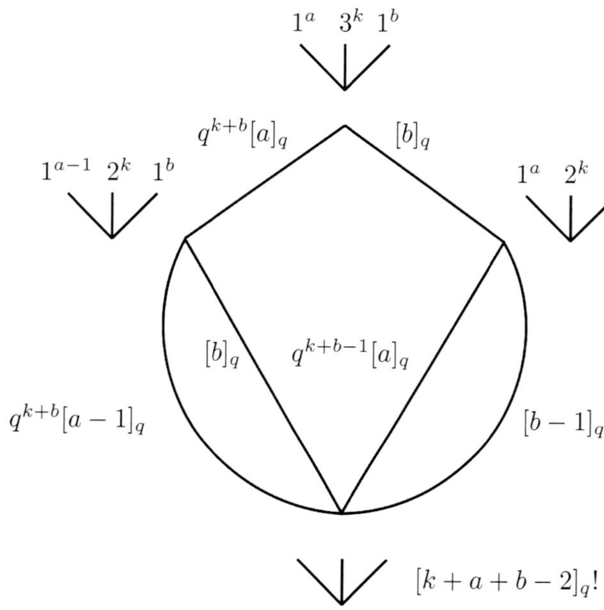

FIGURE 7. Calculation graph for $1^a3^k1^b$. On the bottom we have hedgehog rooted tree $1^{k+a+b-2}$ with the plucking polynomial $[k+a+b-2]_q!$.

REMARK 3.4. Another example of a hedgehog rooted tree with delay function and not unimodal plucking polynomial was given by Wojciech Garstka [**Gar**]. It is a hedgehog rooted tree of type 21412, see Figure 8, with the plucking polynomial

$$q^2(1+q)^2(1+2q+2q^3+q^4) = q^2+4q^3+5q^4+4q^5+5q^6+4q^7+q^8.$$

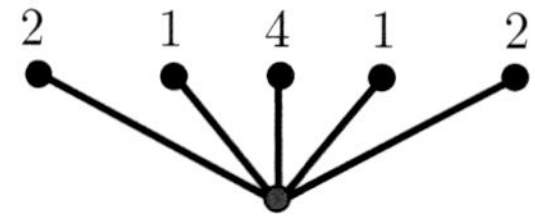

FIGURE 8. The hedgehog rooted tree of type 21412.

Furthermore, type 214412 is also not unimodal. However, types 1214121, 112141211, 2114112 and 211141112 are unimodal.

4. Future work

The plucking polynomial is of particular interest to knot theorists given its connections to the Kauffman bracket skein relations. Experimental calculation suggests the unimodality of the plucking polynomial with delay function values 1 or 2. Thus, we propose the following conjecture.

CONJECTURE 4.1. *(Unimodality Conjecture) Let* $T = T_{1,1,\cdots,1}$ *be the Hedgehog (star) rooted tree with* n *rays and let* f *be a delay function with values* 1 *or* 2*. Then its plucking polynomial with delay function* $Q(T, f)$*, is unimodal.*

The conjecture has been verified up to 8 leaves.

One could also investigate whether the plucking polynomial with delay function having values $1, 2$, or 3 is always unimodal. As of now, there is not enough data to propose it as a conjecture.

The following open problem relates the plucking polynomial with knot theory via the Kauffman bracket of the lattice crossings (see [**DP**]).

OPEN PROBLEM 4.2. Is the plucking polynomial with delay function unimodal when the delay function is anti-unimodal?

References

[CMPWY1] Z. Cheng, S. Mukherjee, J. H. Przytycki, X. Wang, and S. Y. Yang, *Realization of plucking polynomials*, J. Knot Theory Ramifications **26** (2017), no. 2, 1740016, 9, DOI 10.1142/S0218216517400168. MR3604498

[CMPWY2] Z. Cheng, S. Mukherjee, J. H. Przytycki, X. Wang, and S. Y. Yang, *Strict unimodality of plucking polynomials of rooted trees*, J. Knot Theory Ramifications **27** (2018), no. 7, 1841009, 19, DOI 10.1142/S0218216518410092. MR3835751

[CMPWY3] Z. Cheng, S. Mukherjee, J. H. Przytycki, X. Wang, and S. Y. Yang, *Rooted trees with the same plucking polynomial*, Osaka J. Math. **56** (2019), no. 3, 661–674. MR3982046

[DLP] M. K. Dabkowski, C. Li, and J. H. Przytycki, *Catalan states of lattice crossing*, Topology Appl. **182** (2015), 1–15, DOI 10.1016/j.topol.2014.11.015. MR3305606

[DP] M. K. Dabkowski and J. H. Przytycki, *Catalan states of lattice crossing: an application of plucking polynomial*, Topology Appl. **254** (2019), 12–28, DOI 10.1016/j.topol.2018.12.003. MR3894207

[Gar] W. Garstka, private correspondence with J. H. Przytycki, (Plucking polynomial computational project), December 2022.

[Prz] J. H. Przytycki, *q-polynomial invariant of rooted trees*, Arnold Math. J. **2** (2016), no. 4, 449–461, DOI 10.1007/s40598-016-0053-7. MR3564883

[PBIMW] J. H. Przytycki, R. P. Bakshi, D. Ibarra, G. Montoya-Vega, and D. Weeks, *Lectures in knot theory—an exploration of contemporary topics*, Universitext, Springer, Cham, [2024] ©2024, DOI 10.1007/978-3-031-40044-5. MR4738299

[Sta] R. P. Stanley, *Log-concave and unimodal sequences in algebra, combinatorics, and geometry*, Graph theory and its applications: East and West (Jinan, 1986), Ann. New York Acad. Sci., vol. 576, New York Acad. Sci., New York, 1989, pp. 500–535, DOI 10.1111/j.1749-6632.1989.tb16434.x. MR1110850

[Win] A. Wintner, *Asymptotic distributions and infinite convolutions*, Edvards Brothers, Ann Arbor, Michigan, 1938.

School of Mathematics, Monash University, Australia
Email address: dionne.ibarra@monash.edu

University of California, Berkeley, California 94720
Email address: alexc.landry@berkeley.edu

Department of Mathematics, University of Puerto Rico at Río Piedras, San Juan 00925, Puerto Rico
Email address: gabrielmontoyavega@gmail.com

Department of Mathematics, The George Washington University, Washington DC; and Department of Mathematics, University of Gdańsk, Gdańsk, Poland
Email address: przytyck@gwu.edu

Contemporary Mathematics
Volume **834**, 2026
https://doi.org/10.1090/conm/834/16687

Using Fibonacci numbers and Chebyshev polynomials to express Fox coloring groups and Alexander–Burau–Fox modules of diagrams of wheel graphs

Anthony Christiana, Huizheng Guo, and Józef H. Przytycki

ABSTRACT. In this paper we compute the Reduced Fox Coloring Group of the diagrams of Wheel Graphs which can also be represented as the closure of the braids $(\sigma_1\sigma_2^{-1})^n$. In doing so, we utilize Fibonacci numbers and their properties.

Following this, we generalize our result to compute the Alexander–Burau–Fox Module over the ring $\mathbb{Z}[t^{\pm 1}]$ for the same class of links. In our computation, Chebyshev polynomials function as a generalization of Fibonacci Numbers.

1. Introduction

The initial motivation for our work came from curiosity generated by [**Prz2**, Example 2.15] (see also [**BGMMP**, Example 5.2]), where the determinant of the closure of the braid $(\sigma_1\sigma_2^{-1})^n$ is computed. After successfully finding the group of Fox colorings of these links, we shifted our attention to the Alexander module of these links. In doing so the relation to Burau representation of braids becomes clear; thus we called our modules the Alexander–Burau–Fox modules. We use elementary methods, using properties of Fibonacci numbers and Chebyshev polynomials to obtain a simple closed formula for the group of Fox colorings and the ABF-module of $(\sigma_1\sigma_2^{-1})^n$. The connection to Plans' theorem on branched covers of S^3 along links added motivation to our results.

In this section we introduce the Tait Diagram of a plane graph. We then recall the definition of Fox n-colorings and of the Fox Coloring Group of a link (the universal object for Fox Colorings). Then, we formulate our main result about the structure of Fox Coloring Groups.

In Section 2, we prove our main result about Fox Colorings by working with the matrix of relations for the Fox Group using the properties of Fibonacci numbers.

In Section 3, we recall the definition of the Alexander–Burau–Fox Module over the ring $\mathbb{Z}[t^{\pm 1}]$. This module is a generalization of our Fox Coloring Group. We

2020 *Mathematics Subject Classification.* Primary 57K10; Secondary 57M12, 11B39.
Key words and phrases. Determinants of links, Fox coloring, Alexander–Burau–Fox module, Burau representation, Fibonacci numbers, Chebyshev polynomials, knots and links.
The third author was partially supported by the Simons Collaboration Grant 637794.

describe the structure of the ABF Module for the same family of links considered in section one and two, and express it as the sum of two cyclic modules.

In Section 4, we mention the relation with Plan's Theorem on branch covers of links. We also relate our results to those of Minkus and Mulazzani-Vesnin.

We now present some basic definitions.

DEFINITION 1.1. For a plane graph G, we associate an alternating link diagram as follows:

(1) Every edge is replaced by a crossing as illustrated in Figure 1.
(2) We connect the "loose endpoints" of the crossing along the edges; compare Figure 3.

FIGURE 1. Crossing from an edge

DEFINITION 1.2. We define a *Fox n-coloring* of a diagram D to be a function

$$f : arcs(D) \to \mathbb{Z}_n$$

such that every arc is "colored" by an element of $\mathbb{Z}_n$ with the following condition: for every crossing with arcs a, b, and c, $2b - a - c \equiv 0 \mod n$ for overarc b. That is, each crossing has the relation

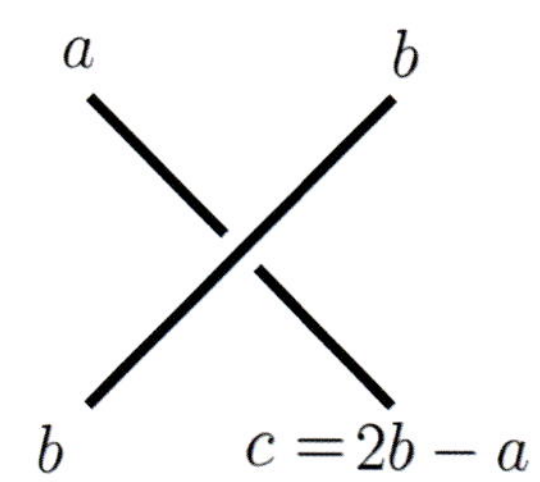

FIGURE 2. Coloring of a crossing

The n-colorings of a diagram D form a group denoted by $Col_n(D)$.

If $f(a_i) = f(a_j)$ for all $a_i, a_j \in arcs(D)$, we call f a *trivial coloring*. These trivial colorings form the subgroup $Col_n^{trivial}(D) \cong \mathbb{Z}_n \in Col_n(D)$ and the quotient $Col_n(D)/Col_n^{trivial}(D)$ is called the Reduced Group of Fox n-colorings denoted by $Col_n^{red}(D)$.

DEFINITION 1.3. An arc is the part of a diagram from undercrossing to undercrossing. We also include in our definition components without a crossing.

The number of arcs is equal to the number of crossing plus the number of trivial components of the diagram.

DEFINITION 1.4. The group $Col(D)$ is the abelian group whose generators are indexed by the arcs of D. The set of arcs is denoted by $arcs(D)$ and the set of generators is denoted by $Arcs(D)$. The relations at each crossing of D are given by $2b - a - c = 0$ where $a, b, c \in Arcs(D)$. That is,

$Col(D) = \{Arcs(D) \mid$ 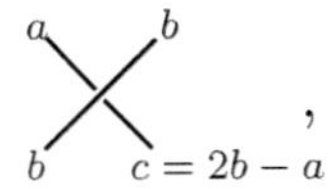

, *where* $a, b, c \in Arcs(D)\}$

The Fox Coloring Group can be also defined for tangles in a similar way. In particular for braids(for n braids treated as n-tangles), the group $Col(T)$ is freely generated by the top arcs of the braid. We will use this later in the proof of Theorem 1.6.

DEFINITION 1.5. Let $Col^{trivial}(D) \leq Col(D)$ be the infinite cyclic subgroup generated by the element $\sum_{a_i \in Arcs(D)} a_i$. This subgroup is isomorphic to $\mathbb{Z}$ and is called the group of trivial colorings of D.

The quotient group $\dfrac{Col(D)}{Col^{trivial}(D)}$ is called the **reduced group of Fox colorings.**[1] We denote it by $Col^{red}(D)$. That is,

$Col^{red}(D) = \{Arcs(D) \mid$ 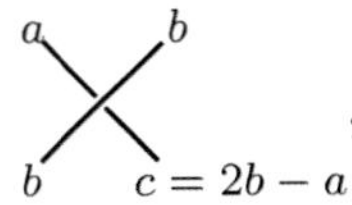

, $\sum_{a_i \in Arcs(D)} a_i = 0\},$

where the sum is taken over all arcs of D.

More information about Fox Colorings can be found in [**Prz1**, **PBIMW**].

We are now ready to formulate the main theorem of the second section, expressing the Reduced Fox Coloring Group using Fibonacci numbers.

THEOREM 1.6. *Let F_k be the Fibonacci sequence defined by*

$$F_0 = 0,\ F_1 = 1 \text{ and } F_{k+2} = F_{k+1} + F_k.$$

Let D_n be the closure of the braid $(\sigma_1\sigma_2^{-1})^n$, that is, $D_n = D(W_n)$ as in Figure 3. Then

$$Col^{red}(D_n) = \begin{cases} \mathbb{Z}_{F_{n-1}+F_{n+1}} \oplus \mathbb{Z}_{F_{n-1}+F_{n+1}} & \text{when } n \text{ is odd,} \\ \mathbb{Z}_{5F_n} \oplus \mathbb{Z}_{F_n} & \text{when } n \text{ is even} \end{cases}$$

In particular, for $n = 2, 3, 4, 5, 6, 7$, we have $\mathbb{Z}_5$, $\mathbb{Z}_4 \oplus \mathbb{Z}_4$, $\mathbb{Z}_{15} \oplus \mathbb{Z}_3$, $\mathbb{Z}_{11} \oplus \mathbb{Z}_{11}$, $\mathbb{Z}_{40} \oplus \mathbb{Z}_8$, and $\mathbb{Z}_{29} \oplus \mathbb{Z}_{29}$, respectively.

2. Fibonacci numbers and the reduced coloring group of wheel graphs, $Col^{red}(D)$

After a few preliminaries, we turn to a proof of Theorem 1.6.

The Tait diagram of a planar graph has a determinant equal to the number of spanning trees of the graph. For wheel graphs in particular, the number of spanning

[1] The group $Col^{red}(D)$ can be interpreted as the first homology of the double branch covering of S^3 branched along D; see [**Prz1**]. The group $Col^{red}(D)$ can also be computed by the Goeritz matrix of the diagram. Using this approach, the determinant of D_n is computed in [**Prz2**].

trees was computed in [**Sed, Mye**]. For our purposes, the determinant is equal to the order of the Reduced Fox Coloring Group. If the determinant is zero, which may only occur in the case of links, then it may result in infinite order.

In the second section of the paper, we obtain concise formulas, using Fibonacci numbers, for the Fox-Coloring Group of diagrams obtained from wheel graphs, which can also be expressed as the closure of 3-braids of the form $(\sigma_1\sigma_2^{-1})^n$ (compare [**DPS**]) where n is the number of spokes in the corresponding wheel graph; compare Figure 3.

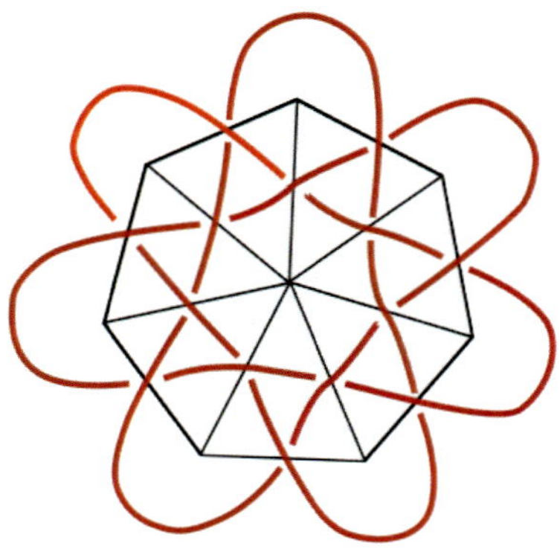

FIGURE 3. Wheel graph W_7 and its Tait diagram, $D_7 = D(W_7)$ representing the closure of the braid $(\sigma_1\sigma_2^{-1})^7$.

The proof is comprised of several propositions and lemmas.

PROPOSITION 2.1. *Consider an arbitrary 3-braid B and label the top arcs by a, b and c and the bottom arcs by a', b', and c', which are uniquely defined by a, b and c.*

(1) *In the Fox Coloring Group of this 3-braid, our arcs satisfy the equation $(a'-a)-(b'-b)+(c'-c)=0$.*
(2) *Recall that a, b and c form a basis of $Col(B)$ and we can change the basis to b, $b-a$ and $b-c$. Then $a'-a$, $b'-b$, and $c'-c$ are linear combinations of $b-a$ and $b-c$.*

PROOF. The proposition is well known and can be proven by the induction on the number of crossings; see, e.g., [**DJP**]. □

To simplify notation, let $x = b-a$ and $y = b-c$. Let $\hat{B}$ be the closure of B, then $Col(\hat{B}) = \{b, x, y \mid a'-a, c'-c\}$. Notice that by Proposition 2.1, $a'-a$ and $c'-c$ is a linear combination of x and y.[2] Let $a'-a = P_B = P_B(x,y) = P_B^x \cdot x + P_B^y \cdot y$ and $c'-c = Q_B = Q_B(x,y) = Q_B^x \cdot x + Q_B^y \cdot y$.

[2]The fact that x and y generate Col^{red} can be easily proved by linear induction. There is a general fact, see e.g. [**DJP**] that if we consider a general n-tangle, T with boundary points denoted by $x_1, x_2, \dots, x_{2n}$ (variable indexed by endpoints) then the relation $\sum_{i=1}^{2n}(-1)^i x_i = 0$ holds in $Col(T)$. In particular elements $x_i - x_{i+1}$ generate $Col^{red}(T)$ group.

The matrix of relations for $Col^{red}(\hat{B})$ is a 2×2 matrix

$$\begin{bmatrix} P_B^x & P_B^y \\ Q_B^x & Q_B^y \end{bmatrix}.$$

Now we work with $B = (\sigma_1\sigma_2^{-1})^n$ and $\hat{B} = D(W_n)$. Therefore, we use notation $P_B = P_n$ and $Q_B = Q_n$.

The following lemmas are illustrated by Figure 4.

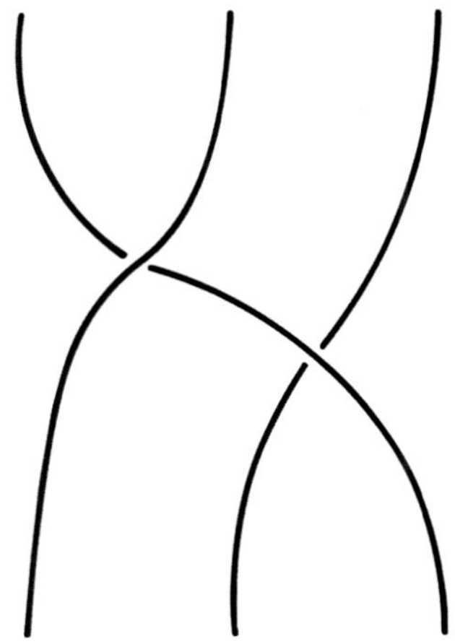

FIGURE 4. Labeling part of $(\sigma_1\sigma_2^{-1})^n$

LEMMA 2.2.

(1) $P_{n+1} = P_n + Q_n + b - a = P_n + Q_n + x$; *thus* $Q_n = P_{n+1} - P_n - x$.

(2) $Q_{n+1} = 2P_{n+1} - P_n + a - c = 2P_{n+1} - P_n + y - x \overset{(1)}{=} P_{n+1} + Q_n + y$; *thus* $P_{n+1} = Q_{n+1} - Q_n - y$.

Thus we can deduce recursive formulas for P_n and Q_n.

LEMMA 2.3.

(Rec1) $P_{n+2} \overset{(1)}{=} P_{n+1} + Q_{n+1} + x \overset{(2)}{=} 3P_{n+1} - P_n + y$.

(Rec2) $Q_{n+2} \overset{(2)}{=} P_{n+2} + Q_{n+1} + y \overset{(1)}{=} P_{n+1} + 2Q_{n+1} + x + y \overset{(2)}{=} 3Q_{n+1} - Q_n + x$.

Then using the notation $P_n = P_n^x \cdot x + P_n^y \cdot y$ and $Q_n = Q_n^x \cdot x + Q_n^y \cdot y$ we recognize P_n^x and Q_n^y as Chebyshev polynomials as follows:

LEMMA 2.4.

(I) $P_{n+2}^x = 3P_{n+1}^x - P_n^x$ *with* $P_0^x = 0$, $P_1^x = 1$, $P_2^x = 3, \ldots$ *Thus* $P_{n+1}^x = S_n(3)$ *where* $S_n(z)$ *denotes the Chebyshev polynomial of the second kind. That is* $S_0(z) = 1, S_1(z) = z$, *and* $S_{n+2}(z) = zS_{n+1}(z) - S_n(z)$. *Compare Section 3.1.* $Q_{n+2}^y = 3Q_{n+1}^y - Q_n^y$ *with* $Q_0^y = 0$, $Q_1^y = 1$, $Q_2^y = 3, \ldots$ *Therefore,* $Q_{n+1}^y = P_{n+1}^x = S_n(3)$.

(II) $P^y_{n+2} = 3P^y_{n+1} - P^y_n + 1$ *with* $P^y_0 = 0,\ P^y_1 = 0,\ P^y_2 = 1, \ldots$ *Similarly,* $Q^x_{n+2} = 3Q^x_{n+1} - Q^x_n + 1$ *with* $Q^x_0 = 0,\ Q^x_1 = 0,\ Q^x_2 = 1, \ldots$ *Therefore we have* $P^y_{n+1} = Q^x_n = P^x_{n+1} - P^x_n - 1 = S_n(3) - S_{n-1}(3) - 1$.

Let F_k be the Fibonacci sequence defined by $F_0 = 0,\ F_1 = 1$ and $F_{k+2} = F_{k+1} + F_k$. It relates to Chebyshev polynomial as follows.

LEMMA 2.5. *We have* $F_{2n} = S_{n-1}(3)$.

PROOF. $F_{2n} = F_{2n-1} + F_{2n-2} = 2F_{2n-2} + F_{2n-3} = 3F_{2n-2} - F_{2n-4}$. Thus F_{2n} and $S_{n-1}(3)$ satisfy the same recursive relation. Noting that $F_2 = 1 = S_0(3)$ and $F_4 = 3 = S_1(3)$ we conclude the lemma. □

The main result of this section can be expressed using Fibonacci number as follows.

THEOREM 2.6. *Let* D_n *be the closure of the braid* $(\sigma_1\sigma_2^{-1})^n$. *Then:*

$$Col^{red}(D_n) = \begin{cases} \mathbb{Z}_{F_{n-1}+F_{n+1}} \oplus \mathbb{Z}_{F_{n-1}+F_{n+1}} & \text{when } n \text{ is odd,} \\ \mathbb{Z}_{5F_n} \oplus \mathbb{Z}_{F_n} & \text{when } n \text{ is even.} \end{cases}$$

The sums $F_{n-1} + F_{n+1}$ are called Lucas numbers denoted by L_n. That is $L_0 = 2, L_1 = 1, L_2 = 3, L_3 = 4, L_4 = 7, dots$ and $L_{n+2} = L_{n+1} + L_n$.

We divide the proof of Theorem 2.6 into three lemmas.

LEMMA 2.7. *The matrix of relations for* $Col^{red}(D_n)$ *is*

$$A_{2n} = \begin{bmatrix} F_{2n} & F_{2n-1} - 1 \\ F_{2n+1} - 1 & F_{2n} \end{bmatrix}.$$

PROOF. From Lemma 2.3, we know

$$Col^{red}(D_n) = \begin{bmatrix} P^x_n & P^y_n \\ Q^x_n & Q^y_n \end{bmatrix} = \begin{bmatrix} P^x_n & P^x_n - P^x_{n-1} - 1 \\ P^x_{n+1} - P^x_n - 1 & P^x_n \end{bmatrix}.$$

Now by Lemma 2.4, $P^x_n - P^x_{n-1} - 1 = S_{n-1}(3) - S_{n-2}(3) - 1 = F_{2n} - F_{2n-2} - 1 = F_{2n-1} - 1$. Therefore $P^x_{n+1} - P^x_n - 1 = F_{2n+2} - F_{2n} - 1 = F_{2n+1} - 1$.

Hence,

$$Col^{red}(D_n) = \begin{bmatrix} F_{2n} & F_{2n-1} - 1 \\ F_{2n+1} - 1 & F_{2n} \end{bmatrix}.$$

We denote this matrix A_{2n}. □

LEMMA 2.8. $Col^{red}(D_n) = \mathbb{Z}_{F_{n-1}+F_{n+1}} \oplus \mathbb{Z}_{F_{n-1}+F_{n+1}}$ *when* n *is odd.*

PROOF.

(1) We replace the first column in A_{2n} by the difference of the first and the second column to get:

$$A_{2n-1} = \begin{bmatrix} F_{2n-2} + 1 & F_{2n-1} - 1 \\ F_{2n-1} - 1 & F_{2n} \end{bmatrix}.$$

(2) We replace the second column in A_{2n-1} by the difference of the second column and the first column to get:

$$A_{2n-2} = \begin{bmatrix} F_{2n-2} + 1 & F_{2n-3} - 2 \\ F_{2n-1} - 1 & F_{2n-2} + 1 \end{bmatrix}.$$

We generalize (1) and (2) to an arbitrary number of column operations (with the same pattern of operations):

After performing k operations of (1) and (2), we get

$$A_{2n-2k} = \begin{bmatrix} F_{2n-2k} + F_{2k} & F_{2n-(2k+1)} - F_{2k+1} \\ F_{2n-(2k-1)} - F_{(2k-1)} & F_{2n-2k} + F_{2k} \end{bmatrix}.$$

Since n is odd, let $k = \frac{n+1}{2}$; then we have:

$$A_{n-1} = \begin{bmatrix} F_{n-1} + F_{n+1} & F_{n-2} - F_{n+2} \\ 0 & F_{n-1} + F_{n+1} \end{bmatrix}.$$

Hence, the gcd of $Col^{red}(D)$ (that is, the gcd of entries of the matix) is $F_{n-1} + F_{n+1}$ and $detA_{n-1} = (F_{n-1} + F_{n+1})^2$. Therefore $Col^{red}(D_n) = \mathbb{Z}_{F_{n-1}+F_{n+1}} \oplus \mathbb{Z}_{F_{n-1}+F_{n+1}}$ for n odd.[3] □

LEMMA 2.9. *$Col^{red}(D_n) = \mathbb{Z}_{5F_n} \oplus \mathbb{Z}_{F_n}$ when n is even.*

PROOF. Let $k = \frac{n}{2}$ and use the proof of Lemma 2.8.

Then we have:

$$A_n = \begin{bmatrix} 2F_n & F_{n-1} - F_{n+1} \\ F_{n+1} - F_{n-1} & 2F_n \end{bmatrix} = \begin{bmatrix} 2F_n & -F_n \\ F_n & 2F_n \end{bmatrix}.$$

Adding twice the last column of the A_n to the first, we get:

$$\begin{bmatrix} 0 & -F_n \\ 5F_n & 2F_n \end{bmatrix}.$$

Here the gcd of $Col^{red}(D)$ is F_n and $|Col^{red}(D)| = 5F_n^2$. Thus we have $Col^{red}(D_n) = \mathbb{Z}_{5F_n} \oplus \mathbb{Z}_{F_n}$ for n even. □

REMARK 2.10. In [**Prz2**] the Goeritz matrix of $D(W_n)$ was reduced to

$$\begin{bmatrix} S_{n-1}(3) & 1 - S_n(3) \\ S_{n-2}(3) + 1 & -S_{n-1}(3) \end{bmatrix}.$$

To show that the group it presents is the same as before, we multiply the last column by -1 and use the equality $S_{n-1}(3) = F_{2n}$, to get:

$$\begin{bmatrix} F_{2n} & F_{2n+2} - 1 \\ F_{2n-2} + 1 & F_{2n} \end{bmatrix} = \begin{bmatrix} F_{2n} & F_{2n+1} + F_{2n} - 1 \\ F_{2n-2} + 1 & F_{2n-1} + F_{2n-2} \end{bmatrix}.$$

After a column operation we get the matrix

$$\begin{bmatrix} F_{2n} & F_{2n+1} - 1 \\ F_{2n-2} + 1 & F_{2n-1} - 1 \end{bmatrix} = \begin{bmatrix} F_{2n-1} + F_{2n-2} & F_{2n} + F_{2n-1} - 1 \\ F_{2n-2} + 1 & F_{2n-1} - 1 \end{bmatrix}.$$

After a row operation we get the matrix

$$\begin{bmatrix} F_{2n-1} - 1 & F_{2n} - 1 \\ F_{2n-2} + 1 & F_{2n-1} - 1 \end{bmatrix},$$

which is the matrix considered in (1) above with a row exchanged. This matrix presents the same abelian group as A_{2n}.

[3] We use the standard fact that for $\mathbb{Z}$ modules given by 2x2 matrix

$$\begin{bmatrix} a & b \\ c & d \end{bmatrix},$$

the related abelian group is equal to $\mathbb{Z}_{d/g} \oplus \mathbb{Z}_g$ where d is the determinant of the matrix and g is $gcd(a, b, c, d)$.

REMARK 2.11. We can deduce several formulas from our column reductions. For example

$$F_{2n} = F_n(F_{n-1} + F_{n+1}).$$

and

$$F_{2n-1} - 1 = \begin{cases} F_n(F_{n-2} + F_n) & \text{for } n \text{ even,} \\ F_{n-1}(F_{n-1} + F_{n+1}) & \text{for } n \text{ odd.} \end{cases}$$

The identities in Corollary 2.11 are well known; e.g. they follow from identities in [**Koshy**] page 97. In particular, $F_{2n} = F_n(F_{n-1} + F_{n+1})$ is exactly identity 54 of [**Koshy**].[4] Thomas Koshy mentions that the identities follow directly from closed forms for F_n and L_n. We present the proof for completeness. We consider roots of the polynomial $x^2 - x - 1 = 0$ say α and β, and note that

$$F_n = \frac{\alpha^n - \beta^n}{\alpha - \beta} \text{ and } L_n = \alpha^n + \beta^n.$$

We have $\alpha = \frac{1+\sqrt{5}}{2}$, $\beta = -\alpha^{-1} = \frac{1-\sqrt{5}}{2}$, $\alpha - \beta = \sqrt{5}$. Our identities can be expressed as product-to-sum formulas (using $F_{-k} = (-1)^{k+1}F_k$):

PROPOSITION 2.12.

$$F_m(F_{n-1} + F_{n+1}) = F_m L_n = F_{m+n} + (-1)^n F_{m-n} = F_{m+n} - (-1)^m F_{n-m}.$$

PROOF.

$$\begin{aligned} F_m L_n &= \frac{(\alpha^m - \beta^m)(\alpha^n + \beta^n)}{\alpha - \beta} \\ &= \frac{\alpha^{m+n} - \beta^{m+n} + \alpha^m\beta^n - \alpha^n\beta^n}{\alpha - \beta} \\ &= \frac{\alpha^{m+n} - \beta^{m+n} + (-1)^n(\alpha^{m-n} - \beta^{m-n})}{\alpha - \beta} \\ &= F_{m+n} + (-1)^n F_{m-n}. \end{aligned}$$

□

3. Alexander–Burau–Fox matrix of a link diagram

In this section we generalize our main result of the second section from the Fox group of colorings to the Alexander–Burau–Fox module over the ring $\mathbb{Z}[t^{\pm 1}]$. Where the Fibonacci numbers appeared in the computation of the the Fox Coloring Group, the analogous combinatorial object appearing in the computation of the ABF Module is the Chebyshev Polynomial of the second kind. The modules we study are Alexander modules of links [**Ale**] in a form which can be deduced from the Burau representation of braids [**Burau, Bur-Z**]. These module directly generalize to the group of Fox colorings where we replace -1 by t.

The formal definition of the ABF Module follows (notice that for $t = -1$, we get Definition 1.4).

[4] There are also the following related identities:

$$\text{Identity 52: } F_{2m+n} - (-1)^m F_n = F_m L_{m+n},$$

$$\text{Identity 53: } F_{2m+n} + (-1)^m F_n = F_{m+n} L_m,$$

DEFINITION 3.1. Let D be an oriented link diagram. The $\mathbb{Z}[t^{\pm 1}]$-module $ABF(D)$ has a presentation where generators are indexed by the arcs of D, denoted by $Arcs(D)$, and whose relations are given by crossings of D (positive and negative) as follows

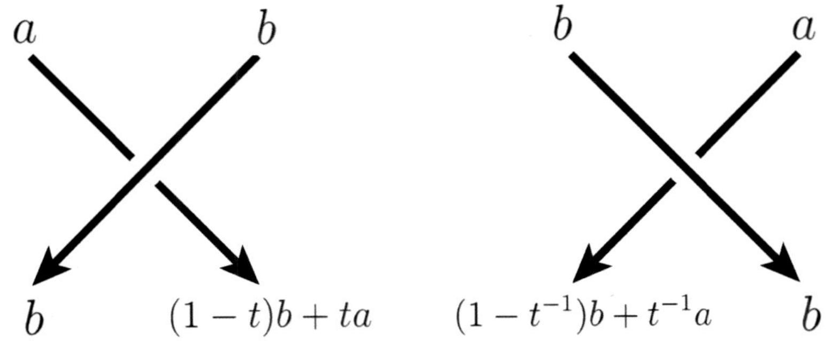

FIGURE 5. ABF Module Relations

$ABF_n(D) = \{Arcs(D) \mid$ 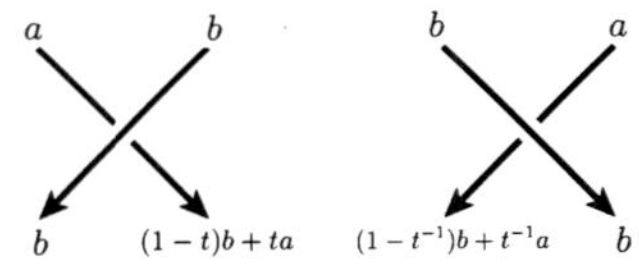 $\text{where } a, b, c \in Arcs(D)\}$

That is,

DEFINITION 3.2. The Reduced Alexander–Burau–Fox Module is the module

$ABF_n^{Red}(D) = \{Arcs(D) \mid$ 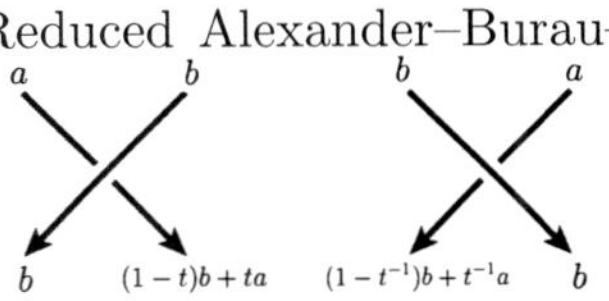 $, \sum_{a_i \in arcs(D)} a_i = 0\}$,

where the sum is taken over all arcs of D. Note that this module eliminates trivial colorings, which is equivalent to setting one arc, say a_s equal to zero (removing one generator and adjusting relations by substituting $a_s = -\sum_{a_i, i \neq s} a_i$).[5]

We now formulate the main result of this section, showing the structure of the ABF module of the diagrams of wheel graphs. We simplify our notation by calling $ABF(D(W_n)) = M_n(D_n)$. Notice that the module decomposes as the sum of cyclic modules (even though $\mathbb{Z}[t^{\pm 1}]$ is not a PID), and that for odd n the module is double and for even n it is "almost" double.

THEOREM 3.3. *The reduced Alexander–Burau–Fox module of diagrams of wheel graphs W_n can be expressed as*

$$M_n^{Red}(D(W_n)) = \begin{cases} \mathbb{Z}[t^{\pm 1}]/(S_{k-1} + S_k) \oplus \mathbb{Z}[t^{\pm 1}]/(S_{k-1} + S_k), & \text{when } n = 2k+1 \\ \mathbb{Z}[t^{\pm 1}]/(S_{k-1}) \oplus \mathbb{Z}[t^{\pm 1}]/((3 - t - t^{-1})S_{k-1}) & \text{when } n = 2k \end{cases}$$

where $z = 1 - t - t^{-1}$ and $S_k(z)$ is the Chebyshev polynomial of the second kind (see Subsection 3.1).

[5]To be precise: the relation $\sum_{a_i \in arcs(D)} a_i = 0$ allows us to replace any other relation so that the coefficient of a_s is equal to zero. In this new presentation a_s can be removed and the presentation has generators $Arcs(D) - a_s$, and in relations we can put $a_s = 0$.

As before, we label the inputs of our braids a, b, c and the outputs a', b', c'; see Figure 6.

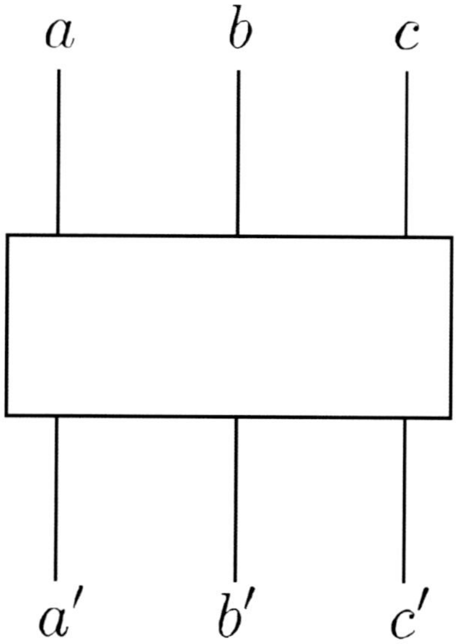

FIGURE 6. Coloring of the tangle

As a special case of the general observation for n-tangles, we have that $t^2(a' - a) + t(b' - b) + (c' - c) = 0$. Thus $b' - b = -t(a' - a) - t^{-1}(c' - c)$, meaning that the middle color can be computed from the first and the third.

Now like in the $t = -1$ case, let the endpoints of $(\sigma_1\sigma_2^{-1})^n$ be $a' = P_n + a$ on the left and $c' = Q_n + c$ on the right. Then $b' - b = -tP_n - t^{-1}Q_n$. As mentioned before, comoputing the Reduced ABF module is equivalent to setting $b = 0$. Thus, we have that $b' = -tP_n - t^{-1}Q_n$.

Writing $P_n = P_n^a \cdot a + P_n^c \cdot c$ and $Q_n = Q_n^a \cdot a + Q_n^c \cdot c$ as before, we see that the matrix of relations for the Reduced Alexander–Burau–Fox module, denoted A_n, is analogous to that of the Reduced Fox Coloring Group:

$$A_n = \begin{bmatrix} P_n^a & P_n^c \\ Q_n^a & Q_n^c \end{bmatrix}.$$

We also produce the following recursive relations, analogous to those found in Section 2. That is,

LEMMA 3.4.

(1) $P_{n+1} = -tP_n - t^{-1}Q_n - a$,
(2) $Q_{n+1} = (1 - t)P_{n+1} + tP_n + a - c$.

PROOF. Considering we work with the Reduced ABF Module, we set $b = 0$. Let b'_n denote the central strand after performing $(\sigma_1\sigma_2^{-1})$ n times. By previous observation, $b'_n = -tP_n - t^{-1}Q_n$. Then we have $P_{n+1} + a = b'_n$, so $P_{n+1} + a = b' = -tP_n - t^{-1}Q_n$. This proves (1) as is shown in Figure 7.

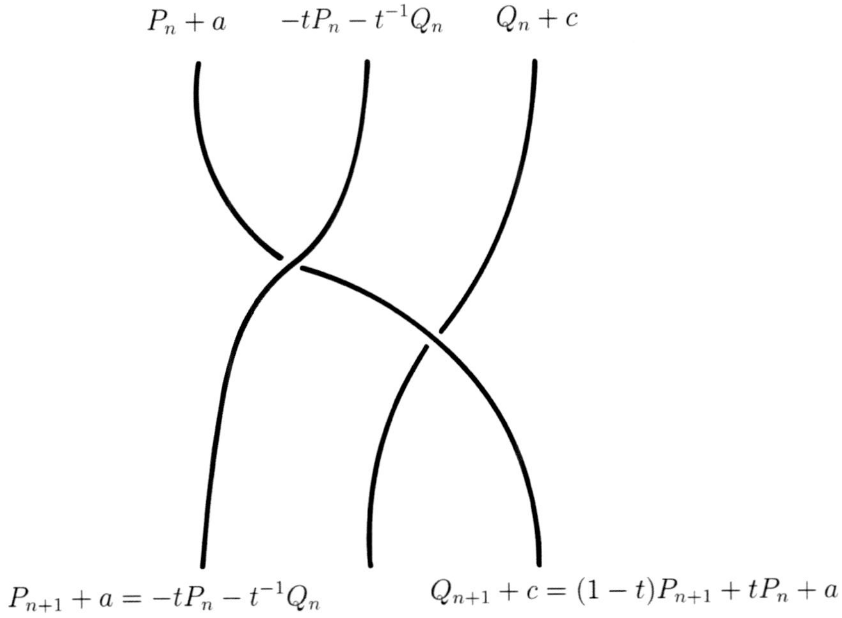

FIGURE 7. Labeling part of $(\sigma_1\sigma_2^{-1})^n$ in ABF Module

Next, we see that

$$\begin{aligned} Q_{n+1} + c &= (1+t)(-tP_n - t^{-1}Q_n) + t(P_n + a) \\ &= -tP_n - t^{-1}Q_n - t^2P_n - Q_n + tP_n + a \\ &= P_{n+1} - t(tP_n - t^{-1}Q_n) + tP_n + a \\ &= P_{n+1} - tP_{n+1} + tP_n + a \\ \Longrightarrow\ Q_{n+1} &= (1-t)P_{n+1} + tP_{n+a}. \end{aligned}$$

Thus, (2) is true. ☐

Immediately from the Lemma 3.8, we can present the matrix of relations A_n in terms of P_n^a and P_n^c by the following lemma.

LEMMA 3.5. *For Q_n^a and Q_n^c as defined above,*

(1) $Q_n^a = (1-t)P_n^a + tP_{n-1}^a + 1$
(2) $Q_n^c = (1-t)P_n^c + tP_{n-1}^c - 1$

PROOF. By Lemma 3.4(2),

$$\begin{aligned} Q_n &= Q_n^a \cdot a + Q_n^c \cdot c \\ &= (1-t)P_{n+1} + tP_n + a - c \\ &= (1-t)P_n^a \cdot a + tP_{n-1}^a \cdot a + a + (1-t)P_n^c \cdot c + tP_{n-1}^c \cdot c - c \\ &= ((1-t)P_n^a + tP_{n-1}^a + 1) \cdot a + ((1-t)P_n^c + tP_{n-1}^c - 1) \cdot c. \end{aligned}$$ ☐

Thus, our Alexander–Burau–Fox matrix is

$$A_n = \begin{bmatrix} P_n^a & P_n^c \\ Q_n^a & Q_n^c \end{bmatrix} = \begin{bmatrix} P_n^a & P_n^c \\ (1-t)P_n^a + tP_{n-1}^a + 1 & (1-t)P_n^c + tP_{n-1}^c - 1 \end{bmatrix}.$$

where $z = 1 - t - t^{-1}$.

The final identity we prove in service of our eventual translation of this matrix into the language of Chebyshev polynomials is the following:

LEMMA 3.6. *For P_n^a and P_n^c we have the following Chebyshev-type recursive relation:*

(1) $P_n^a = zP_{n-1}^a - P_{n-2}^a - (1 + t^{-1})$
(2) $P_n^c = zP_{n-1} - P_{n-2}^c + t^{-1}$

PROOF. Beginning with Lemma 3.8 (1),

$$\begin{aligned} P_n &= P_n^a \cdot a + P_n^c \cdot c \\ &= -tP_{n-1} - t^{-1}Q_{n-1} - a \\ &= -tP_{n-1} - t^{-1}((1-t)P_{n-1} + tP_{n-2} + a - c) - a \\ &= -tP_{n-1} - t^{-1}P_{n-1} + P_{n-1} - P_{n-2} - t^{-1}a + t^{-1}c - a \\ &= (1 - t - t^{-1})P_{n-1} - P_{n-2} - t^{-1}a + t^{-1}c - a \\ &= (zP_{n-1}^a - P_{n-2}^a - t^{-1} - 1) \cdot a - (zP_{n-1}^c - P_{n-2}^c + t^{-1}) \cdot c \end{aligned}$$

□

3.1. Chebyshev Polynomials. We now set about rewriting P_n^a and P_n^c in terms of Chebyshev polynomials using Theorem 3.8, below.

We recall first the definition of Chebyshev Polynomials of the Second Kind.

DEFINITION 3.7. Let

$$\begin{aligned} S_0(z) &= 1, \\ S_1(z) &= z, \text{ and} \\ S_n(z) &= zS_{n-1}(z) - S_{n-2}(z). \end{aligned}$$

Then $S_n(z)$ is the nth polynomial of the Second Kind. The definition of S_n can be extended to negative n by the same recursive relation. For example, $S_{-1} = 0$.

Throughout the paper, we evaluate $S_n(z)$ at $z = 1 - t - t^{-1}$, though we tend to suppress the notation: $S_n(z) \to S_n$.

THEOREM 3.8. *Let the sequence of polynomials $P_n(z, c_i) \in \mathbb{Z}[z, c_i]$ satisfy the recursive relation $P_n = zP_{n-1} - P_{n-2} + c_i$. Then P_n can be expressed using Chebyshev polynomials $S_n(z)$ as follows:*

$$P_n = S_{n-1}P_1 - S_{n-2}P_0 + \sum_{j=0}^{n-2} S_j c_{n-j}.$$

PROOF. We proceed by induction on n. Let $n = 3$ be our base case. Then

$$\begin{aligned}
P_3 &= zP_2 - P_1 + c_3 \\
&= z(zP_1 - P_0 + c_2) - P_1 + c_3 \\
&= z^2 P_1 - zP_0 + zc_2 - P_1 + c_3 \\
&= (z^2 - 1)P_1 - zP_0 + zc_2 + c_3 \\
&= S_2 P_1 - S_1 P_0 + S_0 c_3 + S_1 c_2 \\
&= S_2 P_1 - S_1 P_0 + \sum_{j=0}^{1} S_j c_{n-j}.
\end{aligned}$$

Assume that the proposition holds for $n \geq 3$ up to $n - 1$.

By inductive hypothesis,

$$\begin{aligned}
P_n &= zP_{n-1} - P_{n-2} + c_n \\
&= z(S_{n-2}P_1 - S_{n-3}P_0 + \sum_{j=1}^{n-3} S_j c_{n-1-j}) \\
&\quad - (S_{n-3}P_1 - S_{n-4}P_0 + \sum_{j=0}^{n-4} S_j c_{n-2-j}) + c_n \\
&= (zS_{n-2} - S_{n-3})P_1 - (zS_{n-3} - S_{n-4})P_0 \\
&\quad + z\sum_{j=0}^{n-3} S_j c_{n-1-j} - \sum_{j=0}^{n-4} S_j c_{n-2-j} + c_n \\
&= S_{n-1}P - S_{n-2}P_0 + \sum_{j=0}^{n-3} S_j c_{n-1-j} - \sum_{j=0}^{n-4} S_j c_{n-2-j} + c_n \\
&= S_{n-1}P - S_{n-2}P_0 + zS_0 c_{n-1} + \sum_{j=2}^{n-2} S_j c_{n-j} + c_n \\
&= S_{n-1}P - S_{n-2}P_0 + S_0 c_n + S_1 c_{n-1} + \sum_{j=2}^{n-2} S_j c_{n-j} \\
&= S_{n-1}P - S_{n-2}P_0 + \sum_{j=0}^{n-2} S_j c_{n-j}.
\end{aligned}$$

Thus, our identity holds. □

Thus we can rewrite the recursive relations in Lemma 3.6 with the Corollary 3.9.

COROLLARY 3.9.

(1) $P_n^a(t) = -S_{n-1} - (1 + t^{-1}) \sum_{j=0}^{n-2} S_j(z)$.

(2) $P_n^c(t) = t^{-1} \sum_{j=0}^{n-2} S_j(z)$ *for* $z = 1 - t - t^{-1}$.

PROOF.

(1) By Lemma 3.6(1), $P_n^a = zP_{n-1}^a - P_{n-2}^a - (1+t^{-1})$, so applying Theorem 3.8 with $c = -1 - t^{-1}$ yields the result with initial conditions $P_0^a = 0$ and $P_1^a = -1$.

(2) By 3.6(2), $P^c_n = zP^c_{n-1} - P^c_{n-2} + t^{-1}$ with initial conditions $P^c_0 = 0$ and $P^c_1 = t^{-1}$. □

In the next section, we simplify the equation in Corollary 3.9 using product-to-sum formulas (in fact, we go from sum to product).

3.2. Product-to-sum formulas. Seeking further transformation of A_n, we list the following identities for Chebyshev polynomials in Proposition 3.10, which are analogous to those given for the Fibonacci numbers. They can be found in, for example [**Riv**], [**MH**] or [**BIMPW**], but be aware that [**Riv, MH**] use slightly different notation from ours, which follows the convention introduced to knot theory by R. Lickorish [**Lic**]. Thus let S_n be the Chebyshev polynomial of the second kind (see Definition 3.7), and T_n the Chebyshev polynomial of the first kind. That is, $T_0 = 1, T_1 = x$ and $T_n = xT_{n-1} - T_n$. To obtain closed forms, substite $x = p + p^{-1}$, which yields

$$S_n = \frac{p^{n+1} - p^{-n-1}}{p - p^{-1}}$$

and

$$T_n = p^n + p^{-n}.$$

These closed forms are helpful in the proofs of the five identities of Proposition 3.10, which are sketched for the convenience of the reader.

PROPOSITION 3.10. *For S_n and T_n as defined above,*

(1) $T_m T_n = T_{m+n} + T_{m-n}$.[6]
(2) $S_m T_n = S_{m+n} + S_{m-n}$.
(3) $S_m S_n = S_{m+n} + S_{m+n-2} + dots + S_{m-n+2} + S_{m-n} = \sum_{i=m-n, i\equiv m+n \mod 2}^{m+n} S_i$ *where the sum is taken over i with $i \equiv m+n \mod 2$.*
(4) $S_n(S_n + S_{n-1}) = S_0 + S_1 + dots + S_{2n-1} + S_{2n} = \sum_{i=0}^{2n} S_i$.
(5) $S_n(S_n + S_{n+1}) = S_0 + S_1 + dots + S_{2n-1} + S_{2n} + S_{2n+1} = \sum_{i=0}^{2n+1} S_i$.

PROOF.

(1) $T_m T_n = (p^m + p^{-m})(p^n + p^{-n}) = p^{m+n} + p^{-m-n} + p^{m-n} + p^{n-m} = T_{m+n} + T_{m-n}$.

(2) $S_m T_n = \frac{(p^{m+1} - p^{-m-1})(p^n + p^{-n})}{p - p^{-1}} = \frac{(p^{m+n+1} - p^{-m-n-1}) + (p^{m-n+1} - p^{-m+n-1})}{p - p^{-1}} = S_{m+n} + S_{m-n}$.

(3) We proceed by induction on n, applying (2) repeatedly:
If $n = 0$ we have $S_m S_0 = S_m$ and for $m = 1$, $S_m S_1 = xS_m = S_{m+1} + S_{m-1}$. The inductive step holds as follows: (assuming $m \geq n \geq 2$): $S_m S_n = S_m(S_n - S_{n-2}) + S_m S_{n-2} = S_m T_n + S_m S_{n-2} \overset{(2)}{=} S_{m+n} + S_{m-n} + S_m S_{n-2} \overset{ind}{=} S_{m+n} + S_{m-n} + S_{m+n-2} + S_{m+n-4} + dots + S_{m-n+2}$, as needed.

(4) We use (3) twice for $S_n S_n$ and for $S_n S_{n-1}$.

(5) We use (3) twice for $S_n S_n$ and for $S_n S_{n+1}$. □

The following two corollaries further simplify A_n.

[6]Some authors assume $m \geq n$ or take $|m - n|$ in the formula. It should be noted that this is not necessary as the formula works also for negative numbers.

COROLLARY 3.11.

$$\sum_{j=0}^{n-2} S_j = \begin{cases} S_k(S_k + S_{k-1}) & \textit{when } n = 2k, \\ S_{k-1}(S_{k-1} + S_k) & \textit{when } n = 2k+1. \end{cases}$$

PROOF. This is a restatement of (4) and (5) from Proposition 3.10. □

COROLLARY 3.12.

(1) $S_{2k} = S_k S_k - S_{k-1} S_{k-1} = (S_k - S_{k-1})(S_k + S_{k-1})$
(2) $S_{2k+1} = S_k S_{k+1} - S_{k-1} S_k = S_k(S_{k+1} - S_{k-1})$.

PROOF.

(1) By 3.10(3),

$$\begin{aligned} S_k S_k &= S_0 + S_2 + dots + S_{2k} \\ &= (S_0 + S_2 + \cdots + S_{2k-2}) + S_{2k} \\ &= S_{k-1} S_{k-1} + S_{2k} \end{aligned}$$

So $S_k S_k = S_{k-1} S_{k-1} + S_{2k}$ implies $S_{2k} = S_k S_k - S_{k-1} S_{k-1}$.

(2)

$$S_k S_{k+1} = S_1 + S_3 + dots + S_{2k+1},$$

□

This allows us to reformulate P_n^a and P_n^c.

LEMMA 3.13.

(1) $P_{2k}^a = -S_{k-1}((S_k + S_{k-1}) + t^{-1}(S_{k-1} + S_{k-2}))$
(2) $P_{2k+1}^a = -(S_{k-1} + S_k)(S_k + t^{-1} S_{k-1})$
(3) $P_{2k}^c = t^{-1} S_{k-1}(S_{k-1} + S_{k-2})$
(4) $P_{2k+1}^c = t^{-1} S_{k-1}(S_{k-1} + S_k)$

PROOF.

(1) By Corollary 3.9, $P_{2k}^a(t) = -S_{2k-1}(z) - (1 + t^{-1}) \sum_{j=0}^{2k-2} S_j(z)$, where $z = 1 - t - t^{-1}$, as before. Thus,

$$\begin{aligned} P_{2k}^a &= -S_{2k-1} - (1 + t^{-1}) \sum_{j=0}^{2(k-1)} S_j(z) \\ &= -S_{2k-1} - (1 + t^{-1}) S_{k-1}(S_{k-1} + S_{k-2}) && \text{Cor. 3.11} \\ &= -S_{2(k-1)+1} - (1 + t^{-1}) S_{k-1}(S_{k-1} + S_{k-2}) \\ &= -S_{k-1}(S_k - S_{k-2}) - (1 + t^{-1}) S_{k-1}(S_{k-1} + S_{k-2}) && \text{Cor. 3.12} \\ &= -S_{k-1}(S_k - S_{k-2} + S_{k-1} + S_{k-2} + t^{-1} S_{k-1} + t^{-1} S_{k-2}) \\ &= -S_{k-1}((S_k + S_{k-1}) - t^{-1}(S_{k-1} + S_{k-2})) \end{aligned}$$

(2) By Corollary 3.9, $P^a_{2k}(t)$

$$\begin{aligned}
-P^a_{2k+1} &= -S_{2k+1-1} - (1+t^{-1}) \sum_{j=0}^{2k+1-2} S_j(z) \\
&= -S_{2k} - (1+t^{-1})S_{k-1}(S_{k-1}+S_{k+1}) && \text{Cor. 3.11} \\
&= -(S_k - S_{k-1})(S_k + S_{k-1}) - (1+t^{-1})S_{k-1}(S_{k-1}+S_k) && \text{Cor. 3.12} \\
&= -(S_k + S_{k-1})(S_k - S_{k-1} + S_{k-1} + t^{-1}S_{k-1}) \\
&= -(S_k + S_{k-1})(S_k + t^{-1}S_{k-1})
\end{aligned}$$

(3) & (4) These follow immediately from Cor 3.11.

□

Now, we present the Alexander–Burau–Fox matrix for closures of the braids of type $(\sigma_1\sigma_2^{-1})^n$ in full generality.

PROPOSITION 3.14.

$$A_n = \begin{bmatrix} P^a_n & P^c_n \\ Q^a_n & Q^c_n \end{bmatrix} = \begin{bmatrix} -g_n(g_{n+1} - t^{-1}g_{n-1}) & t^{-1}g_n g_{n-1} \\ t g_n g_{n+1} & -g_n(g_{n+1} - t g_{n-1}) \end{bmatrix}$$

where

$$g_n = \begin{cases} S_{k-1} & \text{when } n = 2k, \\ S_{k-1} + S_k & \text{when } n = 2k+1. \end{cases}$$

PROOF. This follows immediately from Lemma 3.13 and Lemma 3.5. □

3.3. Alexander–Burau–Fox module. With the explicit presentation of the matrix A_n in terms of Chebyshev polynomials from 3.14, we compute the resulting module by row reduction.

Define

$$A'_n = A_n/(-g_n) = \begin{bmatrix} g_{n+1} - t^{-1}g_{n-1} & -t^{-1}g_{n-1} \\ -t g_{n+1} & g_{n+1} - t g_{n-1} \end{bmatrix}.$$

One column operation, $\mathrm{col}_1 - \mathrm{col}_2 \to \mathrm{col}_1$, yields

$$A'_n = \begin{bmatrix} g_{n+1} & -t^{-1}g_{n-1} \\ (-t-1)g_{n+1} + t g_{n-1} & g_{n+1} - t g_{n-1} \end{bmatrix}.$$

We multiply the second column by $-t$:

$$A'_n = \begin{bmatrix} g_{n+1} & g_{n-1} \\ (-t-1)g_{n+1} + t g_{n-1} & -t g_{n+1} + t^2 g_{n-1} \end{bmatrix}.$$

Specifically for $n = 2k+1$, we get

$$A'_{2k+1} = \begin{bmatrix} S_k & S_{k-1} \\ -tS_k & -tS_k + t^2 S_{k-1} \end{bmatrix}.$$

For $n = 2k$, we get

$$A'_{2k} = \begin{bmatrix} S_{k-1} + S_k & S_{k-1} + S_{k-2} \\ -t(S_{k-1} + S_k) & -t(S_{k-1} + S_k) - t^2(S_{k-1} + S_{k-2}) \end{bmatrix}.$$

We will apply the following lemma to A''_n in order to immediately compute the desired module.

LEMMA 3.15.

(1) *The pair* (S_k, S_{k-1}) *can be reduced by Euclidean algorithm (column operations) to* $(S_0, S_{-1}) = (1, 0)$.
(2) *The pair* $(S_k + S_{k-1}, S_{k-1} + S_{k-2})$ *can be reduced by Euclidean algorithm (column operations) to* $(S_1 + S_0, S_0 + S_{-1}) = (z + 1, 1)$.

PROOF.

(1) We notice that $(S_k, S_{k-1}) = (zS_{k-1} - S_{k-2}, S_{k-1})$ thus by taking the first column minus z times the second we get $(-S_{k-2}, S_{k-1})$. Multiplying the first column by -1 we get (S_{k-2}, S_{k-1}) allowing the inductive step.
(2) We notice that $(S_k + S_{k-1}, S_{k-1} + S_{k-2}) = (zS_{k-1} - S_{k-2} + zS_{k-2} - S_{k-3}, S_{k-1}+S_{k-2}) = (z(S_{k-1}+S_{k-2})-(S_{k-2}+S_{k-3}), S_{k-1}+S_{k-2})$, thus by taking the first column minus z times the second we get $(-(S_{k-2} + S_{k-3}), S_{k-1}+S_{k-2})$ and multiplying the first column by -1 we get $(S_{k-2}+ S_{k-3}, S_{k-1} + S_{k-2})$, allowing the inductive step. □

We apply Lemma 3.15 to A'_n. After applying this lemma we get the matrix

$$\begin{bmatrix} 1 & 0 \\ x & y \end{bmatrix} \text{ reduced by row operation to } \begin{bmatrix} 1 & 0 \\ 0 & y \end{bmatrix}$$

where $y = detA'_n$.
Thus we can conclude that our Alexander–Burau–Fox module given by A_n is:

$$\mathbb{Z}[t^{\pm 1}]/(g_n) \oplus \mathbb{Z}[t^{\pm 1}]/(detA'_n \cdot g_n).$$

The $\det(A'_n)$ is computed in the next section.

3.4. Computing determinant of the matrix A'_n. To complete the computation of the Alexander–Burau–Fox module of $(\sigma_1\sigma_2^{-1})^n$, we need the following result.

PROPOSITION 3.16. *For* A'

(1) $detA'_{2k+1} = 1$.
(2) $detA'_{2k} = 3 - t - t^{-1}$.

PROOF.

(1) We have

$$\begin{aligned} detA'_{2k+1} &= (S_k + t^{-1}S_{k-1})(S_k + tS_{k-1}) - S_kS_{k-1} \\ &= S_k^2 + (t + t^{-1})S_kS_{k-1} + S_{k-1}^2 - S_kS_{k-1} \\ &= S_k^2 + (t + t^{-1} - 1)S_kS_{k-1} + S_{k-1}^2 = S_k^2 - zS_kS_{k-1} + S_{k-1}^2 \\ &= S_k^2 - (S_{k+1} + S_{k-1})S_{k-1} + S_{k-1}^2 \\ &= S_k^2 - S_{k-1}S_{k+1} \\ &= 1 \end{aligned}$$

by product-to-sum formulas.

(2) We proceed as in the previous case. We have

$$\begin{aligned}
detA'_{2k} &= \Big((S_k + S_{k-1}) + t^{-1}(S_{k-1} + S_{k-2})\Big)\Big((S_k + S_{k-1})) + t(S_{k-1} + S_{k-2})\Big) \\
&\quad - (S_k + S_{k-1})(S_{k-1} + S_{k-2}) \\
&= (S_k + S_{k-1})^2 + (t + t^{-1})(S_k + S_{k-1})(S_{k-1} + S_{k-2}) \\
&\quad + (S_{k-1} + S_{k-2})^2 - (S_k + S_{k-1})(S_{k-1} + S_{k-2}) \\
&= (S_k + S_{k-1})^2 + (t + t^{-1} - 1)(S_k + S_{k-2})(S_{k-1} + S_{k-2}) \\
&\quad + (S_{k-1} + S_{k-2})^2 \\
&= (S_k + S_{k-1})^2 - z(S_k + S_{k-1})(S_{k-1} + S_{k-2}) + (S_{k-1} + S_{k-2})^2 \\
&= (S_k + S_{k-1})^2 - ((S_{k+1} + S_{k-1}) \\
&\quad + (S_k + S_{k-2}))(S_{k-1} + S_{k-2}) + (S_{k-1} + S_{k-2})^2 \\
&= (S_k + S_{k-1})^2 - (S_{k+1} + S_k)(S_{k-1} + S_{k-2}) \\
&= (S_k S_{k-1} - S_{k+1} S_{k-3}) + (S_k S_k - S_{k+1} S_{k-1}) + (S_{k-1} S_{k-1} - S_k S_{k-2}) \\
&= S_1 + S_0 + S_0 \\
&= z + 2 \\
&= 1 - t - t^{-1} + 2 \\
&= 3 - t - t^{-1}.
\end{aligned}$$

□

Thus, with the computation of $\det(A'_n)$, we complete the proof of the main theorem, Theorem 3.3.

4. Relation to Burau representation

The crossing illustrated in Figure 5 can be interpreted as a 2-braid, and the ABF relation as a linear map: for a positive crossing, $B(a,b) = (b, ta + (1-t)b)$. This map is given in a standard basis $(1,0), (0,1)$ by the matrix:

$$B = \begin{bmatrix} 0 & 1 \\ t & 1-t \end{bmatrix}$$

which is an element of $Gl(2, Z[t^{\pm 1}])$. More generally, for an n-braid we have a homomorphism $B_n \to Gl(n, Z[t^{\pm 1}])$ given on the generators by

$$B(\sigma_i) = \begin{bmatrix} Id_{i-1} & 0 & 0 & 0 \\ 0 & 0 & 1 & 0 \\ 0 & t & 1-t & 0 \\ 0 & 0 & 0 & Id_{n-i-1} \end{bmatrix}$$

This representation is known as the (unreduced) Burau representation [**Burau**]. Because for a negative crossing we have $B(b,a) = ((1-t^{-1})b, t^{-1}a)$, therefore

$$B(\sigma_i^{-1}) = \begin{bmatrix} Id_{i-1} & 0 & 0 & 0 \\ 0 & 1-t^{-1} & t^{-1} & 0 \\ 0 & 1 & 0 & 0 \\ 0 & 0 & 0 & Id_{n-i-1} \end{bmatrix}.$$

Clearly $B(\sigma_i)B(\sigma_i^{-1}) = Id_n$, reflecting the second Reidemeister move. In older literature the role of σ_i and σ_i^{-1} (and t and t^{-1}) is changed (historically negative and positive was not always the same, and until now braid theorists use the opposite convention from knot theorists).

The matrix $B(\gamma) - Id_n$ where $\gamma \in B_n$ gives (with relations in rows) description of the unreduced Alexander–Burau–Fox module. This is, in fact, the matrix used in our calculation in Section 3.

5. Speculations and future directions

To place our result in a broader context, we should mention the classical result of J.Minkus and Mulazzani-Vesnin [**Min, MV**] relating n-fold branch coverings of 2-bridge links with double branch coverings of certain links. In relation to our family of braids, they notice that the n-fold branch covering of S^3 along the figure eight knot is homeomorphic to the double branch covering of S^3 branched along the closure of the braid $(\sigma_1\sigma_2^{-1})^n$. Thus our theorem on the group of Fox colorings can be reformulated in the language of homology of $M_{4_1}^{(n)}$, where $M_L^{(n)}$ denotes the n-fold branch coverings of S^3 branched along the link L. We plan to explore this connection to analyze left orderings of the fundamental groups of some branched coverings; compare [**DaPr**]. We recall the results of [**BGW**], which show that the 2-fold branch covering of S^3 along a non-split alternating link produces a fundamental group which is not left-orderable.

Plans proved in [**Pla**] that an odd branch covering of S^3 along any knot has a double form. That is, there exists G such that

$$H_1(M_k^{(n)}) = G \oplus G \quad \text{for any odd } k.$$

Turaev produced a related result for k even (compare [**Web, DW, Prz4**]). We observe a similar phenomenon in our computation of ABF for $(\sigma_1\sigma_2^{-1})^n$.

Our theorem for the Reduced Fox Coloring Group corresponds well with Plans' and Turaev's results, but what does this mean in the generalization to the ABF module?

Is there some reasonable interpretation of $(\sigma_1\sigma_2^{-1})^\infty$ in relation to ABF module?

References

[A-V] J. W. Alexander, Letter to Oswald Veblen, 1919, Papers of Oswald Veblen, 1881-1960 (bulk 1920-1960), Archival Manuscript Material (Collection), Library of Congress.

[Ale] J. W. Alexander, *Topological invariants of knots and links*, Trans. Amer. Math. Soc. **30** (1928), no. 2, 275–306, DOI 10.2307/1989123. MR1501429

[BIMPW] Józef H. Przytycki, Rhea Palak Bakshi, Dionne Ibarra, Gabriel Montoya-Vega, and Deborah Weeks, *Lectures in knot theory—an exploration of contemporary topics*, Universitext, Springer, Cham, [2024] ©2024, DOI 10.1007/978-3-031-40044-5. MR4738299

[BGMMP] Rhea Palak Bakshi, Huizheng Guo, Gabriel Montoya-Vega, Sujoy Mukherjee, and Józef H. Przytycki, *The generalized Kauffman-Harary conjecture is true*, Algebr. Geom. Topol. **25** (2025), no. 4, 2067–2081, DOI 10.2140/agt.2025.25.2067. MR4950897

[BGW] Steven Boyer, Cameron McA. Gordon, and Liam Watson, *On L-spaces and left-orderable fundamental groups*, Math. Ann. **356** (2013), no. 4, 1213–1245, DOI 10.1007/s00208-012-0852-7. MR3072799

[Burau] Werner Burau, *Über Zopfgruppen und gleichsinnig verdrillte Verkettungen* (German), Abh. Math. Sem. Univ. Hamburg **11** (1935), no. 1, 179–186, DOI 10.1007/BF02940722. MR3069652

[Bur-Z] Gerhard Burde and Heiner Zieschang, *Knots*, 2nd ed., De Gruyter Studies in Mathematics, vol. 5, Walter de Gruyter & Co., Berlin, 2003. MR1959408

[CF] Richard H. Crowell and Ralph H. Fox, *Introduction to knot theory*, Graduate Texts in Mathematics, No. 57, Springer-Verlag, New York-Heidelberg, 1977. Reprint of the 1963 original. MR445489

[DaPr] Mieczysław K. Dąbkowski, Józef H. Przytycki, and Amir A. Togha, *Non-left-orderable 3-manifold groups*, Canad. Math. Bull. **48** (2005), no. 1, 32–40, DOI 10.4153/CMB-2005-003-6. MR2118761

[DPS] A. Di Prisa and O. Savk, *Turk's head knots and links: A survey*, Preprint, `arXiv:2409.20106 [math.GT]`, 2024.

[DW] Pablo Del Val and Claude Weber, *Plans' theorem for links*, Topology Appl. **34** (1990), no. 3, 247–255, DOI 10.1016/0166-8641(90)90041-Y. MR1042282

[DJP] Jan Dymara, Tadeusz Januszkiewicz, and Józef H. Przytycki, *Symplectic structure on colorings, Lagrangian tangles and Tits buildings*, Bull. Pol. Acad. Sci. Math. **68** (2020), no. 2, 169–194, DOI 10.4064/ba201230-2-3. MR4246927

[MH] J. C. Mason and D. C. Handscomb, *Chebyshev polynomials*, Chapman & Hall/CRC, Boca Raton, FL, 2003. MR1937591

[Koshy] Thomas Koshy, *Fibonacci and Lucas numbers with applications*, Pure and Applied Mathematics (New York), Wiley-Interscience, New York, 2001, DOI 10.1002/9781118033067. MR1855020

[Lic] W. B. Raymond Lickorish, *An introduction to knot theory*, Graduate Texts in Mathematics, vol. 175, Springer-Verlag, New York, 1997, DOI 10.1007/978-1-4612-0691-0. MR1472978

[Min] Jerome Minkus, *The branched cyclic coverings of* 2 *bridge knots and links*, Mem. Amer. Math. Soc. **35** (1982), no. 255, iv+68, DOI 10.1090/memo/0255. MR643587

[MV] Michele Mulazzani and Andrei Vesnin, *The many faces of cyclic branched coverings of 2-bridge knots and links*, Atti Sem. Mat. Fis. Univ. Modena **49** (2001), no. suppl., 177–215. Dedicated to the memory of Professor M. Pezzana (Italian). MR1881097

[Mye] B. R. Myers, *Number of Spanning Trees in a Wheel*, IEEE Transactions on Circuit Theory, CT-18 (1971), 280–282.

[Pla] Antonio Plans, *Contribution to the study of the homology groups of the cyclic ramified coverings corresponding to a knot* (Spanish), Rev. Acad. Ci. Madrid **47** (1953), 161–193 (5 plates). MR56923

[Prz1] Józef H. Przytycki, *3-coloring and other elementary invariants of knots*, Knot theory (Warsaw, 1995), Banach Center Publ., vol. 42, Polish Acad. Sci. Inst. Math., Warsaw, 1998, pp. 275–295. MR1634462

[Prz2] Józef H. Przytycki, *From Goeritz matrices to quasi-alternating links*, The mathematics of knots, Contrib. Math. Comput. Sci., vol. 1, Springer, Heidelberg, 2011, pp. 257–316, DOI 10.1007/978-3-642-15637-3_9. MR2777853

[Prz3] Józef H. Przytycki, *Knots and graphs: two centuries of interaction*, Knot theory and its applications, Contemp. Math., vol. 670, Amer. Math. Soc., Providence, RI, 2016, pp. 171–257, DOI 10.1090/conm/670/13449. MR3568528

[Prz4] Józef H. Przytycki, *Plans' theorem for links: an application of* t_k *moves*, Canad. Math. Bull. **31** (1988), no. 3, 325–327, DOI 10.4153/CMB-1988-047-6. MR956364

[PBIMW] Józef H. Przytycki, Rhea Palak Bakshi, Dionne Ibarra, Gabriel Montoya-Vega, and Deborah Weeks, *Lectures in knot theory—an exploration of contemporary topics*, Universitext, Springer, Cham, [2024] ©2024, DOI 10.1007/978-3-031-40044-5. MR4738299

[Riv] Theodore J. Rivlin, *Chebyshev polynomials*, 2nd ed., Pure and Applied Mathematics (New York), John Wiley & Sons, Inc., New York, 1990. From approximation theory to algebra and number theory. MR1060735

[Rol] Dale Rolfsen, *Knots and links*, Mathematics Lecture Series, No. 7, Publish or Perish, Inc., Berkeley, CA, 1976. MR515288

[Sed] Irži Sedlaček, *Teoriya na grafite* (Bulgarian), Nauka i Izkustvo, Sofia, 1967. Translated from the Czech by G. Petrov. MR238725

[Web] Claude Weber, *La démonstration de J. Levine des théorèmes de A. Plans* (French), Algebraic topology, Aarhus 1982 (Aarhus, 1982), Lecture Notes in Math., vol. 1051, Springer, Berlin, 1984, pp. 315–330, DOI 10.1007/BFb0075574. MR764586

Department of Mathematics, The George Washington University, Washington, DC
Email address: ajchristiana@gwmail.gwu.edu

Department of Mathematics, The George Washington University, Washington, DC
Email address: hguo30@gwmail.gwu.edu

Department of Mathematics, The George Washington University, Washington, DC; and Department of Mathematics, University of Gdańsk, Gdańsk, Poland
Email address: przytyck@gwu.edu

Contemporary Mathematics
Volume **834**, 2026
https://doi.org/10.1090/conm/834/16688

Isotopies of links in oriented S^1-bundles over surfaces

Mieczyslaw K. Dabkowski and Sushmita Sinha Roy

Abstract. Links and their invariants in an oriented S^1-bundle M over a compact and connected oriented surface F can be studied using their shadow and arrow diagrams on F. When F is planar or $F = S^2$ and $M = F \times S^1$, there is a natural way to pass from one description to the other. For non-trivial S^1-bundle M, a shadow diagram on F with a marked point can be defined. Although the shadow diagram in F may yield infinitely many non-equivalent (up to arrow moves) arrow diagrams, in an important special case $F = S^2$ both descriptions of links in M are equivalent. This relationship might be used to develop new combinatorial techniques for computing KBSM of 3-manifolds.

1. Introduction

Skein modules of 3-manifolds were introduced by J. H. Przytycki [**15**] in 1987 and later by V. G. Turaev [**19**] in 1988. The Kauffman bracket skein module (KBSM) is one of these modules, defined using the Kauffman bracket skein relation [**9**]. The KBSM was computed for several classes of 3-manifolds in [**16**], [**17**], [**6**], [**7**], [**3**], [**1**], [**2**], [**14**], [**13**], [**11**], and [**12**]. In most cases, methods for computing KBSM rely on representing links in a 3-manifold M, and their ambient isotopy through diagrams and moves on a surface $F \subset M$. For the remainder of this paper, we assume that F is a connected, compact, and oriented surface, and that $p : M \to F$ is an oriented S^1-bundle over F. For such M, links up to ambient isotopy are represented by equivalence classes of arrow diagrams (arcs with arrows and crossings) up to arrow moves (see [**13**]). They can also be described using *shadow diagrams* up to *shadow moves* (see Figure 4.4), as introduced by V. Turaev in [**20**].[1] For a planar surface F or $F = S^2$, shadow diagrams (up to shadow moves) on F are in bijection with ambient isotopy classes of links in M, however shadow diagrams have not yet been used to compute the KBSM of M. When $F \neq S^2$ is non-planar, as noted in [**20**], shadow diagrams (up to shadow moves) on F are no longer in bijection with ambient isotopy classes of links in M (see [**20**, Theorem 3.2]), imposing natural limitations on their application. Nevertheless, we believe that shadow diagrams can be used to study links in Seifert fibered 3-manifolds with base S^2 and to compute the KBSM for this class of 3-manifolds.

The paper is organized as follows. In Section 2, we recall the definitions of shadow and arrow diagrams. In Section 3, we give an argument for Theorem 1

2020 *Mathematics Subject Classification.* Primary 57K31, 57K10.

[1]This idea was based on his earlier observation that links in S^3 can be studied via the Hopf map $p : S^3 \to S^2$ rather than the projection onto $\mathbb{R}^2$.

stated in [**5**] used in [**13**], [**4**] and [**11**]. In Section 4 we recall an action of the first homology of F on links in $p : M \to F$ and we consider special cases of this action in $F \times S^1$, where F is planar or $F = S^2$. Furthermore, in Section 5, for trivial S^1-bundles $p : M \to F$, we discuss relationship between shadow and arrow diagrams. When F is either planar or S^2, shadow diagrams (up to shadow moves) are in bijection with ambient isotopy classes of links in M. In particular, we note that Theorem 1 of [**5**] in this case also follows from [**20**, Theorem 3.2]. We represent a non-trivial S^1-bundle $p : M \to F$ with the Euler number $\chi(p) = \beta$, as a 3-manifold $\hat{M}_\beta$ obtained from $\hat{M} = \hat{F} \times S^1$ by a $(1, \beta)$-Dehn filling, where $\hat{F} = F \setminus D^2_\beta$ and D^2_β is a 2-disk centered at marked point β on F. Links in $\hat{M}_\beta$ can be described using shadow and arrow diagrams on F with the marked point β and we will study the relationships between these two representations. Finally, in Section 6, we briefly discuss the relationship between shadow and arrow diagrams for framed links in the S^1-bundle $p : M \to F$.

2. Arrow and shadow diagrams of links in an oriented circle bundle over oriented surface

Let $p : M \to F$ be an oriented S^1-bundle over F with Euler number $\chi(p) = \beta$. Using terminology introduced in [**20**], a link L of n-components $K_1, K_2, \ldots, K_n$ in M is called *generic* if L is transversal to fibers of p, the immersed curves $p(K_i)$ in F have only double (transversal) crossings, and L intersect F at finitely many points (and L is transversal to F at these points).

Arrow diagrams were introduced in [**13**] to represent generic links in trivial S^1-bundle[2] $M = F \times S^1$. Let $S^1 = \{e^{it} \mid t \in [0, 2\pi]\}$ be the unit circle oriented according to the increasing values of $t \in [0, 2\pi]$, $p : M \to F$ be the projection, and $N = F \times [0, 2\pi]$. Following [**13**], define a map

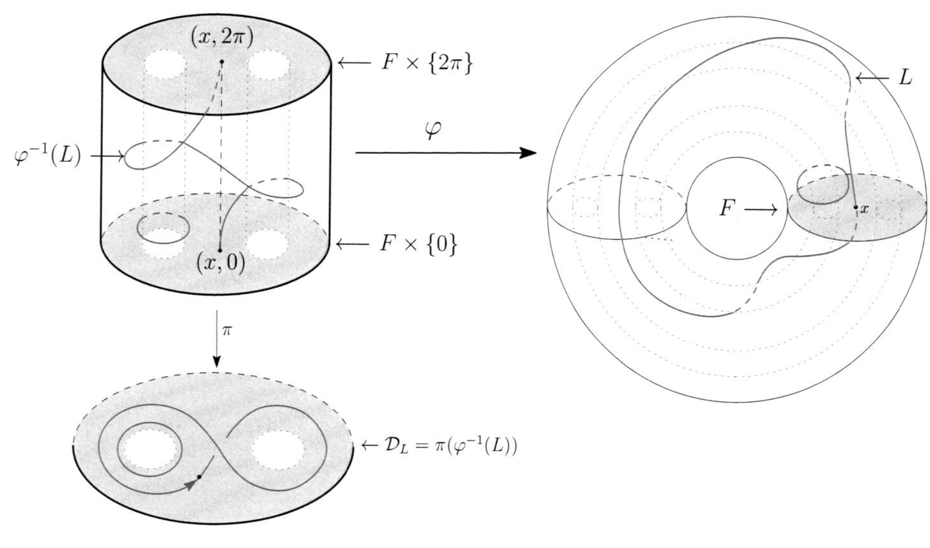

FIGURE 2.1. Arrow diagram for links on F

[2]Arrow diagrams can also be defined for generic links in an oriented S^1-bundle $N \hat{\times} S^1$ over an unoriented, compact and connected surface N (see [**11**] and [**12**]).

$$\varphi : N \to M, \quad \text{by} \quad \varphi(x,t) = (x, e^{it}).$$

For a generic link L in M, its preimage $\varphi^{-1}(L)$ is the relative link in N which consists of finitely many generic knots K_j in the interior of N and/or arcs a_j in N with their boundary points on $F \times \{0\}$ and $F \times \{2\pi\}$ (see Figure 2.1). We identify $F \times \{0\}$ with F and let $\pi : N \to F$, $\pi(x,t) = x$ be the canonical projection onto F. Since for a generic link L in M, each fiber $\pi^{-1}(x)$ over $x \in F$ intersects $\overline{L} = \varphi^{-1}(L)$ in at most two points, the projection $\pi(\overline{L}) = p(L)$ consists of a family of immersed closed curves on F with two types of double points. Either $x \in \pi(\overline{L})$ is a double point of $p(L)$ (i.e., $p^{-1}(x)$ intersects L at two points), or L intersects F at x. In the latter case, we call x a *non-standard double point* of $p(L)$. For the first type of double points of $\pi(\overline{L})$, we assign a crossing, while for each non-standard double point, we assign an arrow according to the rule in Figure 2.2.

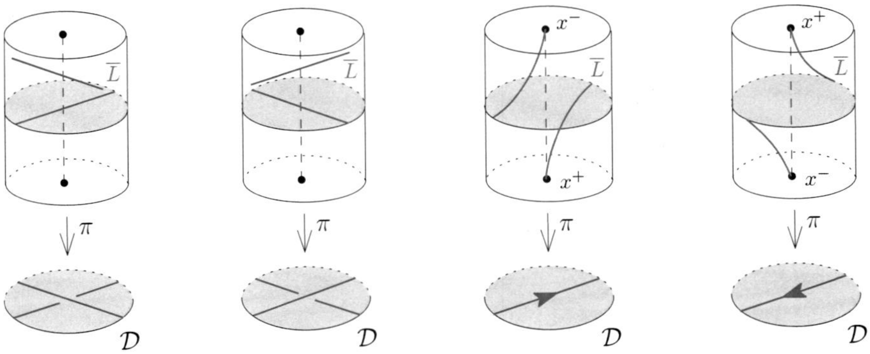

FIGURE 2.2. Arrow Assignment

PROPOSITION 2.1. *For an arrow diagram $\mathcal{D}$ on F, there exists a generic link L in $F \times S^1$ whose arrow diagram is $\mathcal{D}$.*

PROOF. Let $N = F \times [0, 2\pi]$ and $\tilde{\mathcal{D}}$ be the classical diagram on F with non-standard double points on its arcs. Thus, there is a link $\tilde{L}$ in N with diagram $\tilde{\mathcal{D}}$. Let $x \in \mathcal{D}$ be a non-standard double point on an arc of $\tilde{\mathcal{D}}$ and $I_x = \pi^{-1}(x)$ be a fiber over x. Then I_x intersects $\tilde{L}$ at a unique point $\tilde{x}$ (see Figure 2.3). Let x^+ and x^- be ends of two arcs obtained by cutting $\tilde{L}$ at $\tilde{x}$. Depending on the direction of the arrow at x in $\mathcal{D}$, we push along fibers of π the left and right arcs until x^+ and x^- are identified with the points $x \times 0$ and $x \times 2\pi$, as shown in Figure 2.3.

After modifying $\tilde{L}$ by applying one of the two operations shown in Figure 2.3 at each non-standard double point $x \in \tilde{\mathcal{D}}$, we obtain a relative link $\overline{L}$ in N. Clearly, $\overline{L}$ is a link with the arrow diagram $\mathcal{D}$, and hence $L = \varphi(\overline{L})$ is the link in $F \times S^1$ whose arrow diagram is $\mathcal{D}$. □

Let F be a non-planar surface with a non-empty boundary. Then the S^1-bundle over F is trivial, i.e., $M = F \times S^1$ (see Lemma 10.2.1 of [**10**]), and F can be identified with its zero section $s : F \to F \times 0$. Therefore, a generic link L in M has its arrow diagram $\mathcal{D}$ on F, and conversely, by Proposition 2.1, each arrow diagram $\mathcal{D}$ on F corresponds to some generic link $\tilde{L}$ in M.

Assume that F is closed and $\beta \in \mathbb{Z}$. Let $\hat{F} = F \setminus D^2$, where D^2 is an open disk in the interior of F, and $\hat{M} = \hat{F} \times S^1$ be the trivial S^1-bundle over $\hat{F}$ with

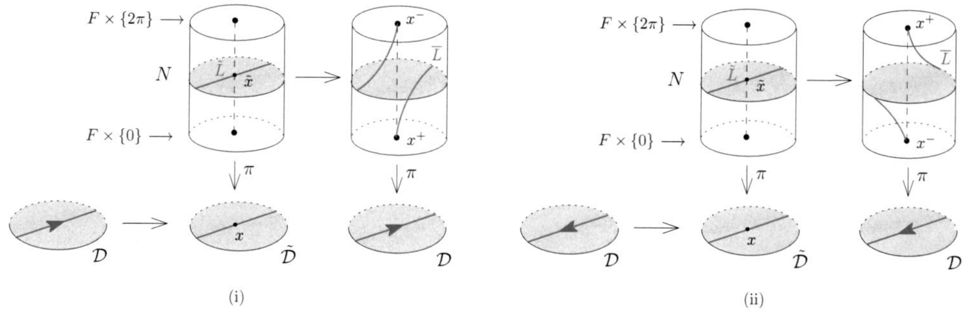

FIGURE 2.3. Links from Arrow Diagrams

a fixed zero section F. We choose meridian (section of p over $\partial\hat{F}$) and longitude on its boundary torus T^2 in a natural way.[3] As shown in the Proposition 10.2.5 and Corollary 10.2.6 of [**10**], $(1,\beta)$-Dehn filling yields the 3-manifold $\hat{M}_\beta$ which is an S^1-bundle over F with Euler number β. Moreover, every oriented S^1-bundle over F with $\beta = \chi(p)$ can be obtained in such a way. Given a link L in $\hat{M}_\beta$ we may assume that $L \subset \hat{M}$. Hence, L can be represented by an arrow diagram $\mathcal{D}$ on $\hat{F}$, and consequently, also on the surface F with a marked point labeled by β. We denote this surface by $\hat{F}_\beta$ and, when p is trivial (i.e., $\beta = 0$), we simply take $\hat{F}_\beta = F$. The above discussion can summarized as a corollary.

COROLLARY 2.2. *Every link L in $\hat{M}_\beta$ has an arrow diagram $\mathcal{D}$ on $\hat{F}_\beta$. Conversely, for every arrow diagram $\mathcal{D}$ on $\hat{F}_\beta$, there exists a generic link $\tilde{L}$ in M whose arrow diagram is $\mathcal{D}$.*

REMARK 2.3. We say that an isotopy of a link L in $\hat{M}_\beta$ is *vertical* if each point of L moves along a fiber of $\hat{M} = \hat{F} \times S^1$. Clearly, generic links L_1 and L_2 with the same arrow diagram on $\hat{F}_\beta$ are vertically isotopic. We will discuss isotopy of links in $\hat{M}_\beta$ in more detail later in Section 3.

In [**20**], V. Turaev introduced a description of generic links L in S^1-bundle $p : M \to F$ with $\chi(p) = \beta$ via shadow diagrams $sh(L)$ on F with $\partial F = \emptyset$. If $\partial F \neq \emptyset$, then $p : M \to F$ is isomorphic with projection $p : F \times S^1 \to F$. As it was mentioned in [**20**, Remark on p. 53] we may choose F as zero section which gives also section over ∂F. Recall, for a generic link L in M its integral shadow diagram $sh(L)$ on F is defined as follows. Assume that $S^1 = \{z \in \mathbb{C} \mid |z| = 1\}$ acts freely on M, orbits of this action coincide with the fibers of p, and the orientation of all fibers is induced by the counterclockwise orientation of S^1. Then $(-1) \in S^1$ acts on M as an involution that sends x to its opposite $(-1) \cdot x$. Let $M' = M/(-1)$ be the quotient of M by this action, then the quotient map $q : M \to M'$ is a double cover of M' (so q restricted to each fiber of p is a map of degree 2) and $p' : M' \to F$ defined by

$$p'(x') = p(x), \text{ for } x \in q^{-1}(x')$$

is an oriented S^1-bundle over F. Clearly, vertically isotopic generic links in M have the same projection on F. Moreover, each generic link L in M can be vertically isotoped to a link L' satisfying the following property:

[3]Boundary of $\hat{F}$ gives the meridian and a fiber over a point $\partial(\hat{F})$ gives its longitude.

PROPERTY 2.4. For every double point $a \in p(L')$, if $x, x' \in p^{-1}(a)$ then $x' = (-1) \cdot x$.

The set $q(L') \subset M'$ is obtained from L' by identifying pairs of points $x, (-1)\cdot x = p^{-1}(a)$ for each double point $a \in p(L') = p(L)$. We note that p' maps $q(L')$ bijectively onto $p(L)$, and its inverse $r : p(L) \to q(L')$ is a section of p' restricted to $p(L)$. Let X be an oriented, connected region of F (with the orientation induced from F) that is cut out by $p(L)$. Consider the restriction $p' \restriction_X$ of p' to X. Since X has a non-empty boundary, we denote by $p'_X : X \times S^1 \to X$ the trivialization of $p' \restriction X$. Let $B = \partial F \cap \partial X$ and $C = \partial X \setminus B$. Let $\tilde{r} : \partial X \to \partial X \times S^1$ be the section of p'_X defined by $\tilde{r}(x) = r(x)$ for $x \in C$ when $B = \emptyset$ and

$$\tilde{r}(x) = \begin{cases} r(x), \text{ if } x \in C \\ q(s(x)), \text{ if } x \in B, \end{cases}$$

When $B \neq \emptyset$ (i.e., $B = B_1 \cup \cdots \cup B_k$ is a disjoint union of $k \geq 1$ circles), then $p'_X = p' \restriction X$ since $M = F \times S^1$. Let $\pi : X \times S^1 \to S^1$ be the projection onto S^1, i.e., $\pi(x, z) = z$. Denote by $\varphi = \pi \circ \tilde{r} : \partial X \to S^1$ the composition of $\tilde{r}$ and the projection π. Let

$$\theta_X = \deg(\varphi)$$

and notice that for $B \neq \emptyset$,

$$\deg(\varphi) = \deg(\pi \circ (\tilde{r} \restriction_C)) + \sum_{i=1}^{k} \deg(\pi \circ (\tilde{r} \restriction_{B_i})),$$

where $(\tilde{r} \restriction_C)$ and $(\tilde{r} \restriction_{B_i})$ are restrictions of $\tilde{r}$ to C and B_i, for $i = 1, 2, \ldots, k$. Clearly, $\pi \circ (\tilde{r} \restriction_{B_i})$ is a constant map, so $\deg(\pi \circ (\tilde{r} \restriction_{B_i})) = 0$ and, therefore

$$\theta_X = \deg(\varphi) = \deg(\pi \circ (\tilde{r} \restriction_C)).$$

As argued in [20], θ_X does not depend on a choice of a trivialization p'_X and a choice of L up to a vertical isotopy. Let ω_X be the number of corners of X. Define *gleam* of X by

$$\mathrm{gl}(X) = (\omega_X - \theta_X)/2.$$

To better understand $\mathrm{gl}(X)$, consider a region X determined by $p(L) = p(L')$, and let $C(X)$ denote the set of corners of X. By the definition of L', each corner $C \in C(X)$ falls into one of the two cases illustrated in Figure 2.4, parts a) and b). It is clear that the contribution of corner C to $\mathrm{gl}(X)$ is 0 in case a) and 1 in case b). Let $C_+(X)$ denote the set of corners that contribute 1 to $\mathrm{gl}(X)$. We refer to $C_+(X)$ as the set of A-regions[4] determined by X. Next, we examine the contribution to $\mathrm{gl}(X)$ made by the arcs bounding X. Consider an arc α connecting two corners, C_1 and C_2, of X, as shown in Figure 2.5. The image $q(\alpha)$ under the quotient map q is shown on the right side of Figure 2.5. Since $\varphi : \partial X \to S^1$ is a composition of $\tilde{r}$ and π, we see that

$$\deg(\varphi) = \Sigma_{C \in C(X)} \deg_C(\varphi) + \deg_A(\varphi),$$

where $\deg_A(\varphi)$ is the contribution to the degree $\deg(\varphi)$ of φ by arcs which bound X. Following the boundary of X in the direction determined by the orientation of X near each corner $C \in C(X)$, we see that the part of $q(L')$ near C is a part of

[4]This choice of terminology will become clear in Section 5, where we compare arrow and shadow diagrams of links in $\hat{M}_\beta$.

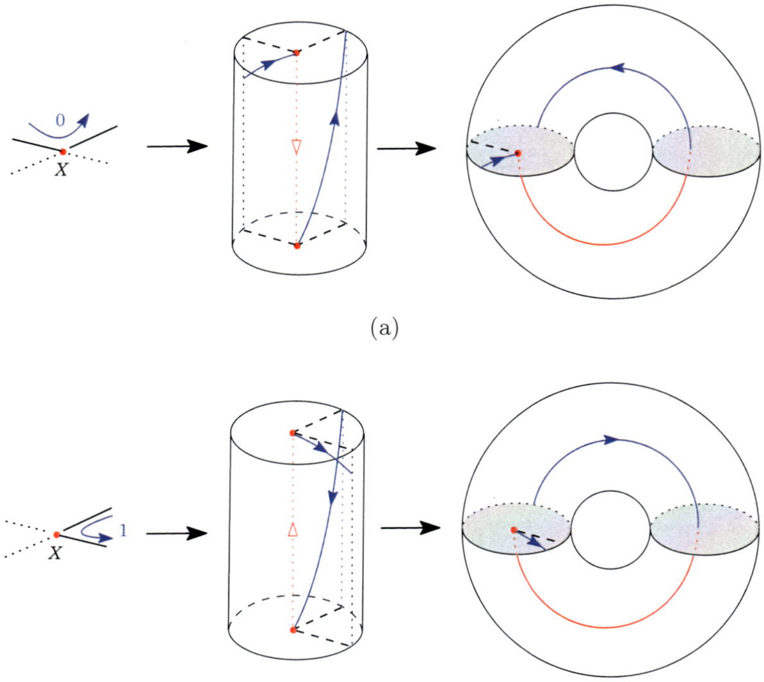

(a)

(b)

FIGURE 2.4. Degree Crossing

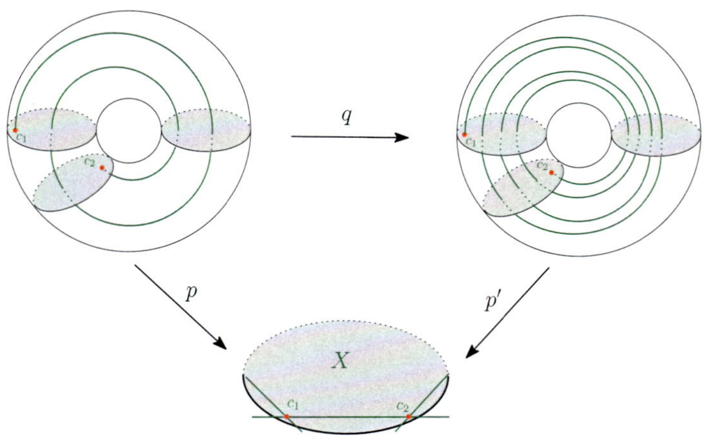

FIGURE 2.5. Degree Crossing Quotient

a curve oriented in either the opposite or the same direction relatively to the fiber S^1 of p' (see Figure 2.4(a) and (b)). Therefore, $\deg_C(\varphi) = -1$ and $\deg_C(\varphi) = 1$, respectively. Hence,

$$\begin{aligned}\frac{1}{2}(\omega_X - \theta_X) &= \frac{1}{2}(\Sigma_{C\in C(X)}(1 - \deg_C(\varphi)) - \deg_A(\varphi))\\ &= \Sigma_{C\in C_+(X)}1 - \frac{1}{2}\deg_A(\varphi)\\ &= |C_+(X)| - \frac{1}{2}\deg_A(\varphi).\end{aligned}$$

Since $\deg(q) = 2$, $\deg_A(\varphi)$ is always even and $\mathrm{gl}(X) = \frac{1}{2}(\omega_X - \theta_X)$ is an integer. We summarize the above in the following proposition.

PROPOSITION 2.5. *Let X be the region determined by the projection $p(L') = p(L)$ of L' onto F. Then* $\mathrm{gl}(X)$ *is an integer given by*

$$\mathrm{gl}(X) = |C_+(X)| - \frac{1}{2}\deg_A(\varphi).$$

Let L be a generic link in M. Its *shadow diagram* $sh(L)$ is the family of immersed curves $p(L)$ together with an integer $\mathrm{gl}(X)$ assigned to each region X cut from F by $p(L)$.

The *total gleam* of $sh(L)$ is defined as the total sum of $\mathrm{gl}(X)$ over all regions X determined by $p(L)$ minus twice the number of double points of $p(L)$ (see [**20**]).

As it was noted in [**20**], the sum of numbers $\omega_X/2$ over all regions X cut from F by $p(L)$ equals twice the number of double points of $p(L)$. Hence, the total gleam of $sh(L)$ equals the sum of $\theta_X/2$ over all regions X which is $-\beta$, where $\beta = \chi(p)$.

3. Isotopy of links in $\hat{M}_\beta$ via moves on arrow diagrams

We say that links L_1 and L_2 in a 3-manifold M^3 are *isotopic* if there exists an *ambient isotopy* of M^3 that deforms L_1 onto L_2. As shown in [**8**, Corollary 6.3], L_1 and L_2 are isotopic in M^3 if and only if they are isotopic by moves. Recall, a *move* is an isotopy h of M^3 with support inside of an open 3-ball B^3 in the interior of M^3, i.e., h is the identity on the complement of B^3. For an oriented S^1-bundle $p : M \to F$, as it was noted in [**20**], each link L in M is isotopic to a generic link. Since in this paper we work with $\hat{M}_\beta$ we will describe isotopies of generic links in this manifold. We consider two types of moves R_0 and R_β in $\hat{M}_\beta$, i.e., one which takes place in a cylinder $N = D^2 \times I$ inside $\hat{M}$ and another one in a 2-handle $H = D^2 \times I$ attached along $(1, \beta)$-curve on the boundary of $\hat{M}$ (see Figure 3.1). We

FIGURE 3.1. R_0 and R_β-moves in $\hat{M}_\beta$

note that H intersects the fiber over the marked point β on $\hat{F}_\beta$, and the R_β-move is also called a 2-handle slide. It might also be helpful to view the R_β-move in two stages, i.e., we first use the R_0-move in the 2-handle $H = D^2 \times I$, and then we attach H along the $(1, \beta)$-curve on the boundary torus of $\hat{M}$ (see Figure 3.2).

Let D^2 be a disk in $\hat{F}$, and let $V = D^2 \times S^1$ be a solid torus in $\hat{M}$. We describe the R_0-move using an arrow diagram on $\hat{F}$. Assume that a generic link L in $\hat{M}$

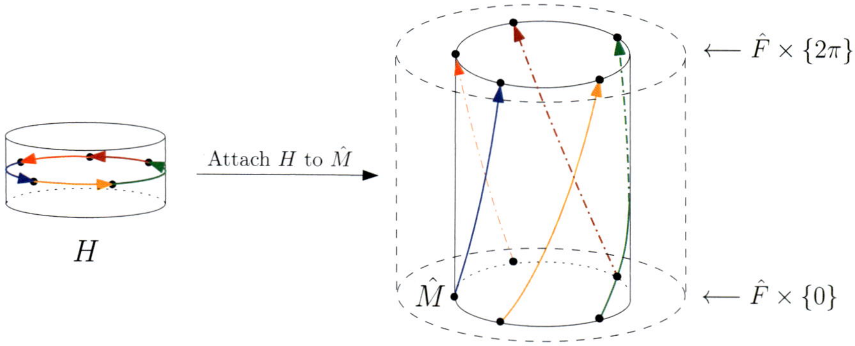

FIGURE 3.2. 2-handle sliding R_β in $\hat{M}_\beta$

intersects V along the arc α shown on the left side of Figure 3.3(a) and (b). Then, there is a 2-handle $N \subset V$ that contains both α and $R_0(\alpha)$ (see Figure 3.3). There are four generic cases (up to symmetry) for the R_0-move, as shown in Figure 3.3. Depending on the positions of the ends of α on the boundary of N (see Figure 3.4 for

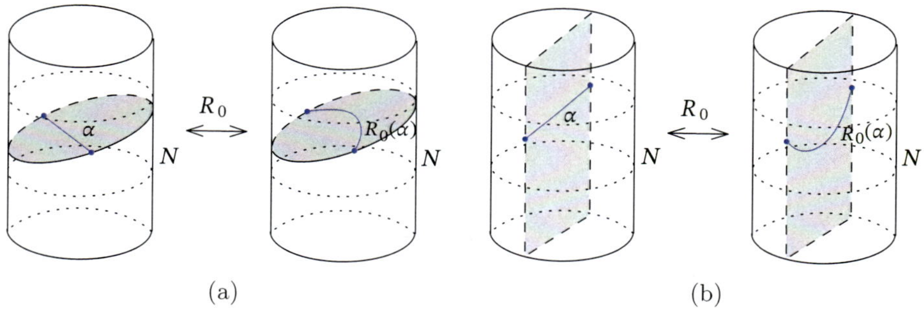

FIGURE 3.3. Generic cases for R_0-move in $\hat{M}$

the different cases corresponding to Figure 3.3(a)), and the position of N relative to $\hat{F} \times \{0\} = \hat{F}$, R_0-move yields corresponding arrow diagrams on $\hat{F}$ shown in Figure 3.5. We clearly see that all moves in Figure 3.5 are consequences of the Q_0 and Q_4-moves on $\hat{F}$.

Now, we describe the R_β-move, which takes place in the 2-handle $H \subseteq \hat{M}_\beta$, using an arrow diagram on $\hat{F}_\beta$. Let $V = D^2_\beta \times S^1$ be a solid torus, where D^2_β is a disk centered at the marked point β on F. Assume that a generic link L in $\hat{M}_\beta$ intersects V along the arc α as shown in Figure 3.6 a) and b). Then there is a 2-handle $H \subset V$ which contains both α and $R_\beta(\alpha)$.[5] Depending on the relative positions of the ends of α in the boundary of H (see Figure 3.6) and the position of H relative to $\hat{F}_\beta \times \{0\} = \hat{F}_\beta$, R_β-move yields the corresponding arrow diagrams on $\hat{F}_\beta$ shown in Figure 3.7.

[5]If the 2-disk with boundary α and $R_\beta(\alpha)$ does not intersect the fiber over β, then R_β-move is same as R_0-move, i.e., we may choose a 2-disk $(D^2)'_\beta$ centered at β so that α and $R_\beta(\alpha)$ are outside of $V = (D^2)'_\beta \times S^1$.

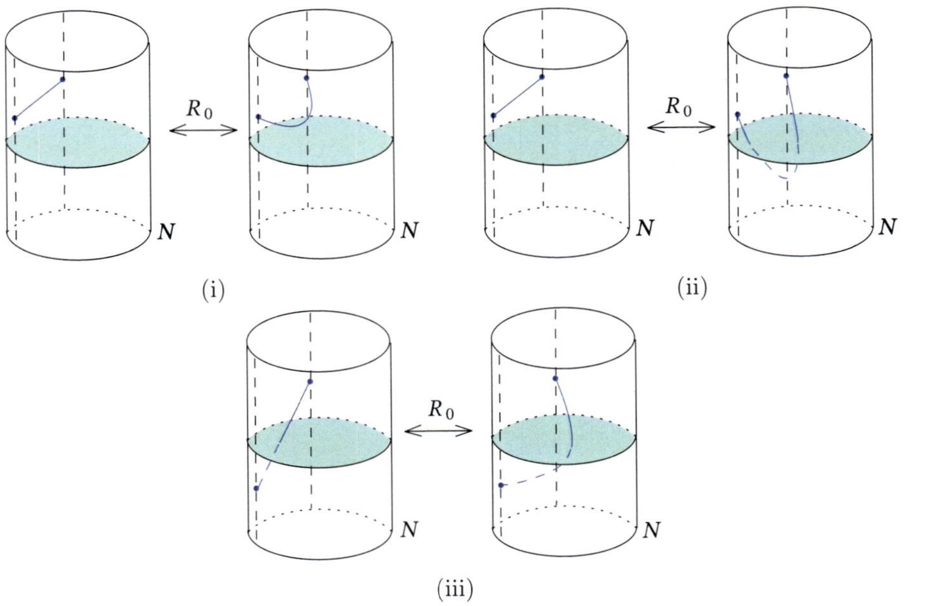

FIGURE 3.4. Different cases for R_0-move in $\hat{M}$

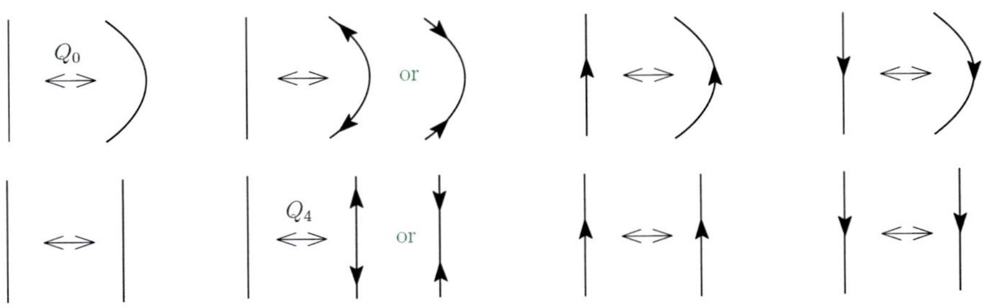

FIGURE 3.5. Arrow diagrams for R_0-move on $\hat{F}$

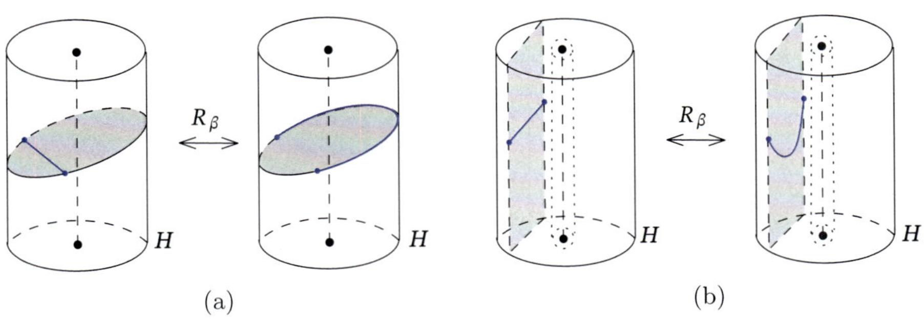

FIGURE 3.6. Generic cases for R_β-move in $\hat{M}_\beta$

We clearly see that all moves in Figure 3.7 are consequences of Q_4 and $Q_{(1,\beta)}$-moves on $\hat{F}_\beta$. In summary, R_0 and R_β-moves in $\hat{M}_\beta$ yield Q_0, Q_4, and $Q_{(1,\beta)}$-moves on $\hat{F}_\beta$.

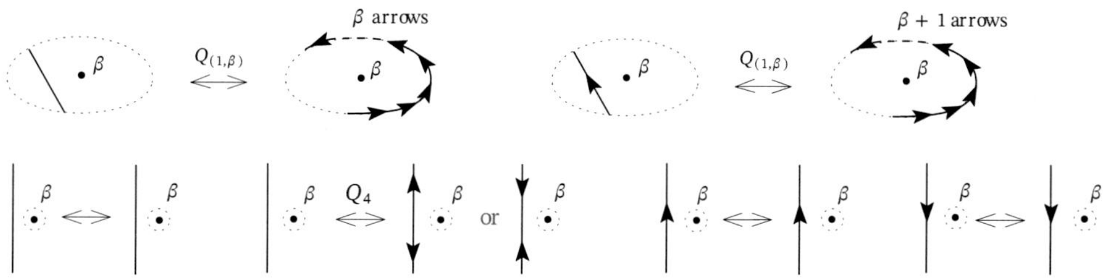

FIGURE 3.7. Arrow diagrams on $\hat{F}_\beta$ corresponding to R_β-move

Let Q_1-Q_5, and $Q_{(1,\beta)}$ be moves in Figure 3.8 and Figure 3.9. Arrow diagrams $\mathcal{D}_1$ and $\mathcal{D}_2$ in $\hat{F}_\beta$ are called *equivalent* if $\mathcal{D}_2$ can be obtained from $\mathcal{D}_1$ by a finite sequence of local moves $Q_1 - Q_5$, and $Q_{(1,\beta)}$.

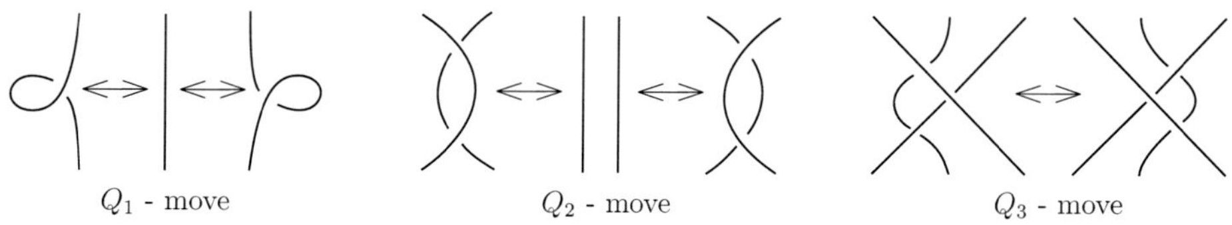

(a) Q_1, Q_2 and Q_3-moves for arrow diagrams on $\hat{F}$

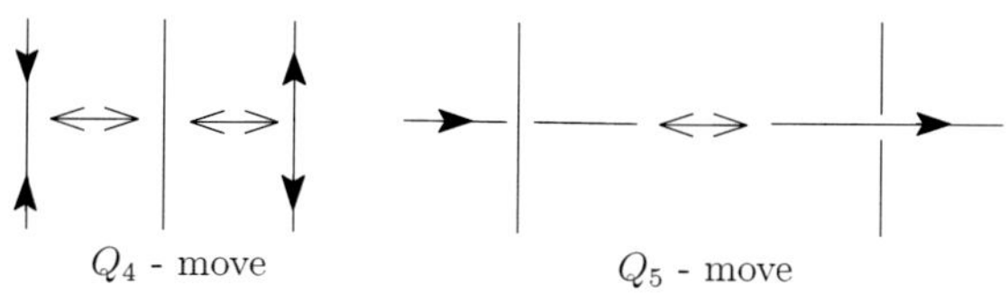

(b) Q_4 and Q_5-moves for arrow diagrams on $\hat{F}$

FIGURE 3.8. $Q_1 - Q_5$-moves for arrow diagrams on $\hat{F}$

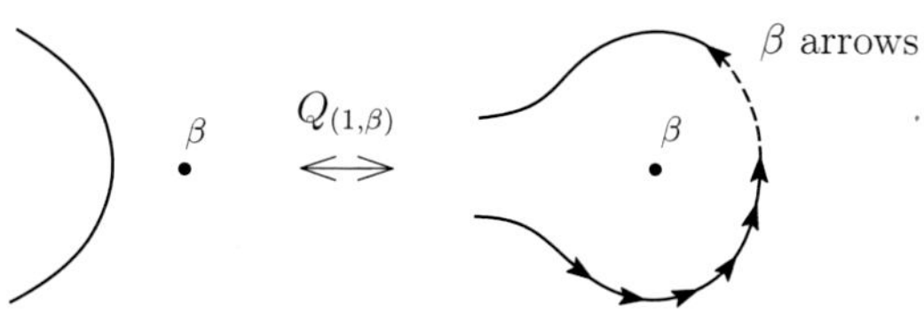

FIGURE 3.9. $Q_{(1,\beta)}$-move for arrow diagrams on $\hat{F}_\beta$

Suppose that L_1 and L_2 are generic links in M which only differ by R_0-move inside 2-handle $N \subseteq \hat{M}$ (respectively 2-handle H in $\hat{M}_\beta$) as shown in Figure 3.10 and Figure 3.11. Therefore, L_1 and L_2 are identical outside of a solid torus containing N (respectively, H). Since R_0 fixes boundary of N (respectively, R_β fixes boundary of H), we may consider tangles T_1 and T_2 with points $\partial(T_1) = \partial(T_2)$ fixed on the boundary of N (respectively, on boundary of H). One argues by using Reidemeister's Theorem [**18**], that all isotopies of tangles in N (respectively H) are generated by R_0, R_β-moves, and one only needs to consider the following three cases as shown in Figure 3.10 a)-c) and Figure 3.11 a)-c). Notice that R_{1_β}, R_{2_β}, and R_{3_β}- moves are the consequences of R_1, R_2, and R_3-moves that take place in $\hat{M}$ and R_β-move that takes place in $\hat{M}_\beta$. Therefore, we only consider R_1, R_2, R_3, and R_β-moves.

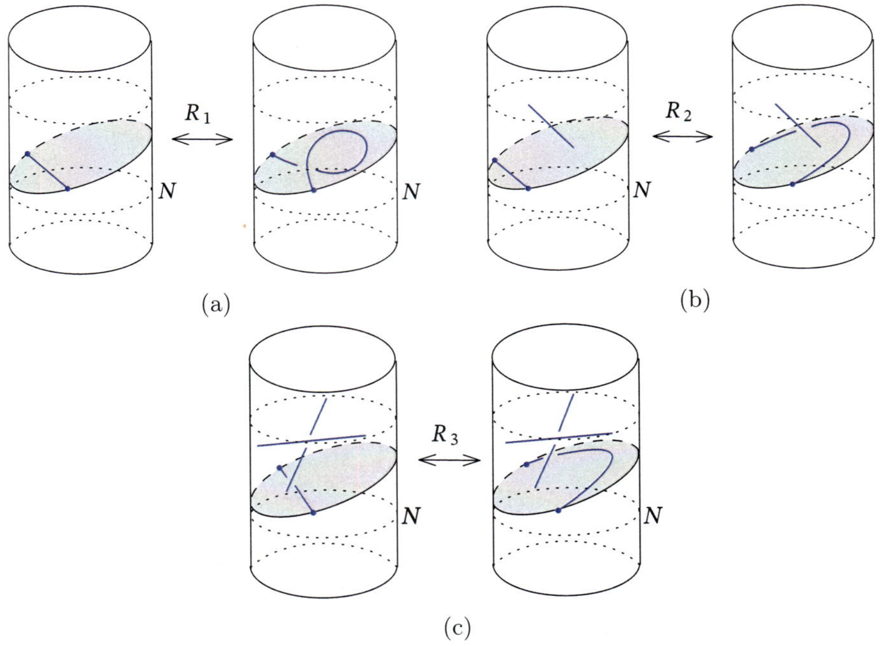

FIGURE 3.10. Different cases for isotopy in 2-handle N

LEMMA 3.1. *Assume that generic links L_1 and L_2 differ only by R_1-move in $N \subset \hat{M}$ (see Figure 3.10). Then their corresponding arrow diagrams $\mathcal{D}_1$ and $\mathcal{D}_2$ are related on $\hat{F}$ by Q_1, Q_4, and Q_5-moves.*

PROOF. Let $T_1 = N \cap L_1$ and $T_2 = N \cap L_2$ be tangles in N. If T_1 and T_2 do not intersect $\hat{F}$ (see Figure 3.10 a)), then their arrow diagrams differ by Q_1-move (see Figure 3.8a). Therefore, we may assume that at least one of T_1 and T_2 intersect $\hat{F}$. We consider the following cases: i) T_1 does not intersect $\hat{F}$ and T_2 intersects $\hat{F}$, ii) T_1 and T_2 both intersect $\hat{F}$.

For i), there are two generic levels for $\hat{F}$, i.e., $\hat{F}$ contains points A_2 and A_3, or A_1 and A_4 (see Figure 3.12). Thus resulting arrow diagrams for T_1 and T_2 are shown on the left of Figure 3.13. Clearly, arrow diagrams of T_2 are obtained from those of T_1 by either applying Q_1-move first followed by Q_4-move, or first by Q_4-move followed by Q_1-move.

For ii), there are three generic levels for $\hat{F}$ to intersect N, i.e., $\hat{F}$ contains A_i, $i = 1, 2, 3$ (see corresponding arrow diagrams of T_1 and T_2 in Figure 3.13). In the first and the last cases, arrow diagrams of T_1 and T_2 differ by Q_1-move on $\hat{F}$ and, in the second case, they differ by Q_5-move followed by Q_1-move. As one checks, in all cases, arrow diagrams of T_1 and T_2 differ by Q_1, Q_4, and Q_5-moves. □

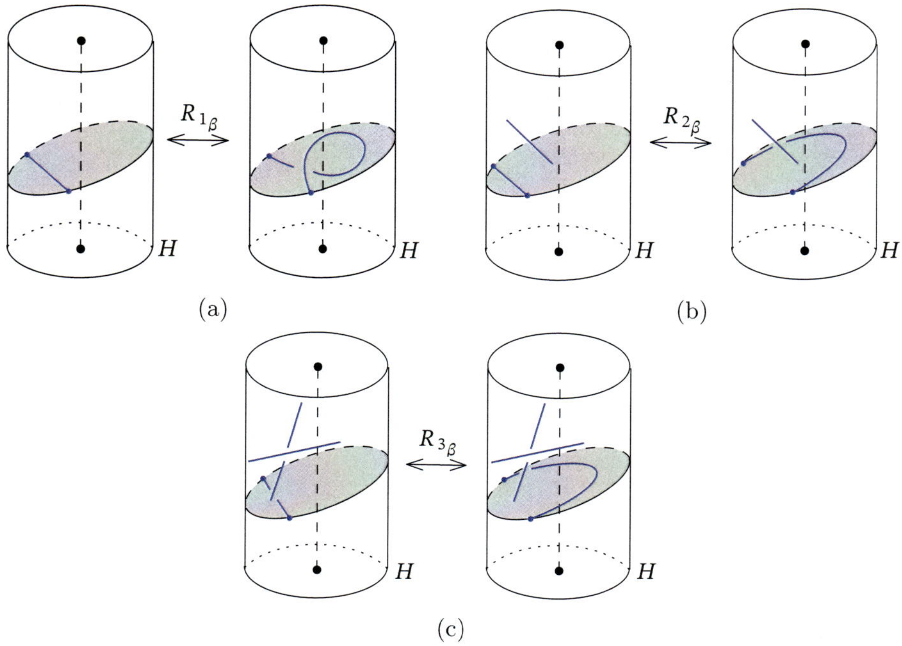

FIGURE 3.11. Different cases for isotopy in 2-handle H

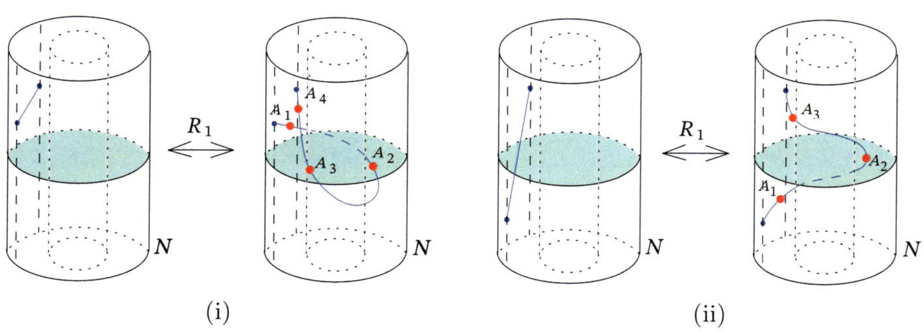

FIGURE 3.12. Different cases for R_1-move in $\hat{M}$

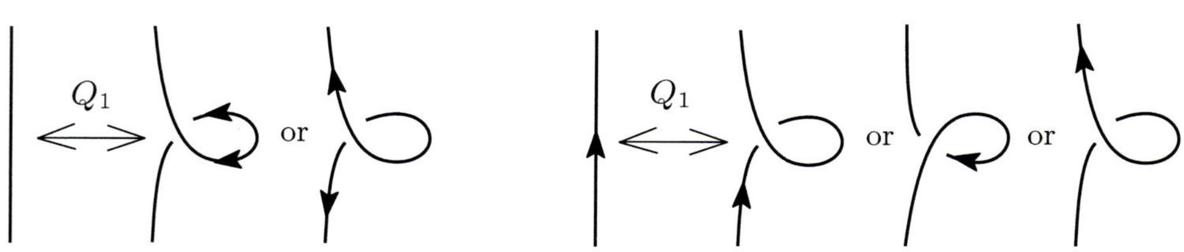

FIGURE 3.13. Arrow diagram on $\hat{F}$ for R_1-move

LEMMA 3.2. *Assume that generic links L_1 and L_2 differ only in $N \subset \hat{M}$ by R_2-move (see Figure 3.10). Then their corresponding arrow diagrams $\mathcal{D}_1$ and $\mathcal{D}_2$ on $\hat{F}$ differ by Q_2, Q_4, and Q_5-moves.*

PROOF. Let $T_1 = N \cap L_1$ and $T_2 = N \cap L_2$ be tangles in $N \subset \hat{M}$. If T_1 and T_2 do not intersect $\hat{F}$ (see Figure 3.10(b)), then arrow diagrams of T_1 and T_2 differ by Q_2-moves (see Figure 3.8(a)). Therefore, we may assume that at least one of T_1 and T_2 intersect $\hat{F}$. We consider the following cases up to symmetry: i) both arcs of T_1 do not intersect $\hat{F}$ and an arc of T_2 involved in R_0-move intersects $\hat{F}$; ii) either an arc of T_1 and T_2 involved in R_0-move intersects $\hat{F}$ or an arc of T_1 and T_2 not involved in R_0-move intersects $\hat{F}$, iii) both arcs of T_1 and T_2 intersects $\hat{F}$ (see Figure 3.14 (i), (ii), and (iii), respectively).

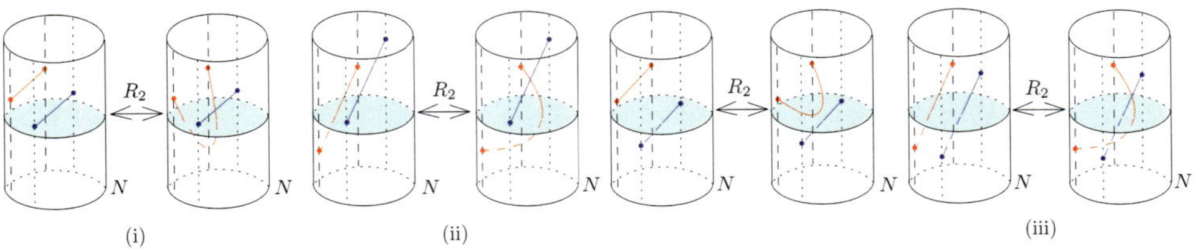

FIGURE 3.14. Different cases for R_2-move in $\hat{M}$

For i), there are two generic positions for $\hat{F}$ shown in Figure 3.14 (i) yielding two arrow diagrams shown in Figure 3.15 (i′). It is easy to see that arrow diagrams of T_2 and T_1 differ on $\hat{F}$ either by Q_4-move followed by Q_2-move or first by two Q_5-moves followed by Q_4-move and then followed by Q_2-move.

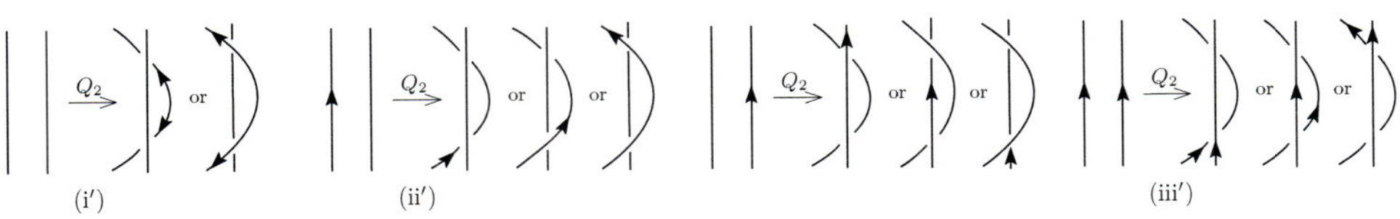

FIGURE 3.15. Arrow diagram on $\hat{F}$ for R_2-move

For ii), there are two cases shown on the left and the right of Figure 3.14 (ii). For the case on the left there are three generic positions for $\hat{F}$ shown in Figure 3.14 (ii) yielding corresponding arrow diagrams for T_2 on the left of Figure 3.15 (ii′). It is easy to see that diagrams of T_2 and T_1 differ by either Q_2-move or Q_5-move followed by Q_2-move. For the case shown on the right of Figure 3.14 (ii), there are also three generic positions for $\hat{F}$ yielding arrow diagrams for T_2 shown on the right of Figure 3.15 (ii′). Clearly, arrow diagrams of T_2 and T_1 differ either by Q_2-move or by Q_5-move followed by Q_2-move.

Finally, for iii), there are three generic positions for $\hat{F}$ shown in Figure 3.14 (iii) yielding three corresponding arrow diagrams for T_2 in Figure 3.15 (iii′). It is easy to see that arrow diagrams of T_2 and T_1 differ by either Q_2-move or two Q_5-moves followed by Q_2-move.

In summary, if L_1 and L_2 differ by R_2-move in $\hat{M}$ then the corresponding arrow diagrams of L_1 and L_2 differ by Q_2, Q_4, and Q_5-moves. □

LEMMA 3.3. *Assume that generic links L_1 and L_2 differ only by R_3-move in N. Then their corresponding arrow diagrams $\mathcal{D}_1$ and $\mathcal{D}_2$ differ on $\hat{F}$ by Q_3, Q_4, and Q_5-moves.*

PROOF. Let $T_1 = N \cap L_1$ and $T_2 = N \cap L_2$ be tangles in N. If T_1 and T_2 do not intersect $\hat{F}$ (see Figure 3.10 c)), then arrow diagrams of T_1 and T_2 differ by Q_3-moves (see Figure 3.8a). Therefore, we may assume that at least one of T_1 and T_2 intersects $\hat{F}$. Depending on the position of $\hat{F}$ relative to T_1 and T_2, there are several cases to consider.

Since all of them yield diagrams of T_1 and T_2 which differ by Q_3 and Q_4-moves, we consider only these cases (up to their symmetry) when resulting diagrams differ by Q_3, Q_4, and Q_5-moves, i.e., i) all arcs of T_1 do not intersect $\hat{F}$ and an arc of T_2 involved in R_0-move intersects $\hat{F}$; ii) an arc of T_1 and T_2 involved in R_0-move intersects $\hat{F}$; iii) all arcs of T_1 and T_2 intersects $\hat{F}$ (see Figure 3.16 (i), (ii), and (iii), respectively).

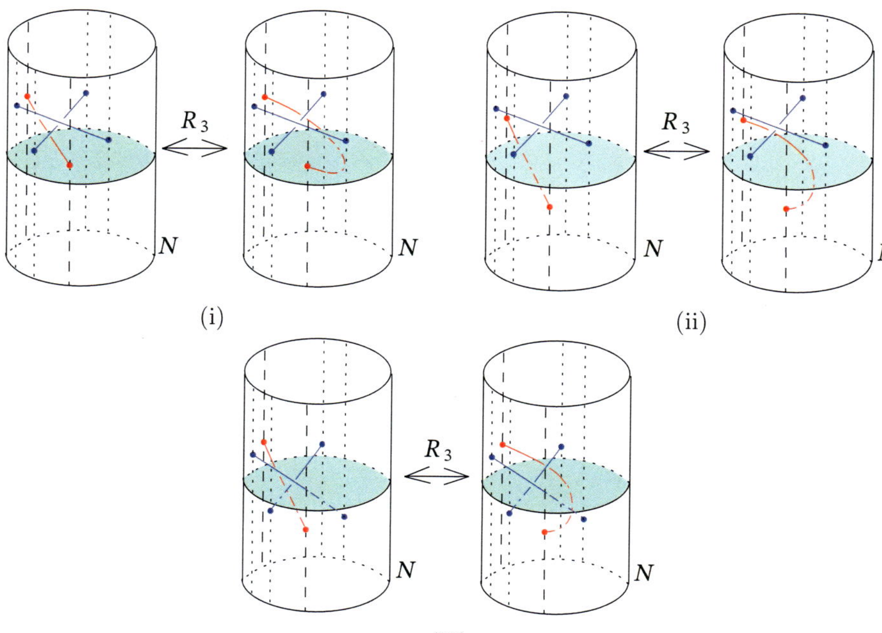

FIGURE 3.16. Different cases for R_3-move in $\hat{M}$

For i), there are two generic positions for $\hat{F}$ shown in Figure 3.16 (i) yielding arrow diagrams shown in Figure 3.17 (i′). It is easy to see that arrow diagrams of T_2 and T_1 differ either by Q_4-move followed by Q_3-move or first by two Q_5-moves followed by Q_4-move then followed by Q_3-move.

For ii), there are three generic positions for $\hat{F}$ shown in Figure 3.16 (ii) yielding arrow diagrams for T_2 on the left of Figure 3.17 (ii′). It is easy to see that arrow diagrams of T_2 and T_1 differ on $\hat{F}$ by either Q_3-move followed by Q_5 or first by Q_5-move followed by Q_3-move.

Finally, for iii), Figure 3.16 (iii) shows one of generic positions for $\hat{F}$ which yields arrow diagram for T_2 in Figure 3.17 (iii′). It is easy to see that arrow diagram of T_2 and T_1 differ by two Q_5-moves followed by Q_3-move and then followed by Q_5-move.

In summary, if L_1 and L_2 differ by R_3-move in $\hat{M}$, then their arrow diagrams differ on $\hat{F}$ by a sequence of Q_3, Q_4, and Q_5-moves. □

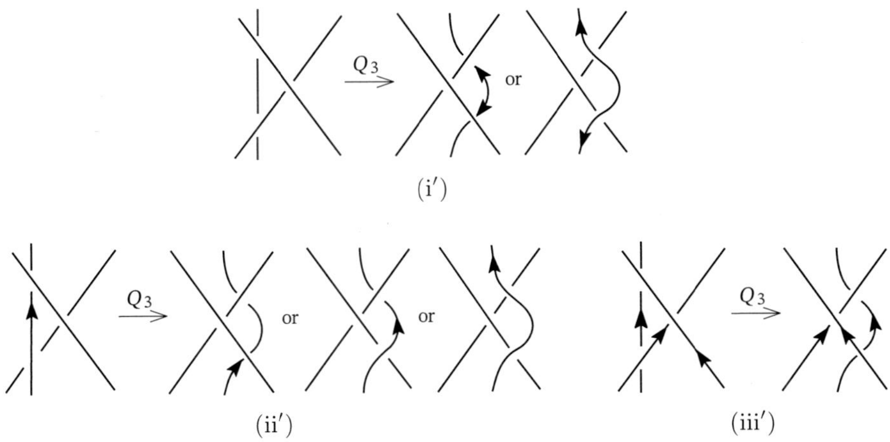

FIGURE 3.17. Arrow diagram on $\hat{F}$ for R_3-move

A version of the following theorem for framed links was used in [**13**], [**11**], and [**12**] without being formally stated. For $M = F \times S^1$, a version of this result can be found in [**5**] (see Theorem 1, pp. 2) and it also stated as Corrolary 3.5 in this paper. Since arrow diagrams have recently been used for computing KBSM of $M = F \times S^1$ over the ring of rational functions $\mathbb{Q}(A)$ in [**4**], we outline its proof below based on Lemma 3.1, Lemma 3.2, and Lemma 3.3.

THEOREM 3.4. *Generic links L_1 and L_2 in $\hat{M}_\beta$ are isotopic if and only if their arrow diagrams $\mathcal{D}_1$ and $\mathcal{D}_2$ differ by a finite number of Q_1-Q_5 and $Q_{(1,\beta)}$-moves on $\hat{F}_\beta$.*

PROOF. Suppose that L_1 and L_2 only differ by one of R_0-R_3-moves in 2-handle $N \subset \hat{M}$. Then, as shown in Figure 3.4, Figure 3.5, Lemma 3.1, Lemma 3.2, and Lemma 3.3, their corresponding arrow diagrams differ by $Q_0 - Q_5$-moves on $\hat{F}$. If L_1 and L_2 differ in 2-handle $H \subset \hat{M}_\beta$ by one of R_β, $R_{1_\beta} - R_{3_\beta}$-moves then, as we noted before, $R_{1_\beta} - R_{3_\beta}$-moves in $\hat{M}_\beta$ are compositions of R_0-R_3 in $\hat{M}$ and R_β-move in $\hat{M}_\beta$. Therefore, corresponding moves $Q_{1_\beta} - Q_{5_\beta}$ between their arrow diagrams on $\hat{F}_\beta$ are compositions of $Q_0 - Q_5$ and $Q_{(1,\beta)}$-moves in $\hat{F}_\beta$. It follows that arrow diagrams of generic links in $\hat{M}_\beta$ related by R_β, $R_{1_\beta} - R_{3_\beta}$-moves differ by $Q_{(1,\beta)}$, Q_{1_β}-Q_{5_β}-moves on $\hat{F}_\beta$.

Assume that arrow diagrams $\mathcal{D}_1$ and $\mathcal{D}_2$ differ on $\hat{F}_\beta$ by one of the local moves $Q_1 - Q_5$ and $Q_{(1,\beta)}$.

If $\mathcal{D}_1$ and $\mathcal{D}_2$ are related by one of $Q_1 - Q_5$-moves on $\hat{F}_\beta$ then, there is a 2-disk on $\hat{F}$, such that solid torus $V = D^2 \times S^1$ intersects L_1 and L_2 along tangles T_1 and T_2, respectively. Using isotopy that proceeds along fibers, we push L_1 to L_2 in $\hat{M}$ (so their arrow diagrams remain same outside of D^2) and tangles T_1 and T_2 have the same ends on boundary of V. Therefore, we may assume that L_1 and L_2 are identical outside V and T_1 and T_2 have the same endpoints on the boundary of V.

If arrow diagrams of T_1 and T_2 differ by Q_1, Q_2, or Q_3-move then T_1 and T_2 are isotopic in V via an isotopy that fixes boundary of V. Therefore, L_1 and L_2 are isotopic in $\hat{M}$, hence in $\hat{M}_\beta$.

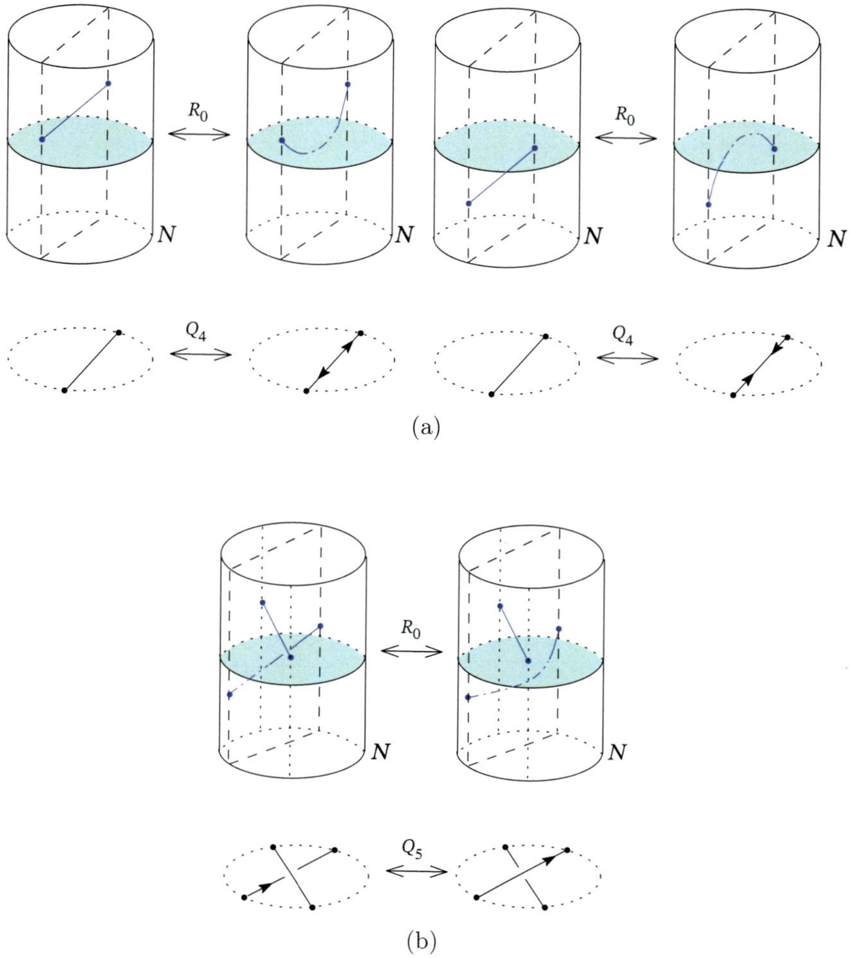

FIGURE 3.18. Isotopy by Q_4 and Q_5-moves

If arrow diagrams of T_1 and T_2 differ by either Q_4-move or Q_5-move then T_1 and T_2 differ by R_0-move in V via an isotopy that fixes boundary of V (see Figure 3.18(a) and (b)). Therefore, again L_1 and L_2 are isotopic in $\hat{M}$, so also in $\hat{M}_\beta$.

If $\mathcal{D}_1$ and $\mathcal{D}_2$ differ on $\hat{F}_\beta$ by $Q_{(1,\beta)}$-move then there is a 2-disk D^2 centered at β, such that solid torus $V = D^2 \times S^1$ intersects L_1 and L_2 along tangles T_1 and T_2. We may assume (by similar arguments that we used before) that boundary of T_1 and T_2 coincide, and L_1 and L_2 are identical outside V. Let α be an arc with ends $\{A, B\}$ on boundary of V. Using R_β-move, we isotope α inside 2-handle H (fixing A and B) to arc $R_\beta(\alpha)$ shown in Figure 3.19 and then we attach H to the torus boundary of $\hat{M}$ along $(1, \beta)$-curve. Therefore, $R_\beta(\alpha)$ becomes a part of $(1, \beta)$ curve on boundary of V containing A and B (see Figure 3.19). Clearly, $R_\beta(\alpha)$ is isotopic to T_2, so consequently, R_β-move in $\hat{M}_\beta$ throws L_1 into L_2. $\square$

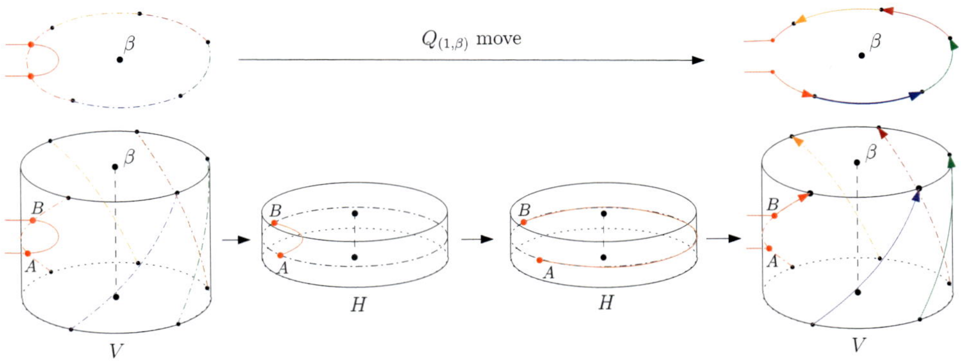

FIGURE 3.19. $Q_{(1,\beta)}$-move on $\hat{F}_\beta$

COROLLARY 3.5. *Links L and L' in $M = F \times S^1$ are isotopic if and only if their arrow diagrams $\mathcal{D}_L$ and $\mathcal{D}_{L'}$ are related by a finite sequence of $Q_1 - Q_5$-moves on F.*

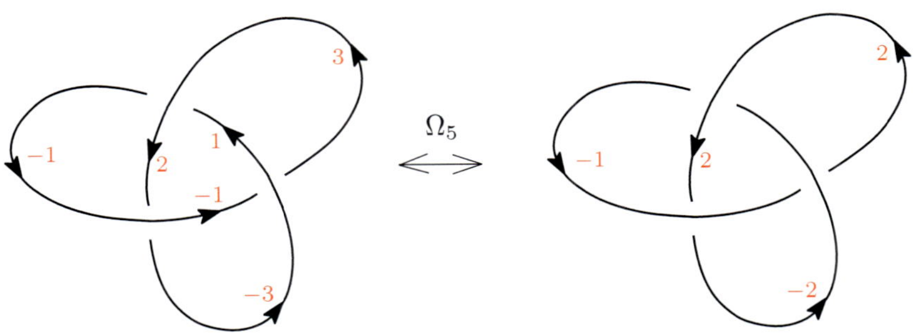

FIGURE 3.20. Reduced Arrow Diagrams

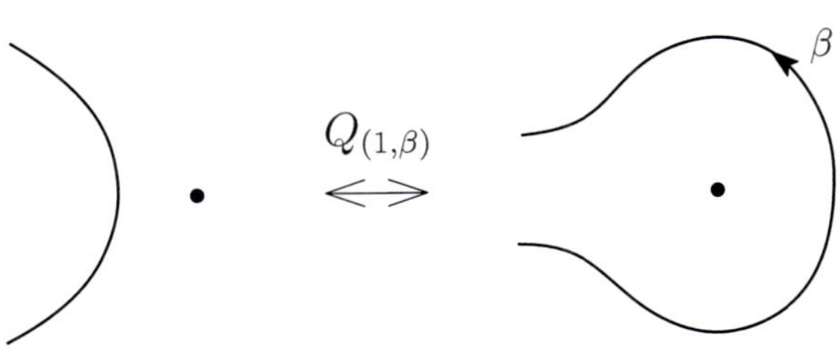

FIGURE 3.21. $Q_{(1,\beta)}$-move

REMARK 3.6. The natural isotopy of generic links in $\hat{M}$ which proceeds along fibers (mentioned first in Section 2) can be obtained as a composition of R_0-moves in Figure 3.3. Clearly, generic links L_1 and L_2 in $\hat{M}$ are related by such an isotopy if and only if their arrow diagrams $\mathcal{D}_L$ and $\mathcal{D}_{L'}$ are related by a finite number of Q_4 and Q_5-moves on $\hat{F}$ (see Figure 3.18). For convenience, we simplify arrow diagram $\mathcal{D}_L$ on $\hat{F}_\beta$ of a link L in $\hat{M}_\beta$ by using the convention that an arrow with an integer n

on an arc α between two successive crossings (see Figure 3.20) stands for $|n|$ arrows on α pointing in the same direction if $n \geq 0$ and in the opposite direction when $n < 0$.

Moreover, we can also represent $Q_{(1,\beta)}$-move on $\hat{F}_\beta$ by arrow diagram in Figure 3.21.

4. Isotopy of links in $\hat{M}_\beta$ via moves on shadow diagrams

An action of the first homology group $H_1(F) = H_1(F, \mathbb{Z})$ of a closed surface F on the set of isotopy classes of links in $p : M \to F$ with Euler number β was defined in [**20**]. We now take a closer look at this action for links in $\hat{M}_\beta$ to describe it using arrow diagrams in $\hat{F}_\beta$. Let L be a generic link in $\hat{M}_\beta$, let γ be a curve in $\hat{F}_\beta$, and let $p : \hat{F} \times S^1 \to \hat{F}$ denote the natural projection. We may assume that γ intersects $p(L)$ transversely at a finite number of points, avoiding the crossings of $p(L)$. Let $\alpha = [a, b]$ be a segment of L such that $p(\alpha)$ intersects γ at exactly one point. Suppose that $p(a)$ is to the left of γ and $p(b)$ is to the right of γ. As in [**20**], we replace the interval α with α', as illustrated on the right side of Figure 4.1 to obtain a new generic link L' from L.

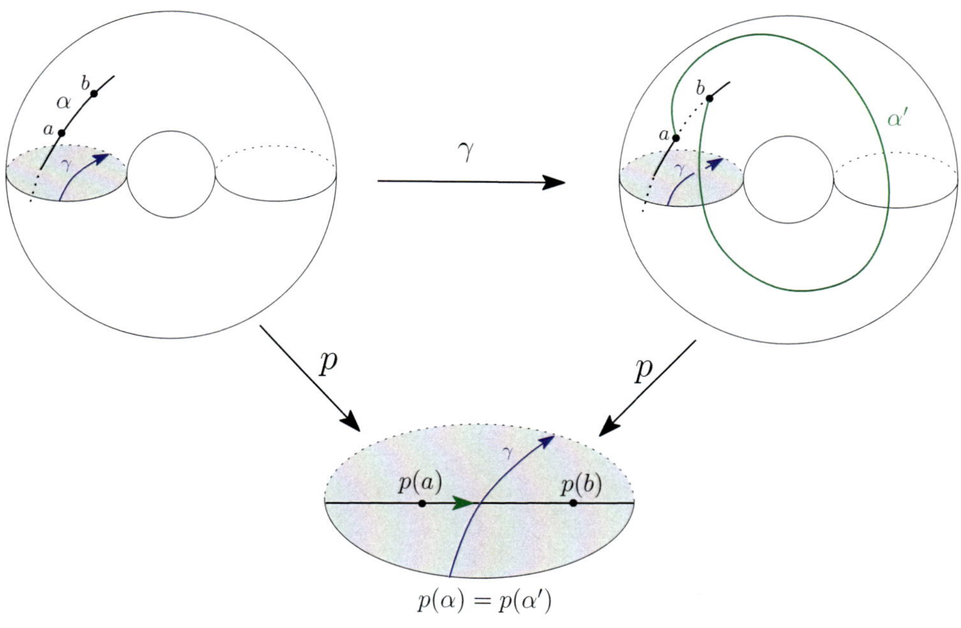

FIGURE 4.1. Action of $H_1(F, \mathbb{Z})$ on isotopy classes of links in $\hat{M}_\beta$

REMARK 4.1. Since arrow diagrams up to $\Omega_1 - \Omega_5$ and $\Omega_{(1,\beta)}$-moves are in bijection with ambient isotopy classes of links in $\hat{M}_\beta$, action of $H_1(F)$ on the generic links in $\hat{M}_\beta$ can be easily described in terms of arrow diagrams in $\hat{F}_\beta$. Let L' be obtained from L by replacing α with α' in $\hat{M}$. As is easy to see, this operation corresponds to adding an arrow to the corresponding arc in the arrow diagram $\mathcal{D}_L$ of L. Thus, at each crossing point c of $\mathcal{D}_L$ and γ, we add (as in Figure 4.1) an arrow directed from $p(a)$ to $p(b)$ (see, e.g., Figure 4.2).

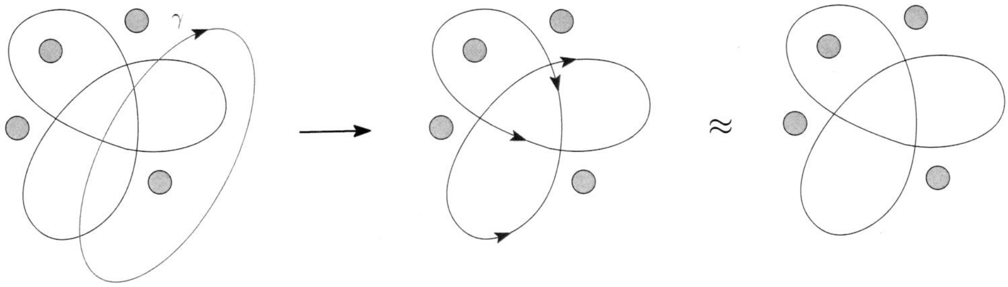

FIGURE 4.2. Action of $H_1(F, \mathbb{Z})$ on links in $\hat{M}$

EXAMPLE 4.2. Consider links L and L' in $M = T^2 \times S^1$ shown in Figure 4.3.

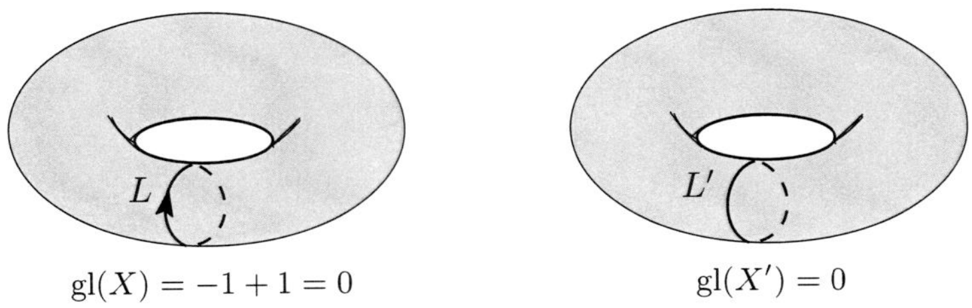

FIGURE 4.3. Non-isotopic links L and L' in $T^2 \times S^1$ with the same gleam

Although arrow diagrams of L and L' are not equivalent by $Q_1 - Q_5$-moves (i.e., L and L' are not isotopic in M), as one may check $\mathrm{gl}(X_L) = \mathrm{gl}(X_{L'}) = 0$. In particular, shadows of L and L' are same on T^2.

For shadow diagrams on F, V. Turaev in [**20**] defined local moves (we will call them *shadow moves* on F) shown in Figure 4.4. We say that two shadow diagrams $s(D_1)$ and $s(D_2)$ on F are *isotopic* if $s(D_1)$ and $s(D_2)$ differ by a finite number of S_1 - S_3-moves. An isotopy class of $s(D)$ on F is denoted by $S(D)$ and we call it a *shadow link*. If two generic links are isotopic in M, then their shadows are isotopic on F (see Lemma 3.1 of [**20**], pp. 48). Since every link L in M is isotopic to a generic link L', we can define shadow link $S(L)$ on F using shadow diagram $sh(L')$. As pointed out in [**20**], $S(L)$ does not depend on the choice of L'.

Example 4.2 shows that a shadow diagram of a link L in $M = F \times S^1$ does not determine its isotopy class. Furthermore, the following result was proved in [**20**].

THEOREM 4.3 ([**20**], Theorem 3.2). *For an oriented S^1-bundle $p : M \to F$ over a closed surface F, the mapping that assigns to each link L in M its shadow diagram $sh(L)$ on F establishes a bijective correspondence between the isotopy types of links in M modulo the action of $H_1(F, \mathbb{Z})$ and the set of integral* shlinks *$S(L)$ on F with total gleam $-\chi(p)$.*

In Remark (part 3) on page 53 of [**20**], V. Turaev mentioned that results of subsection 3 and Theorem 3.2 of his paper remain true for F with non-empty boundary. This is particularly interesting when F is a planar since, in this case, the action of $H_1(F)$ on isotopy classes of links in $F \times S^1$ is actually trivial.

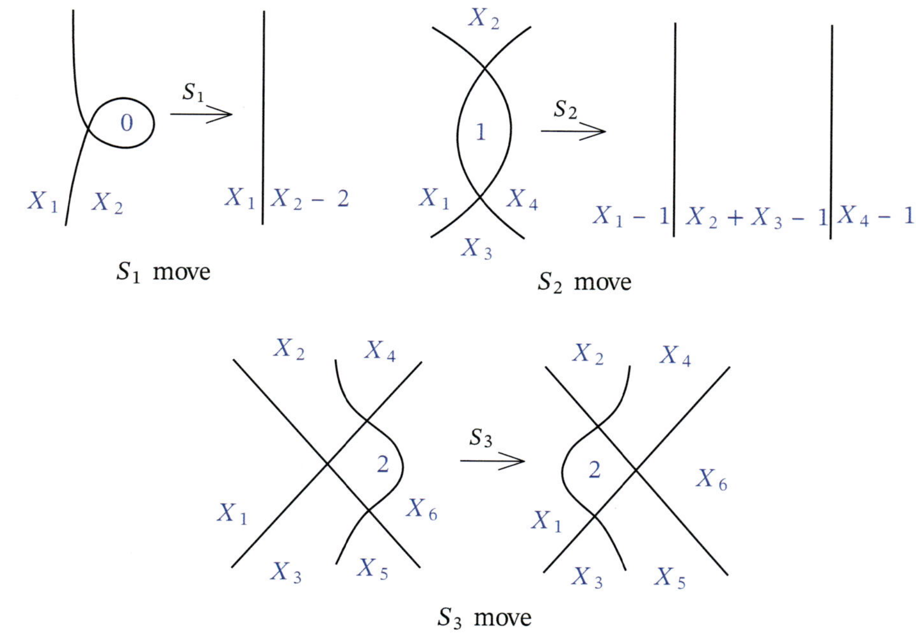

FIGURE 4.4. Shadow moves S_1, S_2 and S_3

LEMMA 4.4. *Let F be a planar and $M = F \times S^1$. Then $H_1(F)$ acts trivially on the set of isotopy classes of generic links in M.*

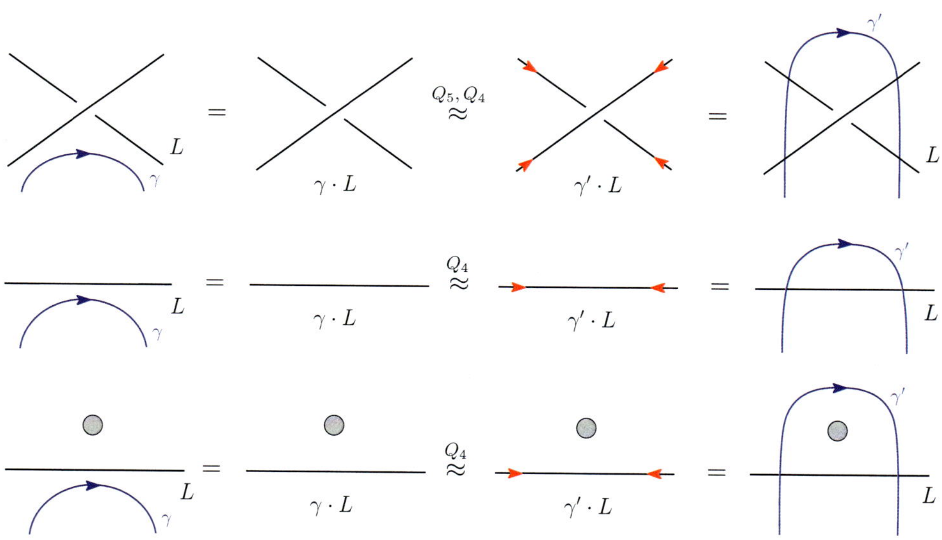

FIGURE 4.5. Action of γ and γ' on L

PROOF. Let L be a link in M with arrow diagram $\mathcal{D}_L$. Using Remark 4.1, we see that for curves γ and γ' shown in Figure 4.5, $\gamma \cdot \mathcal{D}_L = \gamma' \cdot \mathcal{D}_L$.

Applying the operations shown in Figure 4.5, one may modify any curve γ on F into a curve γ' that is disjoint from $\mathcal{D}_L$. Therefore, $\gamma \cdot \mathcal{D}_L = \gamma' \cdot \mathcal{D}_L = \mathcal{D}_L$.

Since the action of $H_1(F)$ is defined is well-defined on the set of isotopy classes of links in M (see argument preceding Theorem 3.2 of [**20**]), $[\gamma] \cdot L = [\gamma'] \cdot L = L$, it follows that $H_1(F)$ acts trivially on the set of ambient isotopy classes of generic links in M. □

PROPOSITION 4.5. *Let F be a planar or $F = S^2$. Then there is a bijection between isotopy classes of links in $M = F \times S^1$ and shadow links on F with total gleam* 0.

We note, that for $F = S^2$ and S^1-bundle $p : M \to F$ the above statement is Corollary 3.3 of [**20**].

PROOF. Assume $F \neq S^2$, then the total gleam of any shadow diagram on F is zero and $H_1(F)$ acts trivially on the isotopy classes of links in M by the Lemma 4.4. Thus, by the Theorem 4.3, there is a bijection between the isotopy classes of links in $F \times S^1$ and the shadow links $S(L)$ on F. □

5. Arrow and shadow diagrams

For $M = F \times S^1$ and a generic link L in M, we choose a generic link L' in M which satisfies Property 2.4. Let $\mathcal{D}'$ be its arrow diagram on F. Consider a 4-valent graph $G(\mathcal{D}') = p(L')$ on F with arrows inherited from $\mathcal{D}'$ (see Figure 5.1) and

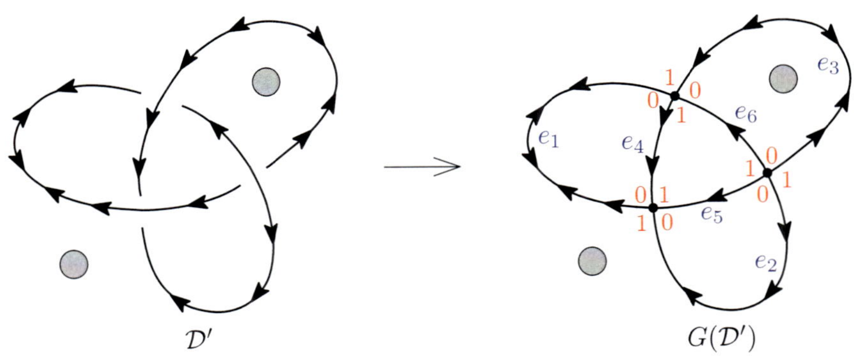

FIGURE 5.1. $G(\mathcal{D}')$ corresponding to $\mathcal{D}'$

integers 1 and 0 assigned to the corners of regions determined by $p(L')$ according to the rule as shown in Figure 5.2.

An edge e of $G(\mathcal{D}')$ is called *separating* if regions on the left and on the right of e are different. Otherwise, we call it *non-separating*. In Figure 5.1, all the edges of $G(\mathcal{D}')$ are separating while both edges of $G(\mathcal{D}')$ in the middle of Figure 5.3 are non-separating. Let X be a region cut from F by edges $e_1, \ldots, e_k$ of $G(\mathcal{D}')$. For each separating edge e_i, we label corresponding line segment on boundary of X by a_i and, for each non-separating edge e_j, we label corresponding two line segments on boundary of X by a_j and a'_j. Using orientation of X induced from F, we assign integers $n(a_i)$ and $n(a'_i)$ to each a_i and a'_i according to the following rule. Let n_i be the sum of positive and negative arrows on e_i. We define $n(a_i) = n_i$ if e_i is separating and we set $n(a_i) = n_i$ and $n(a'_i) = -n_i$ when e_i is non-separating.

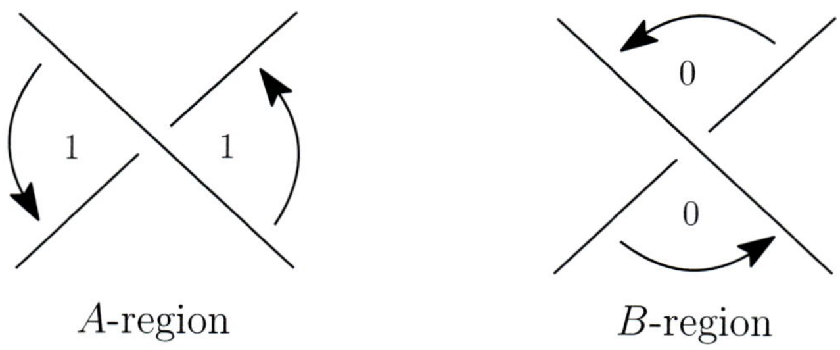

FIGURE 5.2. Convention for A and B-region for $\mathcal{D}'$

Therefore, X comes equipped with integers assigned to each arc segment a_i and a_i' on its boundary and integers 1 or 0 assigned to each of its corners (e.g., see the diagram on the right of Figure 5.3). Let $|C(\mathcal{D}')|$ be the number of vertices of $G(\mathcal{D}')$,

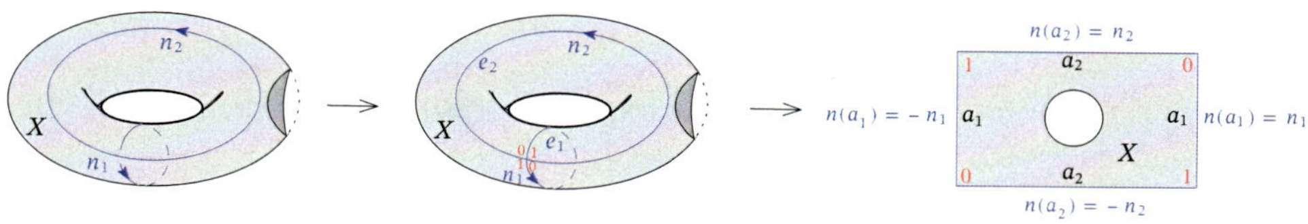

FIGURE 5.3. $s(D')$ obtained from $\mathcal{D}'$ on punctured torus F

$|C_+(X)|$ be the sum of integers assigned to corners of X, and n_X be the sum of all integers $n(a_i)$ and $n(a_i')$. Given an arrow diagram $\mathcal{D}'$, we find its corresponding shadow diagram $s(D')$.

PROPOSITION 5.1. *Let $\mathcal{D}$ be an arrow diagram of a generic link L in $M = F \times S^1$ which satisfies Property 2.4. The gleam $\mathrm{gl}(X)$ assigned to region X cut from F by $p(L)$ in the shadow diagram $sh(L)$ of L is given by*

$$\mathrm{gl}(X) = |C_+(X)| - n_X.$$

Furthermore, the total gleam of the shadow diagram $sh(L)$ is zero.

PROOF. Let φ, ω_X, and θ_X be defined as in Section 2. Clearly, $|C_+(X)|$ is the number of corners C of X with $\deg_C(\varphi) = -1$ and $\deg_A(\varphi) = 2n_X$, where A is the union of arcs in the boundary of X (we can ignore $B = \partial F \cap \partial X$ as $\deg_B(\varphi) = 0$). Therefore, by Proposition 2.5,

$$\mathrm{gl}(X) = |C_+(X)| - \frac{1}{2}\deg_A(\varphi) = |C_+(X)| - n_X.$$

Consequently, we find a shadow diagram $sh(L)$ using its arrow diagram $\mathcal{D}$ by assigning 1 and -1 to the region on the left and on the right for each arrow of $\mathcal{D}$ as shown in Figure 5.4 and assigning either 0's or 1's at each crossing of $\mathcal{D}$ following convention in Figure 5.2. It follows that the gleam $\mathrm{gl}(X)$ of X is the sum of all these numbers. Furthermore, as it is easy to see, the contribution to the total gleam of $sh(L)$ by all arrows of $\mathcal{D}$ is 0 and contribution to the total gleam by all crossings of $\mathcal{D}$ is $2|C(\mathcal{D})|$ (i.e., twice the number of double points of $p(L)$). Thus, the total gleam of the shadow diagram $sh(L)$ is zero. $\square$

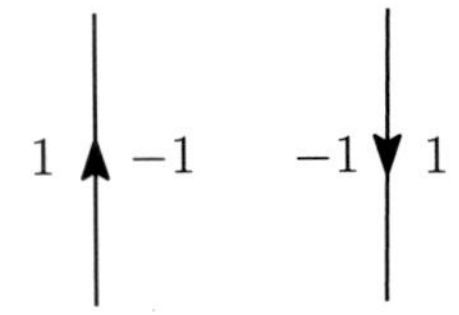

FIGURE 5.4. Contribution by an Arrow

REMARK 5.2. For a link L in $F \times S^1$ with arrow diagram $\mathcal{D}'$ in Figure 5.3, we see that

$$\mathrm{gl}(X) = |C_+(X)| - n_X = 2 - (n_1 + n_2 - n_1 - n_2) = 2$$

and the total gleam of $sh(L)$ is $\mathrm{gl}(X) - 2 = 0$.

EXAMPLE 5.3. Let $\mathcal{D}$ be an arrow diagram of a link L in $S^2 \times S^1$ shown on the left of Figure 5.5. Using algorithm from proof of the Proposition 5.1, we see that

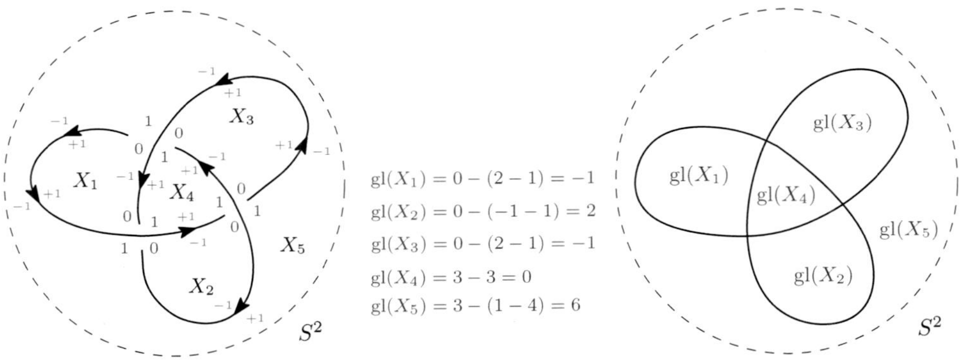

FIGURE 5.5. $s(D)$ obtained from $\mathcal{D}$ on S^2

$\mathrm{gl}(X_1) = -1$, $\mathrm{gl}(X_2) = 2$, $\mathrm{gl}(X_3) = -1$, $\mathrm{gl}(X_4) = 0$, and $\mathrm{gl}(X_5) = 6$. Therefore, the corresponding $sh(L)$ is shown on the right of Figure 5.5. Clearly, its total gleam is zero.

PROPOSITION 5.4. *Let $M = F \times S^1$ be a trivial S^1-bundle. If arrow diagrams $\mathcal{D}$ and $\mathcal{D}'$ differ by a finite sequence of local moves $Q_1 - Q_5$ on F, then the corresponding shadow diagrams $s(D)$ and $s(D')$ differ by a finite sequence of shadow moves $S_1 - S_3$.*

PROOF. Suppose that $\mathcal{D}$ and $\mathcal{D}'$ differ by one of $Q_1 - Q_5$-moves on F. Then, as it is shown in Figure 5.6 a)-e), corresponding shadow diagrams $s(D)$ and $s(D')$ differ on F by $S_1 - S_3$-moves respectively. □

REMARK 5.5. Given an arrow diagram $\mathcal{D}$ on F, we construct a link L in $M = F \times S^1$ using Proposition 2.1. After applying an isotopy which proceeds along the fibers, we choose a link L' in M that satisfies Property 2.4. If $\mathcal{D}'$ is the arrow diagram corresponding to L', then by Proposition 5.1, $sh(L') = s(\mathcal{D}')$. Since by Theorem 3.4, $\mathcal{D}$ and $\mathcal{D}'$ differ by a sequence of $Q_1 - Q_5$-moves. It follows from Proposition 5.4 that $s(D)$ and $s(D')$ differ by a sequence of $S_1 - S_3$-moves. Moreover, by Remark 3.6, $\mathcal{D}$ and $\mathcal{D}'$ differ by Q_4 and Q_5-moves, so in fact $s(D) = s(D')$.

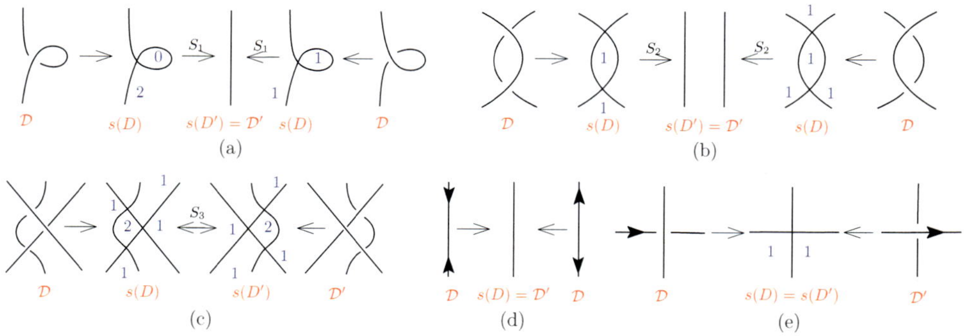

FIGURE 5.6. $Q_1 - Q_5$-moves on $\mathcal{D}$ as consequences of $S_1 - S_3$-moves on $s(D)$

Hence, we may simply use $\mathcal{D}$ to compute $sh(L)$ for a link L in M. In particular, $s(\mathcal{D})$ is well-defined and independent of the choice of $\mathcal{D}$ up to Q_4 and Q_5-moves.

REMARK 5.6. For a shadow diagram $s(D)$ on a non-planar surface $F \neq S^2$ with total gleam zero, $H_1(F)$ acts non-trivially on isotopy classes of links in $M = F \times S^1$. Hence, for a given $s(D)$ there are infinitely many non-isotopic links[6] L in M with $sh(L) = s(D)$ (see e.g., Example 4.2). More precisely, for $[\gamma] \in H_1(F)$ with $[\gamma] \neq 0$ and a link L in M, links $L' = \gamma \cdot L$ and L are not isotopic in M (e.g., $\mathcal{D}_L$ and $\mathcal{D}_{L'}$ for L and L' are clearly not related by $Q_1 - Q_5$-moves), however $sh(L) = sh(L')$.

PROPOSITION 5.7. *Let $M = F \times S^1$ be a trivial S^1-bundle over F and $s(D)$ be a shadow diagram on F whose total gleam is 0. Then, there exists an arrow diagram $\mathcal{D}'$ on F with $s(D') = s(D)$.*

PROOF. Using the argument from the proof of Theorem 3.2 of [**20**], there is a link L in M with $s(D) = sh(L)$. Taking its arrow diagram $\mathcal{D}_L$ on F and applying Proposition 5.1 we obtain corresponding shadow diagram on F which, one may easily check, is the same as $s(D)$. □

REMARK 5.8. Given $s(D)$ on F we may find an arrow diagram $\mathcal{D}$ directly (i.e., without involving part of the proof of Theorem 3.2 of [**20**]) by applying algorithm from our proof of Proposition 5.1 in the opposite direction. Namely, we assign variable n_i to each edge e_i of $s(D)$. Then, for each vertex of $s(D)$, we choose a crossing by assigning 0 and 1 to the regions according to the rule in Figure 5.7. Let X be the region determined by edges $e_{i_1}, e_{i_2}, \ldots, e_{i_k}$ of $s(D)$ and $I(X)$ be the set of i_j corresponding to separating edges e_{i_j}. We also denote $I'(X)$ the set of these i_j's which correspond to the non-separating edges e_{i_j}. Notice that, for the non-separating edges $n(a_{i_j}) = -n(a'_{i_j})$, thus

$$n_X = \sum_{m \in I(X)} n(a_m) + \sum_{m \in I'(X)} (n(a_m) + n(a'_m)) = \sum_{m \in I(X)} n(a_m).$$

[6]There are infinitely many non-equivalent (modulo $Q_1 - Q_5$-moves) arrow diagrams $\mathcal{D}$ which give $s(D)$.

FIGURE 5.7. Convention for a choice of crossing at double points of $s(D)$

Hence, in particular, $n(a_{i_j})$ can be assigned any integer value for each nonseparating edge e_{i_j}. By Proposition 5.1 each region X cut from F by $s(D)$ yields an equation

$$\sum_{m \in I(X)} n(a_m) = |C_+(X)| - \mathrm{gl}(X).$$

If $X_1, \ldots, X_t$ are regions cut from F by $s(D)$, then the above procedure yields the system of t equations with variables $n_1, \ldots, n_r$

$$\sum_{m \in I(X_i)} n(a_m) = |C_+(X_i)| - \mathrm{gl}(X_i),\ 1 \le i \le t, \tag{1}$$

where r is the number of edges of $s(D)$. The system (1) is usually undetermined (e.g., whenever $s(D)$ has non-separating edges). Since by Proposition 5.7, there is an arrow diagram $\mathcal{D}_L$ which yields $s(D)$, (1) has at least one integer solution for $n_1, \ldots, n_r$, hence, also infinitely many whenever $t < r$.

REMARK 5.9. For trivial S^1-bundle $M = F \times S^1$ over a planar surface F or $F = S^2$ by Proposition 4.5, there is a one-to-one correspondence between shadow links $S(D)$ on F and isotopy classes of links in M. Therefore, given $s(D)$, we can find a link L in M with $sh(L) = s(D)$ which is unique up to ambient isotopy of M that proceeds along fibers. In such a case, the above algorithm results in a unique arrow diagram $\mathcal{D}$ (up to Q_4 and Q_5-moves).

EXAMPLE 5.10. Let $F = S^2$ and $s(D)$ be the shadow diagram on the left of Figure 5.8 of a link in $F \times S^1$. We use the method described above to find the arrow

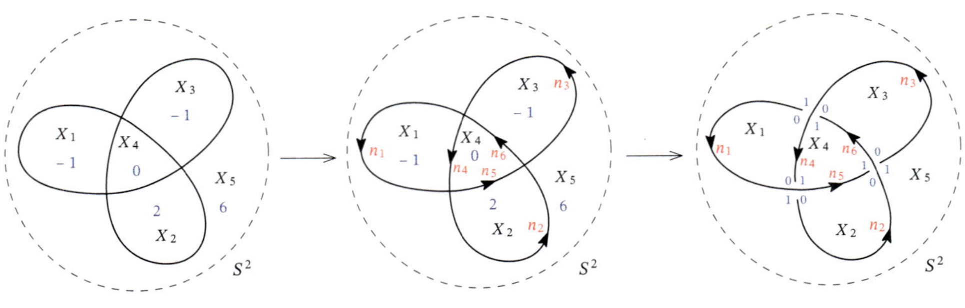

FIGURE 5.8. Arrow diagram $\mathcal{D}$ obtained from shadow diagram $s(D)$ on S^2

diagram $\mathcal{D}$ on S^2 for $s(D)$. Thus, we assign variables $n_1, n_2, \ldots, n_6$ to edges of $s(D)$ and 0 and 1 to corners of each region X_i determined by $s(D)$ using convention from

Figure 5.7. Since $s(D)$ has no non-separating edges, $\mathcal{D}$ is determined uniquely (up to Q_4 and Q_5-moves) from system of equations determined by the regions X_1, X_2, X_3, X_4:

$$
\begin{aligned}
X_1 &: n_1 - n_4 = 0 - \mathrm{gl}(X_1) = 1\\
X_2 &: n_2 - n_5 = 0 - \mathrm{gl}(X_2) = -2\\
X_3 &: n_3 - n_6 = 0 - \mathrm{gl}(X_3) = 1\\
X_4 &: n_4 + n_5 + n_6 = |C_+(X_4)| - \mathrm{gl}(X_4) = 3 - 0 = 3\\
X_5 &: -n_1 - n_2 - n_3 = |C_+(X_5)| - \mathrm{gl}(X_5) = 3 - 6 = -3.
\end{aligned}
$$

Solving the above system gives: $n_1 = 4 - n_5 - n_6$, $n_2 = -2 + n_5$, $n_3 = 1 + n_6$, and $n_4 = 3 - n_5 - n_6$, where $n_5, n_6 \in \mathbb{Z}$. Taking $n_5 = n_6 = 1$ yields $n_1 = 2$, $n_2 = -1$, $n_3 = 2$, and $n_4 = 1$ which, in turn, yields an arrow diagram $\mathcal{D}$ on the left of Figure 5.5.

REMARK 5.11. Notice that arrow diagram $\mathcal{D}$ on the right of Figure 5.8 is equivalent (up to Q_4 and Q_5) to the one on the left of Figure 5.5 as shown in Figure 5.9.

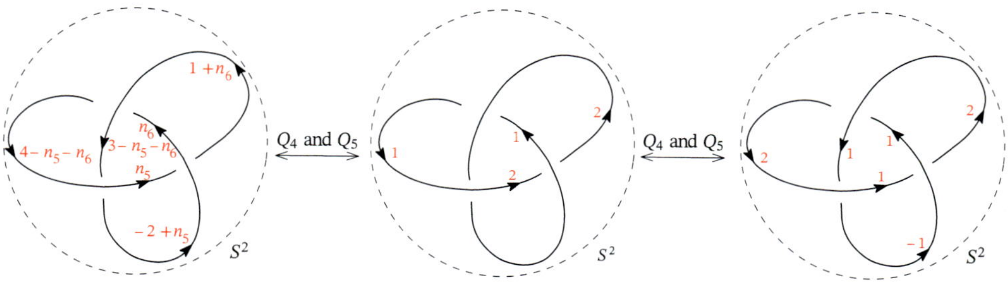

FIGURE 5.9. Arrow diagram $\mathcal{D}$ for $s(D)$ on S^2 up to Q_4 and Q_5-moves

PROPOSITION 5.12. *Let $M = F \times S^1$ be a trivial S^1-bundle over a planar F or $F = S^2$. If shadow diagrams $s(D)$ and $s(D')$ with total gleams 0 differ on F by a finite sequence of shadow moves $S_1 - S_3$ then the corresponding arrow diagrams $\mathcal{D}$ and $\mathcal{D}'$ differ on F by a finite sequence of $Q_1 - Q_5$-moves.*

PROOF. Suppose that $s(D)$ and $s(D')$ differ by one of $S_1 - S_3$-moves on F. By Proposition 4.5, we choose links L and L' in M with $sh(L) = s(D)$ and $sh(L') = s(D')$. Let $\mathcal{D}$ and $\mathcal{D}'$ be their corresponding arrow diagrams on F. Since by Theorem 3.2 of [**20**], links L and L' are isotopic, hence, by Corollary 3.5, $\mathcal{D}$ and $\mathcal{D}'$ differ by a finite sequence of $Q_1 - Q_5$-moves. □

REMARK 5.13. Our proof of Proposition 5.12 does give explicit relationship between shadow moves $S_1 - S_3$ and arrow moves $Q_1 - Q_5$ on F for $M = F \times S^1$, where F is a planar or $F = S^2$. Let $s(D)$ and $s(D')$ be shadow diagrams on F with total gleam 0. If $s(D)$ and $s(D')$ differ by S_1-move, then by the algorithm in Remark 5.8, we find the arrow diagrams $\mathcal{D}$ and $\mathcal{D}'$ for $s(D)$ and $s(D')$. As shown in Figure 5.10, there are two choices for $\mathcal{D}$. However, as shown in Figure 5.10a) and b), in both the cases $\mathcal{D}$ and $\mathcal{D}'$ differ either by Q_1-move or by Q_5 and Q_1-moves.

If $s(D)$ and $s(D')$ differ by S_2-move, then using Remark 5.8, we need to consider four cases shown in Figure 5.11a)-d). For a), as we see in Figure 5.12, $\mathcal{D}$ and $\mathcal{D}'$

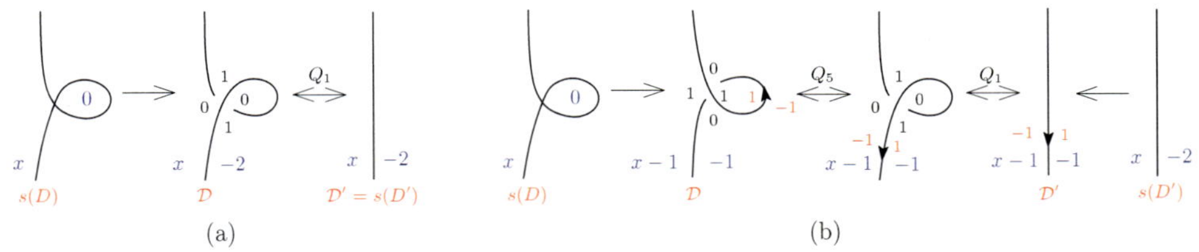

FIGURE 5.10. S_1-move

are related by Q_5 and Q_2-moves and, as one may check, the same is true for all remaining cases b)-d).

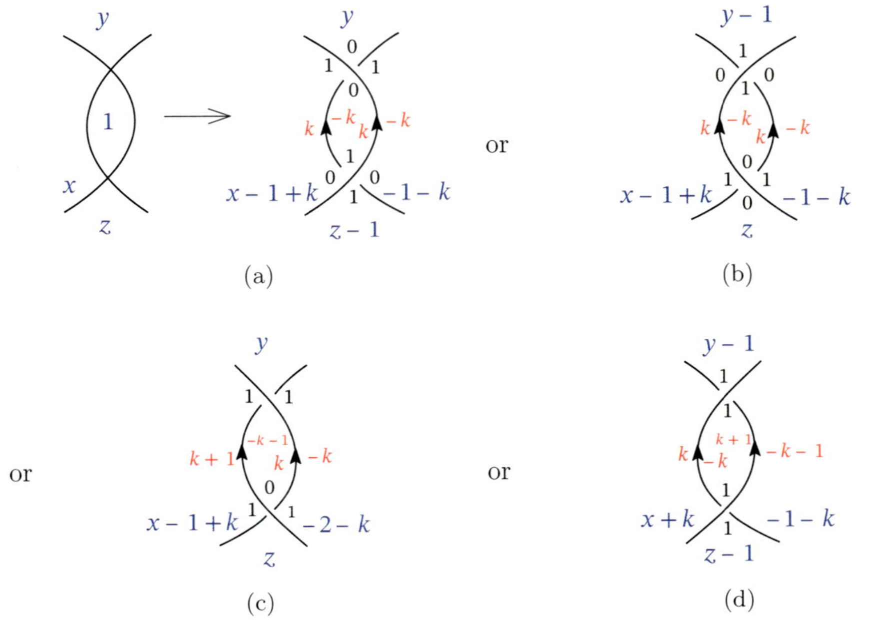

FIGURE 5.11. Different cases of arrow diagrams for S_2-move on $s(D)$

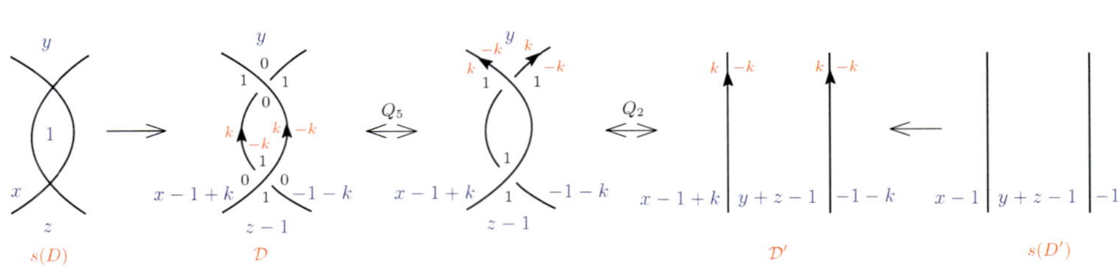

FIGURE 5.12. S_2-move for case (a) in Figure 5.11

Finally, if $s(D)$ and $s(D')$ differ by S_3-move, then using Remark 5.8, we need to consider eight cases a)-h) shown in Figure 5.13.

For a), as we see in Figure 5.14, $\mathcal{D}$ and $\mathcal{D}'$ are related by Q_5 and Q_3-moves. As one may check, this is also true for all remaining cases b)-h). Furthermore, as we

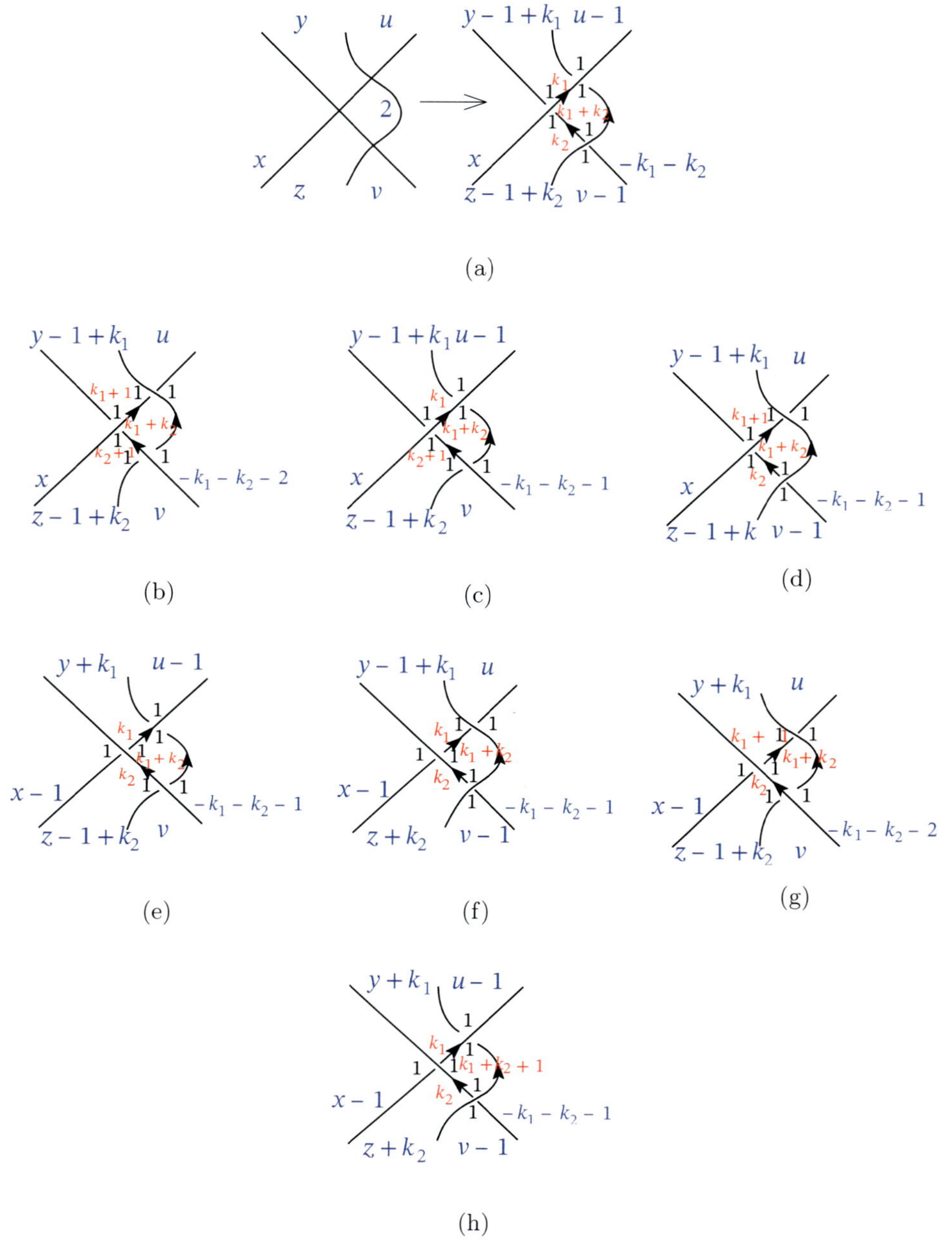

FIGURE 5.13. Different cases of arrow diagrams for S_3-move on $s(D)$

see from the above, $S_1 - S_3$-moves on shadow diagrams do not correspond directly to $Q_1 - Q_3$-moves on arrow diagrams. However, $S_1 - S_3$-moves yield isotopy of links in $F \times S^1$ which is a composition of $R_1 - R_3$-moves and the isotopy that proceeds along fibers (i.e., a composition of R_0-moves shown in Figure 3.18b)).

REMARK 5.14. For a trivial S^1-bundle $M = F \times S^1$ over a planar surface F or $F = S^2$, links in M up to isotopy are described either by their shadow diagrams (up

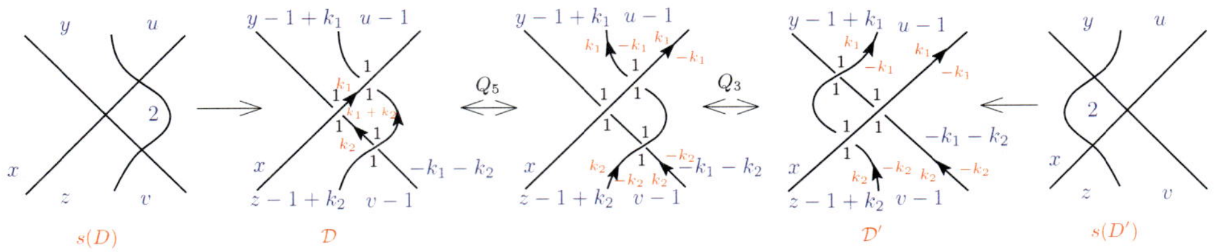

FIGURE 5.14. S_3-move for case (a) in Figure 5.13

to shadow moves) or by their arrow diagrams (up to arrow moves). In particular, Corollary 3.5 is a consequence of Theorem 4.3. However, when $F \neq S^2$ is not planar, Theorem 3.2 of [**20**] neither implies Corollary 3.5 nor Theorem 3.4.

We described above relations between shadow and arrow diagrams of links in a trivial S^1-bundle over F. When F is a closed surface and $\chi(p) = \beta \neq 0$, we consider links in $\hat{M}_\beta$ with their shadow and arrow diagrams on $\hat{F}_\beta$. Since arrow diagrams $\mathcal{D}$ on $\hat{F}_\beta$ up to $Q_1 - Q_5$ and $Q_{(1,\beta)}$-moves are in bijection with ambient isotopy classes of links in $\hat{M}_\beta$, we can use an arrow diagram and arrow moves to define corresponding shadow diagrams and shadow moves on $\hat{F}_\beta$. For $\hat{M} = \hat{F} \times S^1$, let $\pi : \hat{M} \to \hat{F}$ be the canonical projection onto $\hat{F}$ and $\pi(L) = D$ be the 4-valent graph on $\hat{F}$. Therefore, D is also the 4-valent graph on $\hat{F}_\beta$. We denote by X_i regions on $\hat{F}_\beta$ determined by D, $i = 1, \dots, k$. Then $\beta \in X_j$ for some $1 \leq j \leq k$ and for each $i = 1, \dots, k$, and using Proposition 2.5, we define

$$\mathrm{gl}(X_i) = \begin{cases} |C_+(X_i)| - n_{X_i}, & \text{if } i \neq j, \\ |C_+(X_j)| - n_{X_j} - \beta, & \text{if } i = j. \end{cases}$$

The assignment of $\mathrm{gl}(X_i)$ to each region X_i of D defines the shadow diagram $s(D)_\beta$ with total gleam $-\beta$, as one may verify.

PROPOSITION 5.15. *Let $\beta \neq 0$, and L and L' be links in $\hat{M}_\beta$. If L and L' are isotopic, then their corresponding shadow diagrams $sh(L)_\beta$ and $sh(L')_\beta$ differ on $\hat{F}_\beta$ by a sequence of $S_1 - S_3$, and S_β-moves (see Figure* 5.15*).*

PROOF. By Theorem 3.4, L and L' in $\hat{M}_\beta$ are isotopic if and only if their arrow diagrams $\mathcal{D}$ and $\mathcal{D}'$ differ by $Q_1 - Q_5$ and $Q_{(1,\beta)}$-moves. Furthermore, if $\mathcal{D}$ and $\mathcal{D}'$ differ by $Q_1 - Q_5$-moves on $\hat{F}$, then by Proposition 5.4, $s(D)_\beta$ and $s(D')_\beta$ are related by shadow moves $S_1 - S_3$. Therefore, it suffices to determine relation between $s(D')_\beta$ and $s(D)_\beta$, where $s(D')_\beta$ is obtained $\mathcal{D}'$ which differs from $\mathcal{D}$ by $Q_{(1,\beta)}$-move (see Figure 5.15). Let X_i and X_j (here i and j might be same) be regions in $G(\mathcal{D})$ to the left and right of an arc α with $\beta \in X_j$. Let $\mathrm{gl}(X_i) = x$ and $\mathrm{gl}(X_j) = y$. Denote by X_i' and X_j' regions of $G(\mathcal{D}')$ to the left and right of $Q_{(1,\beta)}(\alpha)$ accordingly. Then for $G(\mathcal{D}')$, $\beta \in X_i'$, and $\mathrm{gl}(X_i') = (x + \beta) - \beta = x$, $\mathrm{gl}(X_j') = (y - \beta) + \beta = y$ (see Figure 5.15). Hence, $Q_{(1,\beta)}$-move induces the following S_β-move shown in Figure 5.15. □

It follows from Proposition 5.15, that $sh(L)_\beta$ up to $S_1 - S_3$, S_β-moves is defined for isotopy classes of L in $\hat{M}_\beta$. Hence, we define a shadow link $S(L)_\beta$ on $\hat{F}_\beta$ as an equivalence class of $sh(L)_\beta$ modulo $S_1 - S_3$ and S_β-moves.

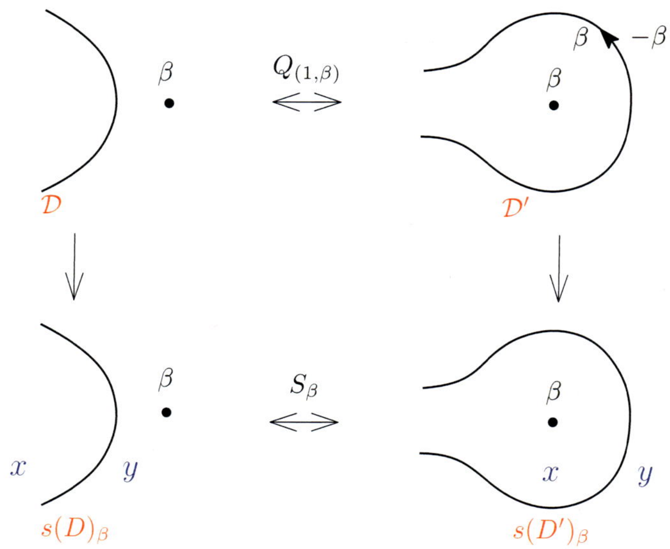

FIGURE 5.15. Shadow move S_β induced by $Q_{(1,\beta)}$-move on $\hat{F}_\beta$

As noted previously, for an S^1-bundle $p : M \to F$ over a closed surface $F \neq S^2$, shadow links $S(D)$ on F are not in bijection with isotopy classes of links in M, even when p is trivial. Using the prove of Theorem 3.2 of [**20**] for $p : M \to F$ with $\chi(p) = \beta$, given a shadow diagram $s(D)$ on F with total gleam $-\beta$, there is a link L in M with $sh(L) = s(D)$. An analogous statement holds for $s(D)\beta$ on $\hat{F}\beta$. Indeed, to construct a link $L \subset \hat{M}_\beta$, one simply modifies the algorithm described in Remark 5.8. That is, if X_i $(1 \leq i \leq t)$ are the regions determined by $s(D)_\beta$ on $\hat{F}_\beta$, with $\beta \in X_j$, then, using the notation from Remark 5.8, each region X_i with $i \neq j$ yields the equation

$$\sum_{m \in I(X_i)} n(a_m) = |C_+(X_i)| - \mathrm{gl}(X_i)$$

while the region X_j yields the equation

$$\sum_{m \in I(X_j)} n(a_m) = |C_+(X_j)| - (\mathrm{gl}(X_i) + \beta).$$

The system of equations above always has a solution. Indeed, given a shadow diagram $s(D)\beta$ with total gleam $-\beta$ on $\hat{F}_\beta$, we define a shadow diagram $s(D)$ on $\hat{F}$ by assigning $\mathrm{gl}(X_i)$ for $i \neq j$ and $\mathrm{gl}(X_j) + \beta$ for $i = j$. Since the total gleam of $s(D)$ is zero, Proposition 5.7 ensures the existence of a link $L \subset \hat{M} = \hat{F} \times S^1$ such that $sh(L) = s(D)$. We then view L as a link in $\hat{M}_\beta$, and, using its arrow diagram on $\hat{F}_\beta$, we find that $sh(L)_\beta = s(D)_\beta$. Consequently, the following lemma, analogous to Proposition 5.7, holds.

LEMMA 5.16. *The mapping that assigns to each link L in $\hat{M}_\beta$ its shadow diagram $sh(L)_\beta$ on $\hat{F}_\beta$ with total gleam $-\beta$ is surjective.*

Form the above lemma, we also naturally conclude a version of Theorem 3.2 of [**20**] for our model $\hat{M}_\beta$ using our terminology.

PROPOSITION 5.17. *The mapping that assigns to each link L in $\hat{M}_\beta$ its shadow diagram $sh(L)_\beta$ on $\hat{F}_\beta$ establishes a bijective correspondence between isotopy types of links in $\hat{M}_\beta$ modulo action of $H_1(F,\mathbb{Z})$ and the set of integral* shlinks *$S(L)_\beta$ on $\hat{F}_\beta$ with total gleam $-\beta$.*

PROOF. By Lemma 5.16, for shadow diagram $sh(L)_\beta$ with total gleam $-\beta$ on $\hat{F}_\beta$, there is link L in $\hat{M}_\beta$.

The remaining part of our proof essentially entails modifying the proof of Theorem 3.2 of [**20**] for $p: M \to F$ by adopting our notions for $\hat{M}_\beta$.

Indeed, let us consider links L and L' in $\hat{M}$ with arrow diagrams $\mathcal{D}$ and $\mathcal{D}'$ on $\hat{F}_\beta$ respectively, and corresponding shadow diagrams $sh(L)_\beta = sh(L')_\beta$. We show that there is a curve γ on $\hat{F}_\beta$, such that $\gamma \cdot L'$ is isotopic to L in $\hat{M}_\beta$. Let $\pi: \hat{M} \to \hat{F}$ be projection onto $\hat{F} \subseteq \hat{F}_\beta$. Given that $sh(L)_\beta = sh(L')_\beta$, it follows that $\pi(L) = \pi(L')$. Consequently, the corresponding graphs $G(\mathcal{D})$ and $G(\mathcal{D}')$ on $\hat{F}_\beta$ have identical vertices and edges. Since Q_4 and Q_5-moves preserve $sh(L)_\beta$ and $sh(L')_\beta$, we may assume that $\mathcal{D}$ and $\mathcal{D}'$ also have the same crossings. Hence, $G(\mathcal{D})$ and $G(\mathcal{D}')$ have the same vertices, edges and all assignments of integers 0 and 1 to their corners. Therefore, $G(\mathcal{D})$ and $G(\mathcal{D}')$ can only differ by number of arrows on their corresponding edges. Let X be a region cut from $G(\mathcal{D})$ and X' be the corresponding region cut from $G(\mathcal{D}')$. Since $\mathrm{gl}(X) = \mathrm{gl}(X')$ and $|C_+(X)| = |C_+(X')|$, it follows that $n_X = n_{X'}$. Let $e_1, \ldots, e_k$ be edges of X and $e'_1, \ldots, e'_k$ be the corresponding edges of X'. Define $\delta_i = n(e_i) - n(e'_i)$, where $n(e_i)$ and $n(e'_i)$ are the numbers of arrows on e_i and e'_i, $1 \leq i \leq k$. Clearly, adding δ_i arrows to edges e'_i of $G(\mathcal{D}')$ results in $G(\mathcal{D})$. Since $n_X = n_{X'}$, $\sum_{i=1}^k \delta_i = 0$. Hence, we are always able to connect points corresponding to additional arrows on arcs of $\mathcal{D}'$ by oriented arcs inside X'. Repeating this procedure for each region X' of $G(\mathcal{D}')$ yields a system of oriented loops γ on $\hat{F}_\beta$ such that $\gamma \cdot L'$ is isotopic to L in $\hat{M}_\beta$ via an isotopy that proceeds along the fibers of $\hat{M}$.

Finally, we show that if $sh(L')$ differs from $sh(L)$ by one of the local moves $S_1 - S_3$ and S_β on $\hat{F}_\beta$, then $\gamma \cdot L$ is isotopic to L' in $\hat{M}_\beta$ for some curve γ on $\hat{F}_\beta$. Let $\mathcal{D}$ and $\mathcal{D}'$ be arrow diagrams for L and L' in $\hat{M}_\beta$, and $s(D)_\beta$ and $s(D')_\beta$ be their corresponding shadow diagrams on $\hat{F}_\beta$. Suppose that $s(D')_\beta$ is obtained from $s(D)_\beta$ by one of shadow moves $S_1 - S_3$ and S_β. Then there is an arrow diagram $\mathcal{D}''$ such that $\mathcal{D}'$ is obtained form $\mathcal{D}''$ by a finite number of $Q_1 - Q_5$ and $Q_{(1,\beta)}$-moves and $s(D'') = s(D)$ (see Remark 5.13). Let L'' be a link in $\hat{M}_\beta$ with arrow diagram $\mathcal{D}''$. Then by Theorem 3.4, links L' and L'' are isotopic in $\hat{M}_\beta$. Since $s(D'') = s(D)$, there is curve γ on $\hat{F}_\beta$, such that $\gamma \cdot L$ is isotopic to L''. Consequently, $\gamma \cdot L$ and L' are isotopic in $\hat{M}_\beta$. □

REMARK 5.18. For S^2 with a marked point β, the isotopy classes of links L in $\hat{M}_\beta$ correspond one-to-one with the shadow links $S(L)_\beta$ on S^2. For all other closed surfaces F, this correspondence does not hold. We demonstrated the cases when $F = S^2$ and $F = T^2$ with a marked point β in Examples 5.19 and 5.20.

EXAMPLE 5.19. Let $\hat{F}_\beta$ be S^2 with a marked point β and L be a link in $\hat{M}_\beta$ with arrow diagram $\mathcal{D}$ in Figure 5.16. Let $s(D)_\beta$ on $\hat{F}_\beta$ be the shadow diagram obtained from $\mathcal{D}$. Since $s(D)_\beta$ has no non-separating edges, we compute the number

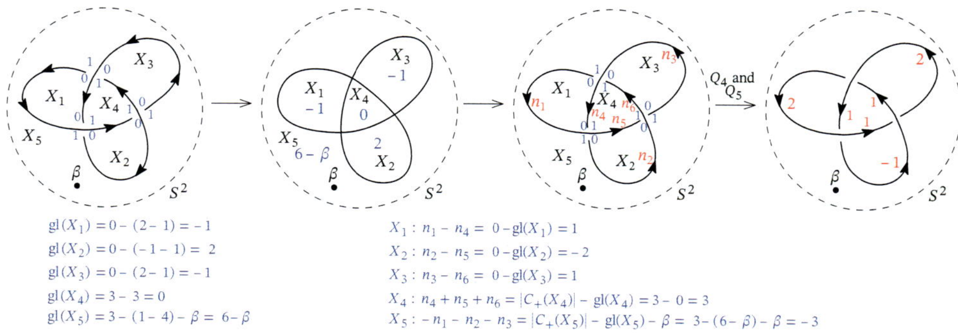

FIGURE 5.16. Arrow and shadow diagrams on $\hat{F}_\beta$

of arrows n_i, for $i = 1, \ldots, 6$, by solving the system of equations determined by the five regions X_i of $s(D)_\beta$, as given by Lemma 5.16. Note that the arrow diagram $\mathcal{D}'$ on $\hat{F}_\beta$, obtained from $s(D)_\beta$, is equivalent to $\mathcal{D}$ after applying the Q_4 and Q_5-moves.

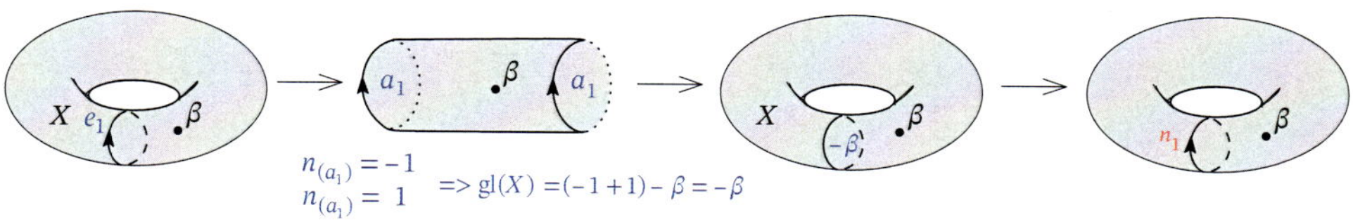

FIGURE 5.17. Non-isotopic links L and L' with $sh(L)_\beta = sh(L')_\beta$

EXAMPLE 5.20. Consider $\hat{M}_\beta$, where $\hat{F}_\beta$ is 2-torus T^2 with marked point β and let L be a link in $\hat{M}_\beta$ with arrow diagram $\mathcal{D}$ in Figure 5.17. Let $s(D)_\beta$ be shadow diagram obtained form $\mathcal{D}$. We construct arrow diagram $\mathcal{D}'$ from $s(D)_\beta$ by applying Lemma 5.16. Since e_1 is a non-separating edge of $s(D)_\beta$ and $s(D)_\beta$ has no separating edges, we may simply assign any integer $n_1 = n(e_1)$ for the number of arrows on its corresponding arrow diagram $\mathcal{D}'$. Let L' be a link in $\hat{M}_\beta$ with arrow diagram $\mathcal{D}'$. When $n_1 \neq 1 \mod \beta$, L and L' are clearly not isotopic in $\hat{M}_\beta$. However, taking for γ a disjoint union of $|n_1| \pm 1$ copies of $(0,1)$-curve on T^2 oriented accordingly to the sign of n_1, $\gamma \cdot L$ and link L' become isotopic.

6. Framed links in oriented S^1-bundle over an oriented surface

Let $p : M \to F$ be S^1-bundle over F with $\chi(p) = \beta$. Recall, by a *framed link* L in M we mean L together with a non-singular normal vector field on L considered up to homotopy of normal non-singular vector fields on L. As noted in [**20**], every framed link in M is isotopic to a unique generic link with framing given by the unit vector field tangent to fibers of p and co-oriented with fibers. We will call such a framed link a *generic framed link* in M. Consequently, a generic framed link L in $\hat{M}_\beta$ is defined as a generic link with framing given by the unit vector field tangent to fibers of $\pi : \hat{M} \to \hat{F}$ and is co-oriented with its fibers. An arrow diagram $\mathcal{D}_L$ for a generic framed link L is same as for unframed L since framing is perpendicular to $\hat{F} \subset \hat{F}_\beta$ (see Figure 6.1(a) and also Figure 2.1). By modifying our reasoning used in proof of Proposition 2.1 for framed links, one proves analogous result for framed links.

PROPOSITION 6.1. *Let $\mathcal{D}$ be an arrow diagram on $\hat{F}_\beta$, then there is a generic framed link L in $\hat{M}_\beta$ whose arrow diagram is $\mathcal{D}$.*

Isotopies of generic framed links in $\hat{M}_\beta$ are composition of framed versions of R_0 and R_β moves (see Figure 3.1) which takes place in a 2-handle N or H. In addition to R_0, R_β, R_2, R_3-moves shown in Figure 3.3, Figure 3.6, Figure 3.10 b), c), and Figure 3.11 b), c), there is a move R_1' shown in Figure 6.1.

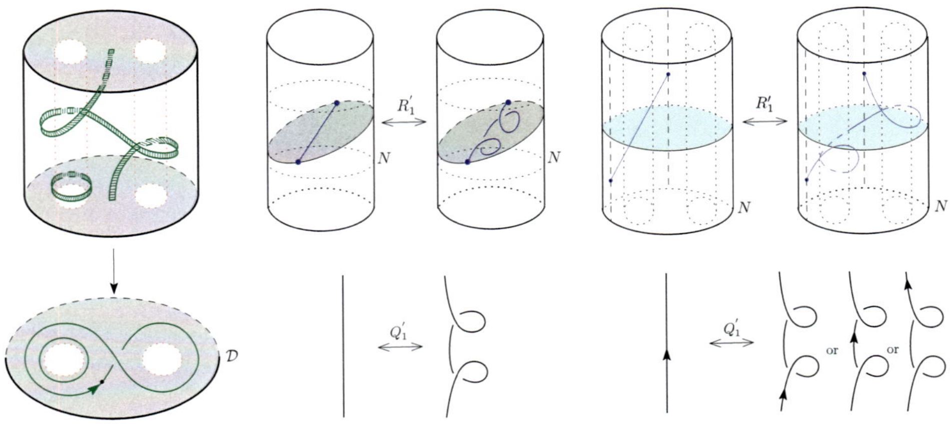

(a) Framed Link (b) Q_1' move (c) Isotopy by Q_1' move

FIGURE 6.1. Generic framed link and isotopy of framed links in $\hat{M}_\beta$ via R_1'-move

As we argued in Section 3, R_0, R_β, R_2, R_3, and R_1'-moves yield Q_2-Q_5, $Q_{(1,\beta)}$, and Q_1'-moves on arrow diagrams.

By modifying arguments from our proof of Theorem 3.4, one also proves analogous result for generic framed links. This result was stated in [**5**] and used in [**11**] and [**12**], we only outline its proof in this paper.

THEOREM 6.2. *Generic framed links L_1 and L_2 in $\hat{M}_\beta$ are isotopic if and only if their arrow diagrams $\mathcal{D}_1$ and $\mathcal{D}_2$ differ on $\hat{F}_\beta$ by a finite number of Q_1', Q_2-Q_5 and $Q_{(1,\beta)}$-moves.*

PROOF. The argument for Q_1-Q_5 as in our proof of Theorem 3.4. For Q_1'-move, we only need to note that if L_1 and L_2 differ by R_1'-move in a cylinder N shown in Figure 6.1(b), i.e., corresponding tangle T_1 intersect $\hat{F}$, then intersection of T_1 and T_2 with $\hat{F}$ yields corresponding arrow diagrams shown in Figure 6.1(c). It is clear that arrow diagrams of T_1 and T_2 are related either by Q_1'-move (for the first and third cases) or by two Q_5-moves followed by Q_1'-move (for the second case). Conversely, as in our proof of Theorem 3.4, we argue that if arrow diagrams of L_1 and L_2 differ by Q_1'-move (see Figure 6.1(b)), then generic framed links L_1 and L_2 differ by a composition of three R_0-moves (see Figure 3.3). □

COROLLARY 6.3. *Let $M = F \times S^1$, where F is a planar or $F = S^2$. Generic framed links L and L' in M are isotopic if and only if their arrow diagrams $\mathcal{D}$ and $\mathcal{D}$ differ on F by a finite sequence of local moves Q_1', Q_2, Q_3, Q_4, and Q_5.*

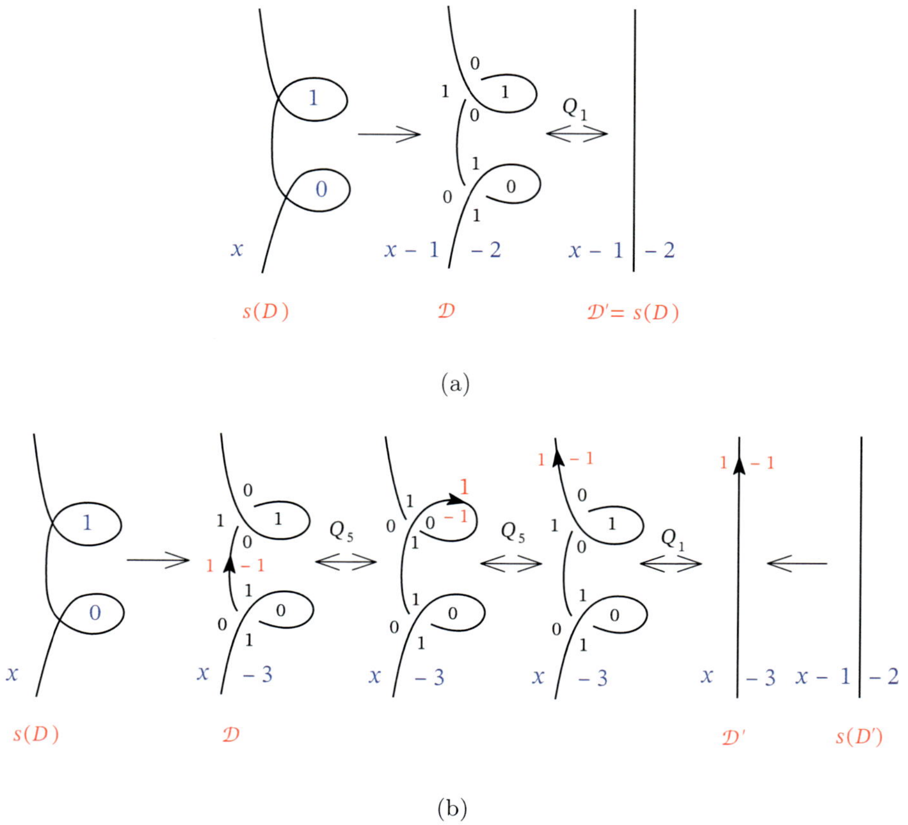

FIGURE 6.2. S_4 move

As it was argued in [20], given a generic framed link L in M, we can define its shadow diagram $sh(L)$ by simply forgetting its framing.

PROPOSITION 6.4. *If arrow diagrams D_L and $D_{L'}$ of framed generic links L and L' differ by a finite sequence of local moves Q_1', Q_2, Q_3, Q_4, and Q_5, then corresponding shadow diagrams $sh(L)$ and $sh(L')$ differ by a finite sequence of shadow moves S_2, S_3 and S_4.*

PROOF. We have already shown the cases when $sh(L)$ and $sh(L')$ differ by S_2-move and S_3-move in our proof of Proposition 5.4. Thus, we only need to check the case when $sh(L)$ and $sh(L')$ differ by S_4 -move. Using Remark 5.8, we need to consider two cases shown in Figure 6.2. For a), shadow move S_4 is a consequence of Q_1'-move on corresponding arrow diagram and Proposition 5.1. In the later case, S_4 is a consequence of two Q_5-moves followed by Q_1'-move and Proposition 5.1. □

PROPOSITION 6.5. *Let $M = F \times S^1$, where F is a planar or $F = S^2$. If shadows diagrams $s(D)$ and $s(D')$ differ by a finite sequence of shadow moves $S_2 - S_4$ then* *corresponding arrow diagrams $\mathcal{D}$ and $\mathcal{D}'$ differ on F by a finite sequence of Q_1', $Q_2 - Q_5$-moves.*

PROOF. We have already shown in Proposition 5.12 that local moves $Q_2 - Q_4$, and Q_5 are consequences of $S_2 - S_3$ shadow moves. Let L and L' differ by Q_1'-move. Using Proposition 5.1, we find $sh(L)$ and $sh(L')$ corresponding to L and L'. As shown in Figure 6.3, Q_1' is then a consequence of S_4-move. □

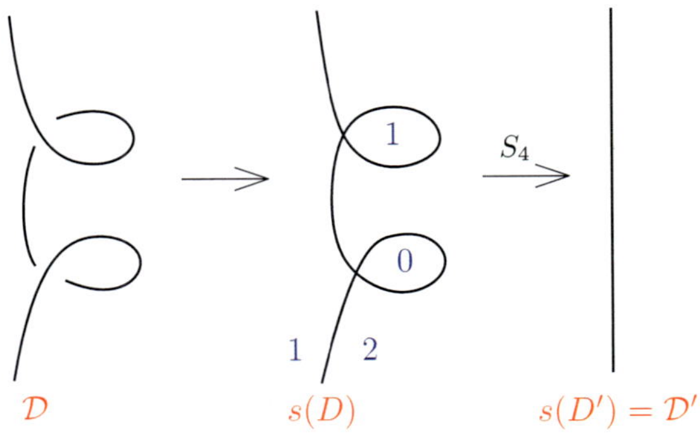

FIGURE 6.3. Q_1'-move as consequences of S_4-move

Finally, we define shadow diagram of a framed generic link in $\hat{M}_\beta$ using its arrow diagram in the same way as for unframed links in $\hat{M}_\beta$. Hence, there is also a version of Proposition 5.17 (i.e., Theorem 3.2 of [**20**]) for generic framed links in $\hat{M}_\beta$.

PROPOSITION 6.6. *The mapping that assigns to each generic framed link L in $\hat{M}_\beta$ its shadow diagram $sh(L)_\beta$ on $\hat{F}_\beta$ establishes a bijective correspondence between isotopy types of framed links in $\hat{M}_\beta$ modulo action of $H_1(F, \mathbb{Z})$ and the set of integral* shlinks $S(L)_\beta$ *on $\hat{F}_\beta$ with total gleam $-\beta$.*

Acknowledgment

Authors would like to thank Professor Józef H. Przytycki and Dr. Cheyu Wu for all valuable discussions and suggestions.

References

[1] D. Bullock, *The $(2,\infty)$-skein module of the complement of a $(2, 2p+1)$ torus knot*, J. Knot Theory Ramifications **4** (1995), no. 4, 619–632, DOI 10.1142/S0218216595000260. MR1361084

[2] D. Bullock, *On the Kauffman bracket skein module of surgery on a trefoil*, Pacific J. Math. **178** (1997), no. 1, 37–51, DOI 10.2140/pjm.1997.178.37. MR1447403

[3] D. Bullock, *Rings of* $\mathrm{SL}_2(\mathbf{C})$*-characters and the Kauffman bracket skein module*, Comment. Math. Helv. **72** (1997), no. 4, 521–542, DOI 10.1007/s000140050032. MR1600138

[4] R. Detcherry and M. Wolff, *A basis for the Kauffman skein module of the product of a surface and a circle*, Algebr. Geom. Topol. **21** (2021), no. 6, 2959–2993, DOI 10.2140/agt.2021.21.2959. MR4344875

[5] B. Gabrovšek and M. Mroczkowski, *Link diagrams in Seifert manifolds and applications to skein modules*, Algebraic Modeling of Topological and Computational Structures and Applications, 2017, pp. 117–141. E-print: `arXiv:1802.03645 [math.GT]`.

[6] J. Hoste and J. H. Przytycki, *The $(2,\infty)$-skein module of lens spaces; a generalization of the Jones polynomial*, J. Knot Theory Ramifications **2** (1993), no. 3, 321–333, DOI 10.1142/S0218216593000180. MR1238877

[7] J. Hoste and J. H. Przytycki, *The Kauffman bracket skein module of $S^1 \times S^2$*, Math. Z. **220** (1995), no. 1, 65–73, DOI 10.1007/BF02572603. MR1347158

[8] J. F. P. Hudson, *Piecewise linear topology*, W. A. Benjamin, Inc., New York-Amsterdam, 1969. University of Chicago Lecture Notes prepared with the assistance of J. L. Shaneson and J. Lees. MR248844

[9] L. H. Kauffman, *State models and the Jones polynomial*, Topology **26** (1987), no. 3, 395–407, DOI 10.1016/0040-9383(87)90009-7. MR899057

[10] B. Martelli, *An Introduction to Geometric Topology*, Createspace Independent Publishing Platform, 2016.

[11] M. Mroczkowski, *Kauffman bracket skein module of the connected sum of two projective spaces*, J. Knot Theory Ramifications **20** (2011), no. 5, 651–675, DOI 10.1142/S0218216511008905. MR2806337

[12] M. Mroczkowski, *Kauffman bracket skein module of a family of prism manifolds*, J. Knot Theory Ramifications **20** (2011), no. 1, 159–170, DOI 10.1142/S0218216511008681. MR2777022

[13] M. Mroczkowski and M. K. Dabkowski, *KBSM of the product of a disk with two holes and S^1*, Topology Appl. **156** (2009), no. 10, 1831–1849, DOI 10.1016/j.topol.2009.03.045. MR2519219

[14] J. M. H. Patrick M. Gilmer, *On the Kauffman bracket skein module of the quaternionic manifold*, J. Knot Theory Ramifications **16** (2007), no. 1, 103–125. E-print:`arXiv:math/0406152 [math.GT]`.

[15] J. H. Przytycki, *Skein modules of 3-manifolds*, Bull. Polish Acad. Sci. Math. **39** (1991), no. 1-2, 91–100. MR1194712

[16] J. H. Przytycki, *Fundamentals of Kauffman bracket skein modules*, Kobe J. Math. **16** (1999), no. 1, 45–66. MR1723531

[17] J. H. Przytycki, *The Kauffman bracket skein algebra of a surface times the interval has no zero divisors*, Proceedings of the 46th Japan Topology Symposium at Hokkaido University, July 26-29 (1999), 52–61.

[18] K. Reidemeister, *Elementare Begründung der Knotentheorie* (German), Abh. Math. Sem. Univ. Hamburg **5** (1927), no. 1, 24–32, DOI 10.1007/BF02952507. MR3069462

[19] V. G. Turaev, *The Conway and Kauffman modules of a solid torus* (Russian, with English summary), Zap. Nauchn. Sem. Leningrad. Otdel. Mat. Inst. Steklov. (LOMI) **167** (1988), no. Issled. Topol. 6, 79–89, 190, DOI 10.1007/BF01099241; English transl., J. Soviet Math. **52** (1990), no. 1, 2799–2805. MR964255

[20] V. G. Turaev, *Shadow links and face models of statistical mechanics*, J. Differential Geom. **36** (1992), no. 1, 35–74. MR1168981

Department of Mathematical Sciences, The University of Texas at Dallas, Richardson, Texas 75080

Email address: `mdab@utdallas.edu`

Department of Mathematics, Florida Gulf Coast University, Fort Myers, Florida 33965

Email address: `ssinharoy@fgcu.edu`

Contemporary Mathematics
Volume **834**, 2026
https://doi.org/10.1090/conm/834/16689

Are Sławianowski and Kijowski theories good models for TQFTs of Schwarz type?

Robert Owczarek

This paper is dedicated to the memory of Professor Jan Jerzy Sławianowski, my informal mentor and friend, on his 80th birthday, and to Professor Jerzy Kijowski, on his 80th birthday, with wishes of long life.

ABSTRACT. The seminal paper by Schwarz [**Sch77**] opened new possibilities of constructing topological invariants using topological quantum field theories on manifolds, which can be formulated without using any metric. The approach has been popularized and gave deep results by the famous work of Witten [**Wit89**] where not only topological invariants of manifolds but also of knots in manifolds, all in $3D$ case, were formulated. It is curious this has not really opened a search for metric independent field theories that would allow for defining other topological quantum field theories leading to new topological invariants. Recently, we initiated with Hanna Makaruk [**MO25**] research in just this direction, by proposing using in such role the Sławianowski theory, which is formulated without using any metric on the manifold. Here I review another theory of such kind, authored by Kijowski [**Kij78**], and shortly review also the Sławianowski theory, for completeness. There are no new results here, the aim of the paper is to attract attention to possible new area of research, and give some hints where the research may go from here.

1. Introduction

First, a timeline...

1.1. A timeline.

- (1978)
  - I was a primary school student.
  - Jerzy Kijowski published the paper [**Kij78**] on affine formulation of General Relativity (GR)
  - Albert Schwarz published the paper [**Sch77**], probably the first paper on Topological Quantum Field Theory
- (1985)
  - I was a student of physics at Warsaw University
  - Vaughn Jones defined new knot invariant, causing an avalanche of new mathematics, especially in topology, but also in physics

2020 *Mathematics Subject Classification.* Primary 57K16.
Key words and phrases. Differential geometry, topology.

  - Jan Jerzy Sławianowski published his paper [**Sła85**] on a theory based on frame bundle, which, as the paper [**Kij78**] had not used any metric on the base manifold
- (1988)
  - I was finishing my MS thesis at the department of physics of Warsaw University
  - At the department of physics happened a conference on Quantum Field Theory if I remember correctly
    * I attended a lecture by Professor Kijowski on his affine formulation of GR
    * I attended a lecture by Professor Schwarz who talked about his first TQFT (in fact, the $U(1)$ Chern–Simons theory formulation of topological invariants of $3D$ manifolds)
  - Edward Witten published the paper [**Wit88**] on his approach to Donaldson invariants of $4D$ manifolds, where the metric is explicitly present
  - I started working at the Institute of Fundamental Technological Research of Polish Academy of Sciences, and switched to research on superfluid helium, where later I used TQFT
- (1989)
  - Edward Witten published the second of his very influential papers on TQFT, this time on TQFT formulation of the Jones polynomial of knots and links and more [**Wit89**], concerning $3D$ manifolds. The theory has not involved any metric in its formulation, it was a general Chern–Simons theory
- (1993)
  - I finished my PhD, in which Witten's approach to knot invariants was used in explanation of the superfluid phase transition in helium and making certain predictions on behavior of near critical (close to phase transition) superfluid helium (see papers [**Owc93a**], [**Owc91**], [**Owc93b**], [**Owc93c**], [**Owc94**], [**Owc95a**], [**Owc95b**], [**Owc96**], [**Owc96b**]). The work of Kijowski [**Kij78**] and Sławianowski [**Sła85**] were a great inspiration for me, though not explicitly present in the thesis
- (2022-...)
  - Perhaps now is the time that these grains mentioned above may give some fruit. I collaborate on that program with Hanna Makaruk [**MO25**].

1.2. The work of A. Schwarz. We know from the paper of Schwarz [**Sch77**] that it is possible to use classical field theory which is independent from the metric to construct via corresponding (topological) quantum field theory (TQFT) invariants of the base manifold. In particular, Schwarz proposed action for his field theory that describes Ray–Singer torsion [**RS71**] to be as follows. On the general n-dimensional manifold M, the action considered by Schwarz reads

$$S = \int_M \omega^r \wedge d\omega^{n-r-1} \tag{1.1}$$

where ω^r is an r-differential form on M, and ω^{n-r-1} is an $n-r-1$-differential form on M. For M which is 3-dimensional, and a 1-form ω in the role of both forms in the formula, this is the standard Chern–Simons action.

$$S = \int_M \omega \wedge d\omega \tag{1.2}$$

There are also generalized invariants, which are proved to be modifications of Ray–Singer torsion, defined in the paper.

1.3. Witten's approach to knot invariants and $3D$ manifold invariants. The invariants defined by Witten are connected with principal bundles over manifolds, in this case $3D$ manifolds, with structure groups which are compact Lie groups. The basic example of such a group is $G = SU(2)$, or, also considered in the paper, $SU(n)$ groups. The action that leads to invariants of the manifolds is defined for the gauge field A, which is a 1-form with values in the Lie algebra $\mathcal{G} = su(2)$ of the group $SU(2)$ (or respective generalization to Lie algebra $su(n)$)

$$A \in C^\infty(T^*M) \otimes \mathcal{G} \tag{1.3}$$

In coordinates x^μ in the manifold M, and introducing a basis $e_a \in su(2)$,

$$A = A^a_\mu dx^\mu \otimes e_a \tag{1.4}$$

For this field, the action that can be defined on a $3D$ manifold that does not depend on the metric on the manifold M is the Chern–Simons action

$$S(A) = \int_M Tr(A \wedge dA + \frac{2}{3} A \wedge A \wedge A) \tag{1.5}$$

where the differential d acts on the 1-form portion of the gauge field A, while Tr is the Killing metric acting on Lie algebra component of the field, which for Lie algebras of compact groups happens to be proportional to Kronecker delta, thus a bit misleading notation Tr. As a result, the manifold invariant is a partition function for the corresponding quantum field theory, which may be written as

$$Z = \int DA\; e^{i\frac{k}{4\pi}S(A)} \tag{1.6}$$

where the factor $\frac{k}{4\pi}$ takes care of certain topological properties of the gauge group itself, with $k \in \mathbb{Z}$.

The goal of the Witten's paper was rather to define knot invariants and prove they are equivalent to the Jones polynomial in case of the manifold M being a $3D$ sphere. The goal was achieved, though using not rigorous arguments (after all, the measures used in defining partition functions are not well defined mathematical objects, so no rigorous proves can be achieved while using them). In this paper we focus on defining manifold invariants rather than invariants of certain embedded submanifolds like knots and links, so we will limit to manifold invariants only. The issue of importance further is the fact that with the Chern–Simons action, the set of fields satisfying the variational principle is the set of flat connections.

2. Standard general relativity

Before presenting Kijowski's theory, we need to review shortly the standard General Relativity. Let (M, g) be a $4D$ smooth manifold M, equipped with a Lorentzian metric g, with signature $+,-,-,-$ or $-,+,+,+$. Let us introduce a connection, at first independently of the metric, on the tangent bundle of the manifold M.

Let $C^\infty(TM)$ denote the set of smooth sections of the tangent bundle TM of the manifold M, which means the set of smooth vector fields over the manifold M, which is a module over $C^\infty(M)$, consisting of the smooth real valued functions on M.

DEFINITION 2.1. A connection ∇ on the tangent bundle TM of the smooth manifold M is a linear map

$$\nabla : C^\infty(TM) \otimes C^\infty(TM) \to C^\infty(TM) \tag{2.1}$$

such that, with notation $\nabla(X \otimes Y) =: \nabla_X(Y)$ the following is true for any $X \in C^\infty(TM)$, $Y \in C^\infty(TM)$, $f \in C^\infty(M)$

(1)
$$\nabla_{fX}(Y) = f\nabla_X(Y)$$

(2)
$$\nabla_X(fY) = f\nabla_X(Y) + (Xf)Y$$

where $Xf = \mathcal{L}_X f$ is the Lie derivative of the function $f \in C^\infty(M)$ along $X \in C^\infty(TM)$

REMARK 2.2. Due to the second of the above conditions, which is a kind of Leibniz rule, connections are not tensor fields, but we can associate with them certain important tensor fields.

DEFINITION 2.3. The Riemann curvature tensor *Riem* of a connection ∇ on TM is defined as follows $\forall X, Y \in C^\infty(TM)$ $Riem(X,Y) \in End(C^\infty(TM))$ such that $\forall Z \in C^\infty(TM)$

$$Riem(X,Y)(Z) := \nabla_X(\nabla_Y Z) - \nabla_Y(\nabla_X Z) - \nabla_{[X,Y]}Z \tag{2.2}$$

This way, the so defined Riemann curvature tensor is a tensor of valence

$$\begin{pmatrix}1\\3\end{pmatrix}.$$

DEFINITION 2.4. The torsion S of a connection ∇ on TM is defined as follows $\forall X, Y \in C^\infty(TM)$

$$S : C^\infty(TM) \otimes C^\infty(TM) \to C^\infty(TM)$$
$$S(X,Y) := \nabla_X Y - \nabla_Y X - [X,Y]$$

REMARK 2.5. Clearly, the torsion S is a tensor field of valence

$$\begin{pmatrix}1\\2\end{pmatrix}.$$

It is useful sometimes to look at local description, using coordinates on the manifold M. Let us introduce then, locally, a system of coordinates x^μ. Then the local basis connected with this coordinates is $\frac{\partial}{\partial x^\mu}$, so the local expression of a vector field (a smooth section of the tangent bundle TM) is

$$X = X^\mu \frac{\partial}{\partial x^\mu}$$

with standard summation convention, allowing skipping the sum sign.

Using coordinates and related components of the vector fields, we can introduce coefficients of the connection ∇ in this basis. We define symbols $\Gamma^{\rho}_{\mu\nu}$ by the following formula

$$\nabla_{\partial_\mu}(\partial_\nu) =: \Gamma^{\rho}_{\mu\nu}\partial_\rho$$

where the abbreviation

$$\partial_\mu := \frac{\partial}{\partial x^\mu}$$

is used.

In this basis, the torsion coefficients are

$$S(\partial_\mu, \partial_\nu) = \nabla_{\partial_\mu}\partial_\nu - \nabla_{\partial_\nu}\partial_\mu - [\partial_\mu, \partial_\nu] = (\Gamma^{\rho}_{\mu\nu} - \Gamma^{\rho}_{\nu\mu})\partial_\rho$$

REMARK 2.6. As a result, the zero torsion condition $S = 0$ in this particular basis would mean

$$\Gamma^{\rho}_{\mu\nu} = \Gamma^{\rho}_{\nu\mu}$$

Therefore, the connection with zero torsion has symmetric coefficients, and thus is, somewhat misleadingly, called a symmetric connection.

Similarly, one can calculate components of the Riemann tensor in the coordinate basis. The results can be summarized in these expressions of components of the torsion and the Riemann curvature tensors

$$S^{\rho}_{\mu\nu} = \Gamma^{\rho}_{\mu\nu} - \Gamma^{\rho}_{\nu\mu}$$
$$R^{\rho}_{\sigma\mu\nu} = \Gamma^{\rho}_{\nu\sigma,\mu} - \Gamma^{\rho}_{\mu\sigma,\nu} + \Gamma^{\alpha}_{\nu\sigma}\Gamma^{\rho}_{\mu\alpha} - \Gamma^{\alpha}_{\mu\sigma}\Gamma^{\rho}_{\nu\alpha}$$

DEFINITION 2.7. Ricci tensor *Ric* is a trace of the Riemann tensor

$$tr(X \mapsto Riem(X, Y)Z) \tag{2.3}$$

or, more frequently defined in components as

$$R_{\mu\nu} = R^{\rho}_{\mu\rho\nu}$$

REMARK 2.8. (notation) The components of the Riemann tensor, Ricci tensor, and Ricci scalar use just one letter "R", but the number and position of the indices unambiguously identify which of the tensors we deal with.

All objects up to the last definition were introduced without using any metric on the base manifold M. Only now, the metric, which means a

$$\binom{0}{2}$$

valence tensor g is introduced and, in order to define a Riemannian (or pseudo-Riemannian) geometry, certain condition will be introduced on the connection ∇.

DEFINITION 2.9. The connection ∇ on the manifold M equipped with a metric g is called metric if $\forall X, Y, Z \in C^\infty(TM)$

$$Z(g(X, Y)) = g(\nabla_Z X, Y) + g(X, \nabla_Z Y)$$

In (pseudo)Riemannian geometry one assumes additionally that the connection is torsion-less, $S = 0$, a condition independent from the metric. The crucial Levi–Civita theorem states that these two conditions describe in a unique way the connection.

THEOREM 2.10. *Riemannian manifold possesses a unique connection that is symmetric and metric.*

PROOF. The essence of the proof is that writing the condition $S = 0$ and the metricity explicitly allows for expressing the connection in unique way, using the metric tensor and its derivatives. Namely,

$$\overset{L.C.}{\Gamma^{\rho}_{\mu\nu}} = \frac{1}{2} g^{\rho\sigma}(g_{\nu\sigma,\mu} + g_{\mu\sigma,\nu} - g_{\mu\nu,\sigma}) \qquad \square$$

It is important to keep in mind that the Levi–Civita connection is expressed using components of the metric and their first derivatives.

$$\overset{L.C.}{\Gamma} = \Gamma(g, g,)$$

The standard general relativity uses the Levi–Civita connection and the scalar curvature, defined as

$$R = g^{\mu\nu} R_{\mu\nu}$$

therefore the metric appears explicitly.

REMARK 2.11. Observe the following facts:

- Riemann tensor depends on $\overset{L.C.}{\Gamma}$ (quadratically) and $\partial \overset{L.C.}{\Gamma}$, the first derivatives of $\overset{L.C.}{\Gamma}$, which on its own depends on the metric and its first derivatives
- As a result, the Riemann curvature tensor depends on up to second derivatives of the metric
- This property descends to the Ricci curvature...
- ...and then to the Ricci scalar curvature

The classical field theories, to which general relativity belongs, are characterized by Lagrangians, which are scalar densities built from the fields constituting the theory. The simplest Lagrangian built from the metric, due to stated above, is

$$\mathcal{L}_g = R\sqrt{|detg|}. \tag{2.4}$$

REMARK 2.12. A historical remark is in place:

- Historically, this variational approach is due to Hilbert, who published so-called Einstein equations shortly before Einstein did.
- Einstein in fact derived his equations earlier but hesitated to publish them because he made an assumption breaking general covariance, $|detg| = 1$. Einstein then discussed the problem with Hilbert, who used the variational principle and very quickly managed to derive the equations in the general case. Einstein was afraid of losing priority and published his derivation soon after Hilbert.

If we define the Einstein tensor as

$$G_{\mu\nu} := R_{\mu\nu} - \frac{1}{2} R g_{\mu\nu},$$

then the Einstein equations are

$$G_{\mu\nu} = 0$$

A proper theory has to include matter interacting via gravity. In standard general relativity, an assumption of so called 'minimal coupling' is made, due to which the Lagrangian of the system consisting of matter interacting via gravitational forces, is just a sum of Lagrangians describing the gravitational field, the Hilbert Lagrangian introduced above and of the matter Lagrangian

$$\mathcal{L} = \mathcal{L}_g + \mathcal{L}_{matt} \tag{2.5}$$

The assumption is made ad hoc. It is not obvious nor it is justified, which weakens this approach to general relativity.

In many cases, though not all, models of matter involve metric, but do not involve connection. As an example, lets look at a scalar field matter. The field is represented by a real-valued function on the manifold, ϕ. The Lagrangian for the scalar field is

$$\mathcal{L}_{matt} = (g^{\mu\nu}\phi_{,\mu}\phi_{,\nu} - m^2\phi^2)\sqrt{|detg|} \tag{2.6}$$

Variation of the ϕ field gives the equation for ϕ

$$0 = \frac{\delta\mathcal{L}_{matt}}{\delta\phi} = \frac{\partial\mathcal{L}_{matt}}{\partial\phi} - \frac{\partial}{\partial x^\mu}\Big(\frac{\partial\mathcal{L}_{matt}}{\partial\phi_{,\mu}}\Big) \tag{2.7}$$

which is a Klein–Gordon-like equation

$$\Big(g^{\mu\nu}\nabla_\mu\nabla_\nu + m^2\Big)\phi = 0 \tag{2.8}$$

with the covariant derivative taken with respect to the Levi–Civita connection for the metric g. Observe that the first of the covariant derivatives, acting on a scalar field, is just a usual derivative.

Now, the crucial moment is the derivation of the Einstein equations by varying the metric field g. The metric is in this standard approach present both in the gravitational and the matter portions of the Lagrangian, thus they both contribute to the final form of the equations. As was stated earlier, the gravitational Lagrangian $\mathcal{L}_g$ depends on the metric via the Ricci scalar (and the square root of the determinant, but this part does not contain derivatives of the metric at all), which is a straight descendant of the Riemann tensor. The Riemann tensor, as could be seen above, it built of derivatives of the connection, and in case of Levi–Civita connection, it is built of the metric components and their first derivatives, so the Riemann tensor for the Levi–Civita connection contains second derivatives of the metric. Therefore, calculating the variational derivative of the Lagrangian we need to remember that the gravitational Lagrangian contains derivatives of the metric up to second order, while usually (e.g. for the scalar matter) it contains metric components only algebraically, no derivatives of the metric are present. As a result, the following is true

(2.9)
$$\frac{\delta\mathcal{L}}{\delta g^{\mu\nu}} = \frac{\delta\mathcal{L}_g}{\delta g^{\mu\nu}} + \frac{\delta\mathcal{L}_{matt}}{\delta g^{\mu\nu}} = \frac{\partial\mathcal{L}_g}{\partial g^{\mu\nu}} - \frac{\partial}{\partial x^\rho}\Big(\frac{\partial\mathcal{L}_g}{\partial g^{\mu\nu}{}_{,\rho}}\Big) + \frac{\partial^2}{\partial x^\rho\partial x^\sigma}\Big(\frac{\partial\mathcal{L}_g}{\partial g^{\mu\nu}{}_{,\rho\sigma}}\Big) + \frac{\partial\mathcal{L}_{matt}}{\partial g^{\mu\nu}}$$

The second derivatives present in the Lagrangian are problematic, since this leads to third order equations, in general, while fundamental equations in physics are of second order. However, the second order terms can be represented as a complete derivative, and thus lead to a boundary term in the action functional, which does not contribute to the equations. Defining the so-called energy-momentum tensor (it exists also under other names)

$$T_{\mu\nu} := \frac{2}{\sqrt{|detg|}} \frac{\partial \mathcal{L}_{matt}}{\partial g^{\mu\nu}} \tag{2.10}$$

Then the Einstein equations modified by presence of matter are

$$G_{\mu\nu} := R_{\mu\nu} - \frac{1}{2} R g_{\mu\nu} = T_{\mu\nu} \tag{2.11}$$

REMARK 2.13. Everywhere above we ignored the physical coupling constants, because we illustrate the ideas rather than do systematic presentation of a physical theory.

3. General relativity and Palatini principle

Some weaknesses of the standard formulation of General Relativity can be avoided at the cost of introducing more variables. Instead of imposing the condition of metricity on the connection and effectively making it a function of the metric and its derivatives, what results in just metric components as basic variables (and there is just 10 independent components in a $4D$ symmetric tensor), one considers as independent both the metric and the connection (which adds to 10 components of the symmetric metric tensor 40 components of the symmetric connection) and derives the metricity condition from a variational principle. Lets focus first on the purely gravitational Lagrangian. It will be, as before, given by Ricci scalar, but calculated without the assumption of metricity.

$$\mathcal{L}_g = g^{\mu\nu} R_{\mu\nu}(\Gamma, \Gamma,) \sqrt{|detg|} \tag{3.1}$$

where "Γ," stands as a shorthand for derivatives of the connection Γ. Now, variation of this Lagrangian with respect to the metric does not touch the Ricci tensor factor, because it is now independent of the metric. Namely,

$$\begin{aligned} \frac{\delta \mathcal{L}_g}{\delta g^{\mu\nu}} &= \frac{\delta}{\delta g^{\mu\nu}} \left(g^{\alpha\beta} R_{\alpha\beta}(\Gamma, \Gamma,) \sqrt{|detg|} \right) \\ &= \frac{\delta g^{\alpha\beta}}{\delta g^{\mu\nu}} R_{\alpha\beta}(\Gamma, \Gamma,) \sqrt{|detg|} + g^{\alpha\beta} R_{\alpha\beta}(\Gamma, \Gamma,) \frac{\delta \sqrt{|detg|}}{\delta g^{\mu\nu}}. \end{aligned} \tag{3.2}$$

Quite obviously,

$$\frac{\delta g^{\alpha\beta}}{\delta g^{\mu\nu}} = \delta^\alpha_\mu \delta^\beta_\nu \tag{3.3}$$

Slightly less obvious but obtainable via standard algebra is

$$\frac{\delta \sqrt{|detg|}}{\delta g^{\mu\nu}} = -\frac{1}{2} \sqrt{|detg|} g_{\mu\nu} \tag{3.4}$$

Therefore,

$$\frac{\delta \mathcal{L}_g}{\delta g^{\mu\nu}} = \left(R_{\mu\nu} - \frac{1}{2} R g_{\mu\nu} \right) \sqrt{|detg|} \tag{3.5}$$

and thus the Einstein equations are recreated in this approach when we equate this variation to zero, though at this moment still for general (symmetric) connection

Γ. However, we have one more trick up our sleeve, the variation of the Lagrangian with respect to the connection. The equations are

$$
\begin{aligned}
(3.6) \qquad 0 = \frac{\delta \mathcal{L}_g}{\delta \Gamma^{\rho}_{\mu\nu}} &= \frac{\partial \mathcal{L}_g}{\partial \Gamma^{\rho}_{\mu\nu}} - \partial_\sigma \Big(\frac{\partial \mathcal{L}_g}{\partial \Gamma^{\rho}_{\mu\nu,\sigma}} \Big) \\
&= \frac{\partial (g^{\alpha\beta} R_{\alpha\beta}(\Gamma, \Gamma,) \sqrt{|detg|})}{\partial \Gamma^{\rho}_{\mu\nu}} - \partial_\sigma \Big(\frac{\partial (g^{\alpha\beta} R_{\alpha\beta}(\Gamma, \Gamma,) \sqrt{|detg|})}{\partial \Gamma^{\rho}_{\mu\nu,\sigma}} \Big) \\
&= g^{\alpha\beta} \sqrt{|detg|} \frac{\partial (R_{\alpha\beta}(\Gamma, \Gamma,))}{\partial \Gamma^{\rho}_{\mu\nu}} - \partial_\sigma \Big(g^{\alpha\beta} \sqrt{|detg|} \frac{\partial (R_{\alpha\beta}(\Gamma, \Gamma,))}{\partial \Gamma^{\rho}_{\mu\nu,\sigma}} \Big),
\end{aligned}
$$

which, after using explicit expression for the Ricci tensor in terms of the connection Γ and its derivatives, and some algebra, can be shown to boil down to

$$
(3.7) \qquad 0 = \nabla_\sigma \Big(g^{\mu\nu} \sqrt{|detg|} \Big)
$$

or, in other words to

$$
(3.8) \qquad 0 = \nabla_\sigma \Big(g_{\mu\nu} \Big),
$$

which is the metricity of the connection Γ with respect to the metric g condition. Combined with symmetry of the connection, using the Levi–Civita theorem, the connection Γ is just the Levi–Civita connection, and thus we recreated the general relativity starting from extra independent variables, the components of the symmetric connection. However, we treated only vacuum case.

What happens when we include matter? In cases the matter Lagrangian does not depend on the connection, like in the example considered above, the variational derivative of the Lagrangian with respect to the connection is zero, so that the connection is still metric. However, in cases the matter Lagrangian does depend on connection, this approach leads to non-metric connection in the areas where the matter is present. The variation with respect to the metric recreates, of course, the standard Einstein equations, with the caveat that connection- dependence of the matter Lagrangian may lead to Einstein tensor that is built of the connection that is not a Levi–Civita connection.

4. Interlude on volume—and volume-like—forms

For understanding the Kijowski approach to General Relativity, as well as to understand the Sławianowski theory, it is useful to look closer at invariance with respect to (orientation-preserving) changes of coordinates of the volume form and the volume-like forms. Lets start with a volume form. On a Riemannian manifold (M, g), in local coordinates x^μ the volume form can be expressed as

$$
(4.1) \qquad vol = \sqrt{|detg|}\; dx^1 \wedge dx^2 \wedge \ldots \wedge dx^n
$$

When we change coordinates, this volume form will be invariant up to the sign of the determinant of the Jacobian matrix of the change of coordinate transformation. On an orientable manifold, when we fix the orientation, that means invariance when we limit to coordinate changes that do not change the orientation. As a result, the volume form is globally defined on the manifold, and in fact ti is independent from the coordinate choice.

Interestingly, even when we do not have a metric on the manifold M, but have defined a

$$\binom{0}{2}\text{-tensor,}$$

say,

$$t = t_{\mu\nu} dx^\mu \otimes dx^\nu \tag{4.2}$$

then the volume-like form

$$\omega_t = \sqrt{|det\ t|}\ dx^1 \wedge dx^2 \wedge \ldots \wedge dx^n \tag{4.3}$$

is invariant with respect to coordinate changes that are orientation-preserving, the same way as for the standard "metric" volume form.

5. Kijowski approach to general relativity

Kijowski theory does not use any metric as a basic field. The only potential for gravitational field is a (symmetric) connection. The Lagrangian should not depend on any metric, but should depend on the connection and its first derivatives (recall the issue with second derivatives of the metric in the standard formulation of General Relativity, which we wish to overcome somehow). General covariance of the equations should also be expressed in the Lagrangian depending on the tensors built from the connection and its first derivatives. Since the torsion is assumed to be zero (the assumption of symmetric connection), the only option is to take Riemann tensor as the building block for the Lagrangian. However, with the tensor of so many indices there might be a problem to built a scalar density. In the light of the consideration from the Interlude section above, the better option is to take just the Ricci tensor, which is of the desirable kind, being a twice covariant tensor. In fact, Kijowski decided to take just its symmetric part

$$K_{\mu\nu} = R_{(\mu\nu)} = \frac{1}{2}(R_{\mu\nu} + R_{\nu\mu}) \tag{5.1}$$

In the light of what was stated in the previous section, the gravitational Lagrangian should be proportional to $\sqrt{|detK|}$,

$$\mathcal{L}(\Gamma, \Gamma,) = \mathcal{L}(\Gamma, K) = C\sqrt{|detK|} \tag{5.2}$$

Now we treat connection components like coordinates in classical mechanics, and then the components of the symmetrized Ricci tensor are like velocities. Lets introduce then analogues of the momenta as

$$\pi^{\mu\nu} = \frac{\partial \mathcal{L}}{\partial K_{\mu\nu}} \tag{5.3}$$

Then we perform a kind of Legendre transform, which allows for transition from coordinates and velocities to coordinates and momenta. What we get we will call a Lagrangian for consistency with the standard General Relativity, though in this approach should be called a Hamiltonian. It depends on connection as coordinates and the above conjugated momenta of the connection. Observe that the momenta $\pi^{\mu\nu}$ are twice contravariant tensorial densities, since the Lagrangian is a scalar density and it is differentiated with respect to a twice covariant tensor. Therefore, there exists a twice covariant tensor, which we call not accidentally $g_{\mu\nu}$, about which we assume to be invertible, with inverse a twice contravariant tensor $g^{\mu\nu}$,

$$g^{\mu\nu} g_{\nu\rho} = \delta^\mu_\rho = g_{\nu\rho} g^{\mu\nu} \tag{5.4}$$

(the order of the indices does not matter, since the tensor $K_{\mu\nu}$ is symmetric, and the property is inherited by both g tensors) such that

$$\pi^{\mu\nu} = \sqrt{|detg|}\ g^{\mu\nu} \tag{5.5}$$

Therefore, there is essentially a $1 - to - 1$ correspondence between $\pi^{\mu\nu}$ and $g^{\mu\nu}$. The new Lagrangian depends on Γ and g, and is defined as follows

$$L(\Gamma, g) = \pi^{\mu\nu}K_{\mu\nu} - \mathcal{L}((\Gamma, \Gamma, (g)) = \sqrt{|detg|}\ g^{\mu\nu}K_{\mu\nu} - \mathcal{L}((\Gamma, \Gamma, (g)) \tag{5.6}$$

The first term is just the standard Lagrangian of General Relativity. What is the second term? We have to find out, in terms of the tensor g, which will be the metric in this theory. If the initial Lagrangian is

$$\mathcal{L}(\Gamma, \Gamma,) = C\ \sqrt{|detK|} \tag{5.7}$$

then we can calculate the momenta

$$\pi^{\mu\nu} = \frac{\partial \mathcal{L}}{\partial K_{\mu\nu}} = \frac{\partial\ (C\ \sqrt{|detK|})}{\partial K_{\mu\nu}} = \frac{C}{2}\sqrt{|detK|}\ K^{\mu\nu} \tag{5.8}$$

where $K^{\mu\nu}$ is the inverse tensor of $K_{\mu\nu}$. By definition of the tensor g

$$\sqrt{|detg|}\ g^{\mu\nu} = \frac{C}{2}\sqrt{|detK|}\ K^{\mu\nu}. \tag{5.9}$$

This is solved by

$$g_{\mu\nu} = \alpha\ K_{\mu\nu} \tag{5.10}$$

with

$$\alpha = sgnC\ \left(\frac{|C|}{2}\right)^{\frac{2}{n-2}}. \tag{5.11}$$

The resulting Lagrangian (or rather Hamiltonian) is then of the following form

$$L(\Gamma, g) = g^{\mu\nu}K_{\mu\nu}\sqrt{|detg|} - \Lambda\sqrt{|detg|} \tag{5.12}$$

with

$$\Lambda = -C\ \left(\frac{|C|}{2}\right)^{\frac{n}{n-2}},$$

so it is General Relativity with a cosmological constant.

REMARK 5.1. For the issue of topological invariants, we ignore matter, but since it plays important role in Kijowski theory, lets mention how things are going with matter included. We use the example of scalar matter, discussed shortly before, just for illustration. In general case, in order to apply the tricks above, one has to rewrite matter Lagrangian in terms of connection and its derivatives only, without invoking a metric. The metric will still appear as an object related to conjugate momentum of the connection, so that

$$g^{\mu\nu}\sqrt{|detg|} = \pi^{\mu\nu} = \frac{\partial \mathcal{L}}{\partial K_{\mu\nu}} \tag{5.13}$$

but now the Lagrangian contains both gravitational and the matter degrees of freedom. For example, in the case of scalar matter the Lagrangian is

$$\mathcal{L}(\phi, \phi, ; \Gamma, \Gamma,) = \frac{2}{m^2\phi^2}\ \sqrt{|det\Big(K_{\mu\nu} - \phi,_{\mu}\phi, \nu\Big)|} \tag{5.14}$$

which give Klein–Gordon equations with metric dependent both on Ricci curvature and on the matter field components and their derivatives, and Einstein equations with energy-momentum tensor as in standard theory, but with metric derived rather than fundamental quantity.

6. Sławianowski theory

In Sławianowski theory [**Sła85**], [**Sła86**], [**SKG$^+$18**], which aimed at another metric-independent description of gravity, the basic field was chosen to be a frame, understood as a (smooth) section of the frame bundle. There are a number of natural bundles associated with a given smooth manifold M. First, there is a tangent bundle TM consisting of smooth vector fields on the manifold. Then, there is a cotangent bundle, T^*M, which consists of 1-forms, which are covector fields over the manifold M, dual to vector fields. Then, tensoring them together, we can have bundles of tensor fields of various grade. Next, in the tangent spaces we can introduce bases at each point of the manifold, changing from point to point in a smooth way. The set of all such choices can be organized into another bundle, called the frame bundle, FM. It is a principal bundle, which means its fibers over each point of the manifold are identifiable with a Lie group, in this case the general linear group $GL(n,\mathbb{R})$. The action is natural, lets briefly describe it. We choose one frame (basis) at a point of M, which is choosing a point in the fiber of the manifold FM over this point. The point is an n-tuple of linearly independent vectors

$$e_A, A = 1, 2, \ldots, n \tag{6.1}$$

An element of $GL(n,\mathbb{R})$ is a matrix $M \in GL(n,\mathbb{R})$ that acts on the right on the chosen element e_A of the fiber

$$e'_A = e_B\ M^B_A, A = 1, 2, \ldots, n, B = 1, 2, \ldots, n \tag{6.2}$$

As usual, the summation convention is applied here. When a metric is introduced on the manifold M, the frame bundle is reduced to an appropriate orthogonal frame bundle, consisting of frames that are orthogonal at each point, so that the structure group is reduced to corresponding orthogonal group (or to its subgroup, when we introduce more structure, e.g. when we choose orientation for orientable manifold, we reduce the bundle further, to the special orthogonal group). Since we want to keep away from introducing any metric on the manifold, we do not reduce the structure group, besides, possibly to $GL^+(n,\mathbb{R})$ when we choose orientation for an orientable manifold. If we choose coordinates, the frames, which means sections of the frame bundle, can be represented locally as vector fields as follows

$$e_A = e^\mu_A \frac{\partial}{\partial x^\mu}, A = 1, 2, \ldots, n, \mu = 1, 2, \ldots, n \tag{6.3}$$

The action of $GL(n,\mathbb{R})$ on the frame bundle translates into action of the same group on the dual bundle of coframes. The coframes are n-tuples of covector field e^A defined by natural (no metric involved) duality between frames and coframes

$$< \cdot, \cdot >: C^\infty(F^*M) \times C^\infty(FM) \to C^\infty(M) \tag{6.4}$$

defined in such a way that

$$< e^A, e_B >= \delta^A_B, A, B = 1, 2, \ldots, n \tag{6.5}$$

The transformation law comes from the following chain of equalities

$$\begin{aligned} \delta^A_B &=< e'^A, e'_B >=< e^C N^A_C, e_D M^D_B > \\ &=< e^C, e_D > N^A_C M^D_B = \delta^C_D N^A_C M^D_B = N^A_C M^C_B \end{aligned} \tag{6.6}$$

with the matrix N to be established from the equation. Apparently

$$N = M^{-1} \tag{6.7}$$

and the dual bundle is also a principal bundle with structure group $GL(n, \mathbb{R})$, just acting with inverse matrices. Similarly, in the local coordinates on the manifold M, the coframes can be expressed as

$$e^A = e^A_\mu dx^\mu, A = 1, 2, \ldots, n, \mu = 1, 2, \ldots, n \tag{6.8}$$

The natural connection associated with a frame is the teleparallelism connection, defined by

$$\nabla_{e_A} e_B = 0, \;\; A, B = 1, 2 \ldots, n \tag{6.9}$$

It is clear that so defined connection has its nonholonomic coefficients equal to zero

$$\Gamma^A_{BC} = 0 \tag{6.10}$$

In the coordinate basis, the coefficients will not be zero. Lets remind the definition

$$\nabla_{\partial_\mu} \partial_\nu = \Gamma^\rho_{\mu\nu} \partial_\rho \tag{6.11}$$

The above condition reads

$$\begin{aligned} 0 = \nabla_{e_A} e_B &= \nabla_{e^\mu_A \partial_\mu} (e^\nu_B \partial_\nu) = e^\mu_A \nabla_{\partial_\mu} (e^\nu_B \partial_\nu) \\ &= e^\mu_A \Big(\partial_\mu (e^\nu_B) \partial_\nu + e^\nu_B \nabla_{\partial_\mu} (\partial_\nu) \Big) \\ &= e^\mu_A \Big(\partial_\mu (e^\nu_B) \partial_\nu + e^\nu_B \Gamma^\rho_{\mu\nu} \partial_\rho \Big) \\ &= e^\mu_A \Big(\partial_\mu (e^\rho_B) \partial_\rho + e^\nu_B \Gamma^\rho_{\mu\nu} \partial_\rho \Big) \\ &= e^\mu_A \Big(e^\rho_{B},_\mu + e^\nu_B \Gamma^\rho_{\mu\nu} \Big) \partial_\rho . \end{aligned} \tag{6.12}$$

Because e_A is a basis, then this is satisfied when

$$0 = e^\rho_{B},_\mu + e^\nu_B \Gamma^\rho_{\mu\nu} \tag{6.13}$$

or

$$e^\nu_B \Gamma^\rho_{\mu\nu} = - e^\rho_{B},_\mu . \tag{6.14}$$

Using the coreper, one can write it as

$$\Gamma^\rho_{\mu\nu} = - e^B_\nu e^\rho_{B},_\mu . \tag{6.15}$$

Alternative form is

$$\Gamma^\rho_{\mu\nu} = e^\rho_A e^A_{\nu,\mu} . \tag{6.16}$$

This connection is flat, which means the Riemann curvature tensor is zero. On the other hand, its torsion is non-zero

$$S^\rho_{\mu\nu} = \Gamma^\rho_{[\mu\nu]} = - e^B_{[\nu} e^\rho_{B},_{\mu]} \tag{6.17}$$

We need to build a second order covariant tensor (a priori not a metric) from the torsion tensor, which will allow us to build a Lagrangian in already well established way. After some thought, one can establish there is a family of such tensors, defined using two free parameters

$$\gamma_{\mu\nu} = a S^\rho_{\mu\sigma} S^\sigma_{\nu\rho} + b S^\rho_{\mu\rho} S^\sigma_{\nu\sigma} . \tag{6.18}$$

The Lagrangian built accordingly to earlier formulated recipe is then

$$\mathcal{L}(e_A, e_{A,}) = \sqrt{|det\gamma|}. \tag{6.19}$$

The field equations coming from variational principle look as follows

$$\nabla_\mu H_\rho^{\nu\mu} = -2H_\rho^{\nu\mu} S^\sigma_{\sigma\mu} \tag{6.20}$$
$$H_\rho^{\nu\mu} = e^A_\rho \frac{\partial \mathcal{L}}{\partial e^A_{\mu,\nu}}.$$

References

[Kij78] J. Kijowski, *On a new variational principle in general relativity and the energy of the gravitational field*, Gen. Relativity Gravitation **9** (1978), no. 10, 857–877, DOI 10.1007/BF00759646. MR512794

[MO25] H. Makaruk and R. Owczarek, *Sławianowski's theory and topological invariants of manifolds*, Inverse problems: In memory of Professor Zbigniew Oziewicz, Contemp. Math., vol. 824, Amer. Math. Soc., [Providence], RI, 2025, pp. 67–72, DOI 10.1090/conm/824/16454. MR4945857

[MO95] H. Makaruk and R. Owczarek, GL(n, **R**)*-invariant tetrad theory and its relation to the purely affine formulation of general relativity*, Proceedings of the XXVII Symposium on Mathematical Physics (Toruń, 1994), Rep. Math. Phys. **36** (1995), no. 2-3, 355–363, DOI 10.1016/0034-4877(96)83631-7. MR1382924

[Owc91] R. Owczarek, *Topological defects in superfluid helium*, Internat. J. Theoret. Phys. **30** (1991), no. 12, 1605–1612, DOI 10.1007/BF00673638. MR1138341

[Owc93a] R. Owczarek, *Knotted vortices and superfluid phase transition*, Modern Physics Letters B, vol. 7, (1993), pp. 1523–1526.

[Owc93b] R. Owczarek, *Knotted vortices and fermionic excitations in bulk superfluid helium*, Modern Physics Letters B, vol. 7, (1993), pp. 1383–1386

[Owc93c] R. Owczarek, *Frames and fermionic excitations of vortices in superfluid helium*, Journal of Physics: Condensed Matter, vol. 5, (1993), 8793.

[Owc94] R. Owczarek, *Non-zero helicity generalization of the Nielsen-Olesen string solution*, Rep. Math. Phys. **34** (1994), no. 3, 305–310, DOI 10.1016/0034-4877(94)90003-5. MR1339467

[Owc95a] R. Owczarek, *Notes on superconductivity by quantum size effect and non-equivalent spin structures*, Journal of Technical Physics, vol. 36, (1995), pp. 181–185.

[Owc95b] R. Owczarek, *On geometric methods in the description of quantum fluids*, Quantization, coherent states, and complex structures (Białowieża, 1994), Plenum, New York, 1995, pp. 265–269. MR1403169

[Owc96] R. Owczarek, *Quantization of ideal fluids: Perturbative or non-perturbative approach*, Acta Phys. Polon. B **27** (1996), no. 10, 2387–2391. New ideas in the theory of fundamental interactions (Zakopane, 1995). MR1457303

[Owc96b] R. Owczarek, *On the role of helicity in stability and quantization problems of fluid flows*, Journal of technical physics, vol. 37, (1996), pp. 539-542

[RS71] D. B. Ray and I. M. Singer, *R-torsion and the Laplacian on Riemannian manifolds*, Advances in Math. **7** (1971), 145–210, DOI 10.1016/0001-8708(71)90045-4. MR295381

[Sch77] A. S. Schwarz, *The partition function of degenerate quadratic functional and Ray-Singer invariants*, Lett. Math. Phys. **2** (1977/78), no. 3, 247–252, DOI 10.1007/BF00406412. MR676337

[Sła85] J. J. Sławianowski, *Field of linear frames as a fundamental self-interacting system*, Rep. Math. Phys. **22** (1985), no. 3, 323–371, DOI 10.1016/0034-4877(85)90028-X. MR872646

[Sła86] J. J. Sławianowski, *Lie-algebraic solutions of affinely-invariant equations for the field of linear frames*, Rep. Math. Phys. **23** (1986), no. 2, 177–197, DOI 10.1016/0034-4877(86)90020-0. MR0894911

[SKG+18] J. J. Sławianowski, V. Kovalchuk, B. Gołubowska, A. Martens, and E. E. Rożko, *Space-time as a structured relativistic continuum*, Math. Methods Appl. Sci. **41** (2018), no. 14, 5404–5422, DOI 10.1002/mma.5087. MR3845919

[Wit88] E. Witten, *Topological quantum field theory*, Comm. Math. Phys. **117** (1988), no. 3, 353–386. MR953828

[Wit89] E. Witten, *Quantum field theory and the Jones polynomial*, Comm. Math. Phys. **121** (1989), no. 3, 351–399. MR990772

Department of Mathematics, University of New Mexico, Los Alamos, New Mexico 87544; and Department of Mechanical Engineering, University of New Mexico, Albuquerque, New Mexico 87106

Email address: rowczare@unm.edu

SELECTED PUBLISHED TITLES IN THIS SERIES

835 **Mee Seong Im and Tony Shaska, Editors,** Artificial Intelligence and Mathematics Research, 2026

834 **Mee Seong Im, Hanna E. Makaruk, Bach Nguyen, and Robert Owczarek, Editors,** Inverse Problems, Geometric and Topological Aspects of Representation Theory, and Low-Dimensional Topology, 2026

832 **Nils Bruin, David Kohel, and Chloe Martindale, Editors,** Arithmetic, Geometry, Cryptography and Coding Theory, 2026

831 **Irena Swanson and Kei-ichi Watanabe, Editors,** Lectures on Tight Closure and Its Applications, 2025

830 **O. Bezushchak, Yu. Drozd, P. Etingof, V. Futorny, R. Grigorchuk, and E. Zelmanov, Editors,** Modern Algebra Volume 2, 2025

829 **O. Bezushchak, Yu. Drozd, P. Etingof, V. Futorny, R. Grigorchuk, and E. Zelmanov, Editors,** Modern Algebra Volume 1, 2025

828 **Hafedh Herichi, Maria Rosaria Lancia, Therese Basa Landry, Anna Rozanova-Pierrat, and Steffen Winter, Editors,** Fractal Geometry in Pure and Applied Mathematics, 2025

827 **Carmen Caprau, J. Scott Carter, Neslihan Gügümcü, and Sam Nelson, Editors,** Algebraic Structures in Knot Theory, 2025

826 **Ahmed Laghribi and André Leroy, Editors,** Algebraic Structures and Applications, 2025

825 **Saurabh Verma, María A. Navascués, and Amit Priyadarshi, Editors,** Advances in Dimension Theory, Fractal Functions and Measures, 2025

824 **Hanna E. Makaruk and Robert M. Owczarek, Editors,** Inverse Problems: In Memory of Professor Zbigniew Oziewicz, 2025

823 **Mikhail Bershtein, Anton Dzhamay, and Andrei Okounkov, Editors,** The Versatility of Integrability, 2025

822 **Ahmad Barhoumi, Roozbeh Gharakhloo, and Andrei Martinez-Finkelshtein, Editors,** Recent Developments in Orthogonal Polynomials, 2025

821 **Bogdan D. Suceavă, Adara M. Blaga, Marian Ioan Munteanu, Cezar Oniciuc, Shoo Seto, Gabriel-Eduard Vîlcu, and Lihan Wang, Editors,** Recent Advances in Differential Geometry and Related Areas, 2025

820 **Sudeb Mitra, Rakesh Kumar Parmar, Saminathan Ponnusamy, and Toshiyuki Sugawa, Editors,** Geometric Function Theory and Related Topics, 2025

819 **Howard S. Cohl, Roberto S. Costas-Santos, and Robert S. Maier, Editors,** Applications and q-Extensions of Hypergeometric Functions, 2025

818 **Howard S. Cohl, Roberto S. Costas-Santos, and Robert S. Maier, Editors,** Classical Hypergeometric Functions and Generalizations, 2025

817 **Mikhail Khovanov, Joshua Sussan, and Anton Zeitlin, Editors,** From Representation Theory to Mathematical Physics and Back, 2025

816 **Luca F. Di Cerbo and Laurenţiu G. Maxim, Editors,** Geometry and Topology of Aspherical Manifolds, 2025

815 **Daniel Orr and Joshua Jeishing Wen, Editors,** Macdonald Theory and Beyond, 2025

814 **Susan Montgomery and Siu-Hung Ng, Editors,** Quantum Groups, Hopf Algebras, and Applications, 2025

813 **Thomas Creutzig, Julia Plavnik, and David Ridout, Editors,** Quantum Symmetries, 2025

812 **Ross Geoghegan, Craig R. Guilbault, and Kim Ruane, Editors,** Topology at Infinity of Discrete Groups, 2025

For a complete list of titles in this series, visit the
AMS Bookstore at **www.ams.org/bookstore/conmseries/**.

FSC
www.fsc.org
MIX
Papier | Fördert
gute Waldnutzung
FSC® C083411

Kolibri 360 GmbH
Ferdinand-Jühlke-Straße 7
99095 Erfurt, Deutschland
produktsicherheit@kolibri360.de